Biologically Active Substances Usable in Food, Pharmaceutical and Agrobiological Fields

This concise text on biologically active substances of the food, pharmaceutical and agricultural industries presents data on natural compounds of vegetable and animal origin. Various nutrients in food, phytochemicals and zoochemicals are discussed, including their uses for prophylactic, metaphylactic and therapeutic purposes in personalized medicine. Along with these compounds, prebiotics isolated by biotechnological methods from plant tissues are reviewed, with the aim of obtaining compounds with an oligoglucide structure. Metabolism of nutrients and the biodegradation of xenobiotics are hot topics and access routes into the human body for the various biologically active substances are covered.

Features:

- Biologically active substances and related chemistry, biochemistry and agrochemistry data are rigorously discussed.
- Data regarding natural compounds of vegetable origin detected from plants present in the spontaneous flora and plants obtained in agricultural crops (medicinal plants, aromatic plants and more) are presented.
- Discusses the natural compounds of animal origin detected in the organisms of some terrestrial and aquatic animals.
- Covers prebiotics isolated by technological and biotechnological methods from plant tissues, with the aim of obtaining compounds with oligoglucide structure.
- Broad audience including all those in biochemistry, the food and pharmaceutical industries and agricultural fields.

Medicinal Plants and Natural Products for Human Health
Series Editor: Christophe Wiart

We are at a point in history where the global population is increasingly interested in medicinal plants, natural products and herbalism. This interest has accelerated with COVID-19 and long COVID symptoms. Rediscovery of the connection between plants and health is responsible for a new generation of botanical therapeutics that include plant-derived pharmaceuticals, multicomponent botanical drugs, dietary supplements, functional foods and plant-produced recombinant proteins. Many of these products can complement conventional pharmaceuticals in the treatment, prevention and diagnosis of diseases. This series is a critical reference for anyone involved in the discovery of biopharmaceuticals for the improvement of human health.

Alternative Medicines for Diabetes Management: Advances in Pharmacognosy and Medicinal Chemistry
Edited by: Varma H. Rambaran and Nalini K. Singh

Therapeutic Perspectives of Tea Compounds: Potential Applications against COVID-19
Authored by: Farnoosh Dairpoosh and Kianoosh Dairpoosh

Biologically Active Substances Usable in Food, Pharmaceutical and Agrobiological Fields
Authored by: Zeno Garban and Gheorghe Ilia

For more information on this series please visit: https://www.routledge.com/Medicinal-Plants-and-Natural-Products-for-Human-Health/book-series/MPNPHH

Biologically Active Substances Usable in Food, Pharmaceutical and Agrobiological Fields

Zeno Garban and Gheorghe Ilia
Department of Biochemistry and
Molecular Biology, University of Life Sciences "King Michael I"
Timisoara and Romanian Academy-Timisoara Branch
Department of Biology-Chemistry,
West University Timisoara

CRC Press
Taylor & Francis Group
Boca Raton London New York

CRC Press is an imprint of the
Taylor & Francis Group, an **informa** business

Front cover image: Trutta/Shutterstock

First edition published 2024
by CRC Press
2385 Executive Center Drive, Suite 320, Boca Raton, FL 33431

and by CRC Press
4 Park Square, Milton Park, Abingdon, Oxon, OX14 4RN

Library of Congress Cataloging-in-Publication Data
Names: Garban, Zeno (Professor of biochemistry), author.
Title: Biologically active substances usable in food, pharmaceutical and agrobiological fields / by Zeno Garban, Department of Biochemistry and Molecular Biology, Romanian Academy, Gheorghe Ilia, Department of Biology-Chemistry, West University.
Description: First edition. | Boca Raton : CRC Press, 2024. |
Series: Medicinal plants and natural products for human health |
Includes bibliographical references and index. |
Identifiers: LCCN 2023057168 (print) | LCCN 2023057169 (ebook) |
ISBN 9781032698618 (hardback) | ISBN 9781032702513 (paperback) |
ISBN 9781032702520 (ebook)
Subjects: LCSH: Bioactive compounds. | Natural products. |
Natural products—Biotechnology.
Classification: LCC QD415 .G285 2024 (print) | LCC QD415 (ebook) |
DDC 572/.69—dc23/eng/20240409
LC record available at https://lccn.loc.gov/2023057168
LC ebook record available at https://lccn.loc.gov/2023057169

ISBN: 978-1-032-69861-8 (hbk)
ISBN: 978-1-032-70251-3 (pbk)
ISBN: 978-1-032-70252-0 (ebk)

DOI: 10.1201/9781032702520

Typeset in Times
by codeMantra

Contents

Preface

Approaching the problem of biologically active substances of food, pharmaceutic and agrobiologic interest has always aroused the interest of specialists concerned with fundamental specific research (e.g., chemical structure, biological activity and mechanisms of action) and with specific application (e.g., technologies/biotechnologies, physiological, pharmacological and pathobiochemical effects).

In this framework, data regarding natural compounds of vegetable origin detected from plants present in the spontaneous flora and plants obtained in agricultural crops (medicinal plants, aromatic plants, etc.) are presented. Also, information about the natural compounds of animal origin detected in the organisms of some terrestrial and aquatic animals is discussed.

Such substances are present next to other nutrients in food or can be obtained as extracts from plant organs and tissues (e.g., leaves, flowers and roots) – in the case of phytochemicals or from animals (e.g., tissues and various products) – in the case of zoochemicals. They are used in the food or pharmaceutical field for prophylactic, metaphylactic and/or therapeutic purposes (implicitly in personalized medicine).

Along with the previously mentioned substances, prebiotics isolated by technological/biotechnological methods from plant tissues, aiming to obtain compounds with an oligosaccharide structure, are discussed. These can be used both in the food and pharmaceutical industry.

Another area (often treated too briefly) that has been addressed in this book includes compounds of agrobiological interest. From this group the plant growth bioregulators are discussed because nowadays they are increasingly used in agro-biology, predilectly for the production/processing of some foods (fruits and vegetables).

In all the above-mentioned cases, addressing the topic of biologically active substances, there are rigorous discussions of not only chemistry/biochemistry/agrochemistry data but also problems related to specific physiological, pharmaceutical and toxicological aspects encountered in various compounds.

In a more extensive framework, the integrative connection with the metabolization of nutrients and the biodegradation of chemical xenobiotics was insured. Also, some information regarding the access routes into the human body for the various biologically active substances under discussion was exposed.

In the preparation of the manuscript for publication, Robert Ujhelyi – chemist at the Calivita Romania Company in Timisoara – also participated.

About the Authors

Zeno Garban – Ph.D., consulting professor of Biochemistry and Molecular biology, senior researcher and currently a corresponding member of the Romanian Academy. He is a member of some national and international scientific associations as well as of the editorial boards of some journals.

His educational background includes engineering studies at the University of "Life Sciences" Timişoara – Faculty of Agronomy, followed by the graduation in chemistry studies at the "Babeş-Bolyai" University Cluj-Napoca – Faculty of Chemistry. Afterward, he defended his doctoral thesis and obtained Ph.D. in Chemistry (Biochemistry).

Prof. Garban has an impressive research activity encompassing a broad spectrum of domains starting with comparative animal physiology (focusing on blood circulation – lactopoiesis, electrophoretic studies); neuroendocrine interrelationships in animal vasectomy: ethanol in teratology (on animal models), etc. He conducted studies in biochemistry and molecular biology, on specific interactions of DNA with various substances (metal ions, cytostatic drugs – in vitro and in vivo); the metabolization of nutrients and biotransformation of chemical xenobiotics of nutritional and pharmaceutical interest; the role of chemical xenobiotics in nutrition and pathobiochemistry. Professor Garban was a doctoral advisor in the domain of chemistry (biochemistry) at the Polytechnic University of Timisoara.

His scientific activity was focused on the domains of food science, approaching the complex problems of metallomics and proteomics. The interdisciplinary character has been put into evidence both in the published articles and in books and treatises.

These achievements resulted in an intense publishing activity with over 380 scientific papers in the country and abroad, 32 books (28 as single author); chapters in books published by Pergamon Press-Oxford (1984), Libbey-London (1997) and Wiley-VCH-Weinheim (2004). His most important works are represented by the books: *Biochemistry: Comprehensive Treatise* – four volumes, published in five editions (in Romanian); *Molecular Biology: Concepts, Methods, Appplications* (in Romanian) – awarded with the "Traian Savulescu" Prize of the Romanian Academy; and *Quo vadis food xenobiochemistry* – three editions (the last one in English).

Professor Garban initiated and organized the International symposium "Metal Elements in Environment, Medicine and Biology" and was a co-editor of the published 10 proceedings tomes. His scientific activity in the last few years included biologically active substances of nutritional and pharmaceutical interest.

Formerly, as full professor, he has also been involved in university teaching at the Faculty of Food Products Technology of the University of Life Sciences, Timişoara. He held courses in biochemistry, molecular biology, nutrition and food xenobiochemistry.

Dr. Eng. Gheorghe Ilia graduated from "Polytechnica" University in Timisoara as Chemical Engineer and earned Ph.D. in Chemistry at the same university. He is a senior researcher at "Coriolan Dragulescu" Institute of Chemistry, Timisoara, part of Romanian Academy and Professor at West University Timisoara, Chemistry Doctoral School as Ph.D. coordinator. He published more than 315 papers in peer-reviewed journals, 11 books and 17 chapter books in domestic and abroad publishing houses. His fields of expertise are phosphorus-containing polymers obtained by polycondensation or radical polymerization and their use in biomedical applications, as flame retardants, or as solid polymer electrolytes for electrochemical devices (Li-polymer batteries or fuel cells); organic–inorganic hybrids based on phosphorus derivatives using sol-gel method or by grafting on inorganic supports with potential applications in catalysis; MOF's based on phosphorus derivatives; green synthesis of phosphorus compounds; biologically active substances, especially plant growth bioregulators. Also he is an associate editor of *Current Green Chemistry* and an editorial board member of *Current Catalysis*.

1 Biologically Active Substances
Conceptual and Applicative Aspects

1.1 INTRODUCTION

Biologically active substances are compounds or mixtures of compounds that have beneficial or harmful effects on organisms. Such substances are found in nature, in plant and animal tissues as well as in microorganisms. They are used directly in culinary foods preparation or in the industrial processing of foods. These substances have also aroused the interest of pharmacists, who – based on pharmacologic and pharmacognostic studies – have sought to obtain extracts (such as tinctures, syrups, ointments, etc.) from plant and animal tissues for therapeutical use. Obviously, cosmetology (including also cosmetic dermatology) has large interest in this domain for the beneficial effects obtained from plant-based cosmetical products (Katz and Weaver, 2003; Liu, 2013; Godlewska-Żyłkiewicz et al., 2020; Hernandez et al., 2021; Bergonzi et al., 2022; Goyal et al., 2022). Last but not least, the problem of these substances has also been of interest in molecular toxicology with applications in nutrition, pharmacology and cosmetology, with peculiar interest in specific pathobiochemical aspects.

Over time, by chemical synthesis in the laboratory, "synthetic compounds" were obtained with a structure identical to natural biologically active substances and with similar properties to them. In the case of extraction and synthesis compounds, their chemical composition and biological effects are controlled in advance (according to specific protocols).

In biochemistry and modern molecular biology, *biologically active substances* are of particular interest in terms of conceptual and applicative aspects in food science, pharmacology, agrobiology, etc. The concept of "biologically active substance" (gr. *bios*-life; lat. *activus*-active, dynamic) was gradually developed, conditioned by the discovery of new substances with (recognized) effects on biochemical interactions in living matter and implicitly on physiological processes. In biochemistry and physiology, the term *bioactive substances* is used frequently.

Broadly speaking, the notion of *biologically active substance* refers to various specific characteristics of some chemical compounds, i.e., *physiologically active* – in the case of compounds that accompany the nutrients in food or enter the organism as food supplements, novel foods, etc.; *pharmacologically active* – in the case of pharmaceutical products – medicines obtained by extraction (starting

DOI: 10.1201/9781032702520-1

from psychopharmacology information) or by organic synthesis; *toxicologically active* – in the case of strictly toxic compounds (biocides) used in phytopathology and medical (veterinary/human) pathology for hygienization.

It is also worth mentioning that the study of biologically active substances is of interest to xenobiochemistry. This interest is explained by the fact that, along with nutrients, various chemical xenobiotics can coexist in ingested food, as well as different pharmaceutical products (the latter also considered xenobiotics) and during the biotransformation of xenobiotics, toxic xenobioderivatives can result.

Bioconstituents of plants and animals as well as of the human body (implicitly biologically active substances) were investigated in terms of chemical structure, biological activity and structure-activity relationship, i.e., structure-activity relationship (SAR), later extended to the quantitative structure-activity relationship (QSAR). In biochemistry, along with *bioconstituents* considered classics (organic and inorganic biomolecules), the *biochemical effectors* involved in metabolic processes have been investigated. The circumscribed evaluation of the *biochemical effectors* concluded that they include distinct groups of compounds: vitamins, enzymes, biochemical messengers (hormones, neurotransmitters and pheromones) and biologically active substances.

In this context, the study of *biologically active substances*, which includes *natural compounds of plant origin* (phytochemicals), *animal origin* (zoochemicals), *prebiotics*, *plant growth bioregulators*, etc., has been expanded, diversified and deepened (Arteca, 1996; Hasler and Blumberg, 1999; Ward and Bruce, 2003; Heller et al., 2004; Bruneton, 2009; Gârban, 2018).

Thus, knowledge of remedies based on the use of biologically active substances, especially in the field of phytochemicals, was also developed. These have been called "herbal remedies", the terminology referring to substances extracted from "herbs" that can be added to the "diet" or suitable for use in the treatment of various diseases, of interest to the so-called *complementary medicine* (Barnes, 2003). Ethnopharmacology has often been used to develop this field.

Using modern genetics, biochemistry, molecular biology and modern biotechnologies, the scientific community developed skills in the design and manufacture of novel foods, food for special medical purposes and food supplements with specific characteristics (Devlin, 1992; Biesalski et al., 2009; Gârban, 2018).

To ensure the beneficial effects of the biologically active substances extracted from phyto- and zoochemicals, it is necessary to evaluate the compounds or mixtures of extracted compounds in order to validate them. Validation is done by *in vitro* and *in vivo* tests on biological entities. For this purpose, the bioavailability and the dose-effect level in various target tissues are followed. In some cases, the use of biomarkers (biological markers) for different biologically active substances is considered (Fung et al., 2000; Crews and Hanley, 1995).

The issue of biologically active substances of food interest has also been addressed in the *Encyclopedia of Food and Culture* (Katz and Weawer, 2003). That volume mentions that the notion of biologically active substances (bioactive) in food refers to *non-essential biomolecules* present in foods that can modulate one or more metabolic processes with improved health status.

In different works dealing with biologically active substances, these are named "extra-nutritional" food constituents because they are present in extremely small quantities both in the plant and animal origin foods (Teodoro, 2019). Sometimes they are also called "non-nutrient" compounds (Liu, 2013).

This book – related to biologically active substances – includes phytochemicals, zoochemicals, prebiotics and plant growth regulators. With reference to them, various aspects were addressed such as structure, activity, natural distribution, beneficial/adverse effects, evaluation methods, etc.

1.2   CHARACTERISTICS OF THE CHEMICAL COMPOSITION OF THE LIVING ORGANISMS

1.2.1   BASIC CONCEPTS

Living organisms carry out complex vital processes during which – according to systems theory – continuous exchanges of *substance, energy and information* with the environment take place. These exchanges are achieved through physicochemical transformations aimed at biodegrading substances (nutrients) that enter the body by food and the biosynthesis of substances necessary for the processes of morphogenesis and energogenesis.

But, along with nutrients, the organism can access chemical xenobiotics (food and/or pharmaceutical interest), which, in turn, will be subjected to biodegrading processes. The main processes of biodegradation and biosynthesis occur, in fact, during the *metabolization of nutrients* and the *biotransformation of xenobiotics* – the fundamental dynamic features of biological processes.

Based on physiological and biochemical criteria, the biologically active substances in this volume were included in the category of biochemical effectors within the bioconstituents of living organisms.

As biologically active substances enter into the body mostly with foods of various sources and participate in the metabolization of nutrients, from a nutritional point of view, they were classified in the category of "other nutrients" alongside the water and dietary fiber.

The previously mentioned statements attest to the need to address related issues to bioconstituents, nutrients and chemical xenobiotics.

1.2.2   BIOCONSTITUENTS, NUTRIENTS AND CHEMICAL XENOBIOTICS

In the living world, there is a morphophysiological specificity of organisms (plants/animals), conditioned by natural bioconstituents. The sources of nutrients and the peculiarities of trophicity are at their origin. In this context, in relation to trophicity, the nutritional requirement, considered, roughly speaking, as a summary of the need for morphogenesis (natural bioconstituents), and the requirement for energogenesis (specific to biochemical interactions) is considered.

In historical terms, during the Renaissance (15th–16th centuries), it is mentioned that the preoccupations of science regarding biological effects also concerned application problems. The works of the Swiss alchemist and physician known as Paracelsus (1493–1541) were noted. He performed the first tests on biological fluids in the body of animals. *Paracelsus* also hypothesized that *Alle Dinge sind giftig, die Dosis ist entscheidend* (in German) (Everything is poisonous, the dose is decisive).

The contribution of Paracelsus led to the emergence, in the 16th and 17th centuries, of "medical chemistry", the so-called "iatrochemistry" (gr. *iatros*-doctor). The issued opinion explains the use of biologically active substances in pharmacology as well as the toxicological effects of some compounds used in too high doses (see Chapter 2).

1.2.2.1   Bioconstituents

Living matter constituents are generically denominated as *bioconstituents*. Topobiochemically, they are located in the ultrastructural elements which are specific to biological systems.

During evolution, the living organism selected from the environment various elements indispensable for the morphogenesis of ultrastructures as well as for own functions. These are known as "bioelements" or "biogenic elements". Considering the quantitative criterion, there are macro-, oligo- and microbioelements (Figure 1.1).

> *Macrobioelements* or *macrobiogenic elements* – represent around 99.70% of the constituents of the living matter. Among these, quaternary bioelements are included: oxygen, carbon, hydrogen and nitrogen (96.20%), plus calcium and phosphate (2.50%) together composing the *group of biomacroelements*.
>
> *Oligobioelements* – are found in low proportions (0.05%–0.75%). They are present in the structure of certain bioorganic and/or bioinorganic compounds, in a dissociated or nondissociated state.
>
> *Microbioelements* – are present in extremely low quantities in living matter, being sometimes referred to as "trace elements". They are classified as *invariable (indispensable) microbioelements* present in all living organisms. Among these, Fe, Zn, Cu, Co, Mn, Mo, F, I, etc. – *variable microbioelements* – are only present in some organisms. Such elements are Ni, Se, Si, B, etc. The bioelements present in the organism are associated with various chemical combinations. These are characterized by well-defined compositions and structures, which ensure chemical "biocompatibility" and involvement in physicochemical transformations occurring in the organism.

Biomolecules may be classified as organic biomolecules, inorganic biomolecules and biochemical effectors. Such molecules are found in all organisms. In fact, they are contained in living acellular matter (viruses) and cellular matter (prokaryotic and eukaryotic cells). The differences in composition, chemical structure and biologic activity explain the *molecular basis* of diversity.

1.2.2.1.1   Organic Biomolecules

According to the classical concept – currently used in biochemistry treaties – this group of biomolecules includes the main classes of compounds: carbohydrates, lipids and proteins.

> *Carbohydrates* – are natural chemical compounds found in the vegetal and animal regna.
>
> *Lipids* – are compounds with heterogeneous structure and varied chemical properties, with two essential characteristics: hydrophobicity and water

Bioconstituents

- Organic biomolecules

- Carbohydrates
  - Oses
    - Aldoses: dioses, trioses, tetroses, pentoses...
    - Ketoses: trioses, tetroses, pentoses...
  - Osides
    - Holosides (holosaccharides)
      - Oligosaccharides
      - Polisaccharides
    - Heterosides (glycosides)

- Lipids
  - Simple
    - Acylglycerides
    - Sterides
    - Cerides
    - Etholides
  - Complex
    - Glycerophospholipids
    - Sphingolipids

- Proteins
  - Amino acids (monopeptides)
  - Peptides
    - Oligopeptides
    - Polypeptides
  - Proteins
    - Holoproteins
    - Heteroproteins

- Inorganic biomolecules

- Water in the cellular environment and in biological fluids

- Biomineral compounds
  - With cationic character
    - Macroelements: Ca
    - Oligoelements: Na, K, Mg
    - Microelements: Fe, Cu, Zn, Mn
  - With anionic character
    - Macroelements: P
    - Oligoelements: S, Cl
    - Microelements: I, F, Se, Si

- Biochemical effectors

- Vitamins
  - Liposoluble
    - Vitamin A (retinols)
    - Vitamin D (calciferols)
    - Vitamin E (tocopherols)
    - Vitamin K (phylloquinone/menaquinone/menadione)
  - Hydrosoluble
    - Vitamin B Complex: thiamine (B_1), riboflavin (B_2), nicotinamide (B_3), pantothenic acid (B_5), pyridoxine (B_6), biotin (B_7), folic acid (B_9), cobalamin (B_{12})
    - Vitamin C (ascorbic acid)

- Enzymes
  - a) Oxidoreductases (EC 1.); b) Transferases (EC 2.); c) Hydrolases (EC 3.); d) Lyases (EC 4.); e) Isomerases (EC 5.); f) Ligases (EC 6.)

- Biochemical mediators
  - Endoactives
    - Hormones
      - Peptide
      - Steroid
    - Neurotransmiters
  - Ectoactives
    - Pheromones

- Biologically active substances
  - Phytochemicals
    - 1) Organic acids; 2) Alkaloids; 3) Phytoncides; 4) Heterosides; 5) Plant sterols; 6) Natural plant pigments; 7) Tannins; 8) Essential oils
  - Zoochemicals
    - 1) Polyunsaturated fatty acids; 2) Colostrinin; 3) Immunoglobulins; 4) Casein; 5) Coenzyme Q; 6) Carnitin; 7) Choline; 8) Natural animal pigments; 9) Animal alkaloids a.o.
  - Prebiotics
    - 1) Fructo-oligosaccharides; 2) Galacto-oligosaccharides; 3) Isomalto-oligosaccharides; 4) Xylo-oligosaccharides etc.
  - Plant growth regulators
    - 1) Auxines; 2) Gibberellins; 3) Ethylene; 4) Cytokinins; 4) Abscisic acid etc.

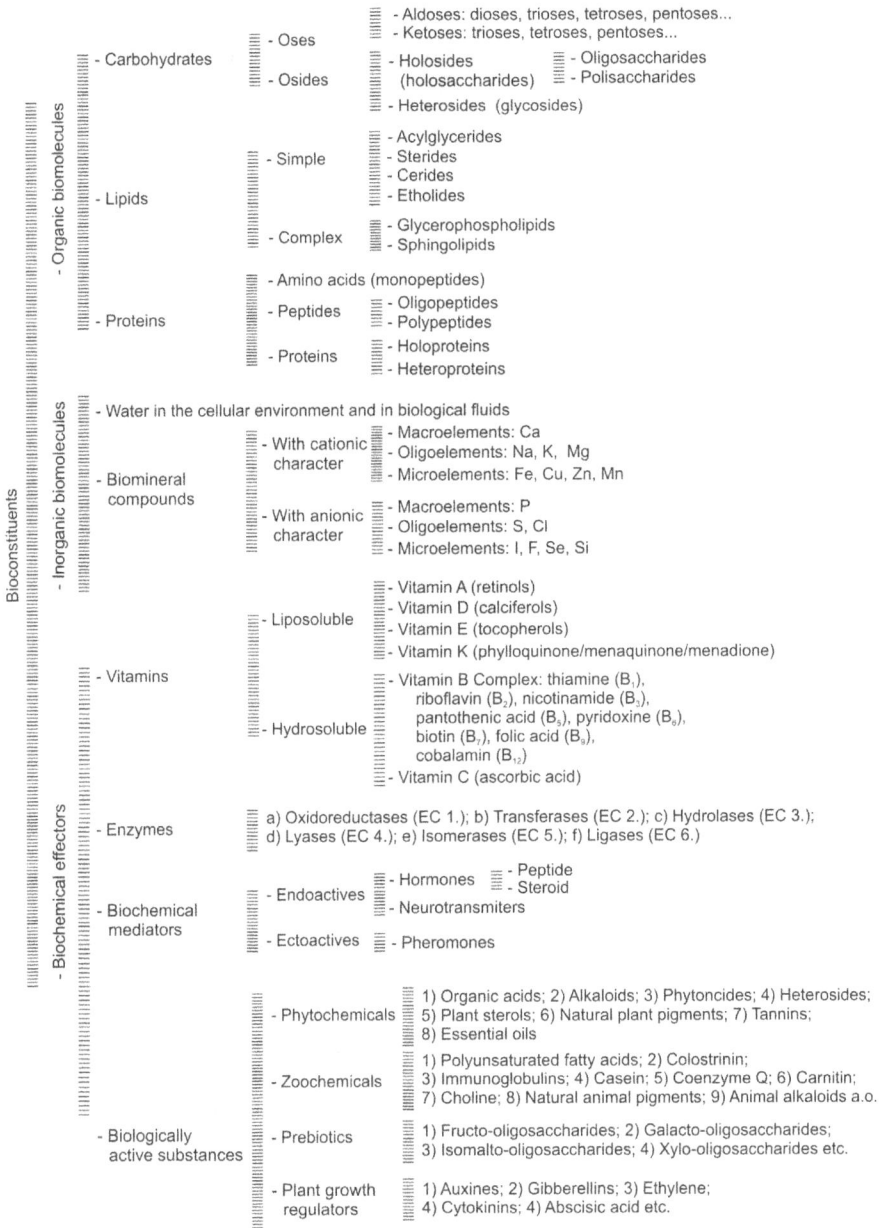

FIGURE 1.1 Bioconstituents identified in living organisms (overview) – Gârban, 1999- updated.

insolubility. Lipids are found in various vegetal and animal food products and, obviously, in the human body.

Proteins – Represent a large class of bioconstituents ubiquitary distributed in the living matter. Depending on the chemical composition, proteins are classified into three groups with different structural complexity: (i) amino acids – quaternary substances containing carbon, oxygen, hydrogen and nitrogen; (ii) peptides – with amino acids as basic structural units released by chemical (acid, alkaline) or biochemical (enzymatic) hydrolysis; and (iii) proteins which also contain, together with C, O, H, N other elements: phosphor, sulfur, metallic oligo- and microelements (Mg, Fe, Zn, Cu, Mn), etc.

1.2.2.1.2   Inorganic Biomolecules

This category includes water and biomineral compounds.

Water – Water represents the chemical component present in the highest proportion in living organisms. The quantum is between 40% and 94%, with regnum, species, gender and age-related variations.

Thus, the living organism may be seen as a dispersed heterogeneous system with water as the dispersion environment. Water can come from exogenous (foods) or endogenous (biochemical reactions) sources.

Biomineral compounds – Biomineral compounds found in living organisms (often referred to as "mineral salts") represent 3%–5%. These compounds are mainly represented by substances, which may generate anions and cations by electrolyte dissociation.

1.2.2.1.3   Biochemical Effectors

Depending on the structure, the *natural biochemical effectors* – which represent a separate chapter of modern biochemistry – are *vitamins, enzymes, chemical mediators* and *biologically active substances*, etc.

Vitamins – are exogenous biochemical effectors (with some exceptions such as vitamin D) represented by the organic compounds required in very small quantities for various biochemical pathways of metabolic processes in the organism.

Enzymes – are mainly endogenous biochemical effectors – resulting during the biosynthesis of enzyme-proteins, but also exogenous – especially resulting from vegetal foods.

Biochemical mediators – this group of substances includes *hormones, neurotransmitters and pheromones* – represented by varying different compounds in terms of chemical structure as well as biological activity. The diversity of natural distribution and, especially the complexity of the structure-activity relation in various classes of biologically active compounds have only recently retained the attention of biochemists.

Biologically active substances – represented mostly by phytochemicals having at their origin various plants, zoochemicals – present in the organism of some animals, prebiotics and plant growth regulators – are essential for the plant growth phases.

To have a more conclusive image of the biologically active substances, Figure 1.2 is presented, in which the substances mentioned above were included.

```
                                    ≡ - Organic acids
                                    ≡ - Alkaloids
                                    ≡ - Phytoncides
                                    ≡ - Heterosides
            ≡ - Phytochemicals     ≡ - Plant pigments
            ≡                       ≡ - Tannins
            ≡                       ≡ - Essential oils

            ≡                       ≡ - Polyunsaturated fatty acids
            ≡                       ≡ - Conjugated linoleic acid
            ≡                       ≡ - Animal alkaloids
            ≡                       ≡ - Colostrinin
            ≡                       ≡ - Immunoglobulins (IgG; IgM; IgA)
            ≡ - Zoochemicals        ≡ - Casein
            ≡                       ≡ - Carnitine
            ≡                       ≡ - Coenzyme Q
            ≡                       ≡ - Choline
            ≡                       ≡ - Lipoic acid
            ≡                       ≡ - Animal pigments

            ≡                       ≡ - Fructo-oligosaccharides
            ≡                       ≡ - Galacto-oligosaccharides
            ≡ - Prebiotics         ≡ - Isomalto-oligosaccharides
            ≡                       ≡ - Xylo-oligosaccharides

            ≡                       ≡ - Auxins
            ≡                       ≡ - Gibberellins
            ≡ - Plant growth        ≡ - Cytokinins
            ≡   regulators          ≡ - Ethylene
            ≡                       ≡ - Abscisic acid
```

(vertical label: Biologically active substances)

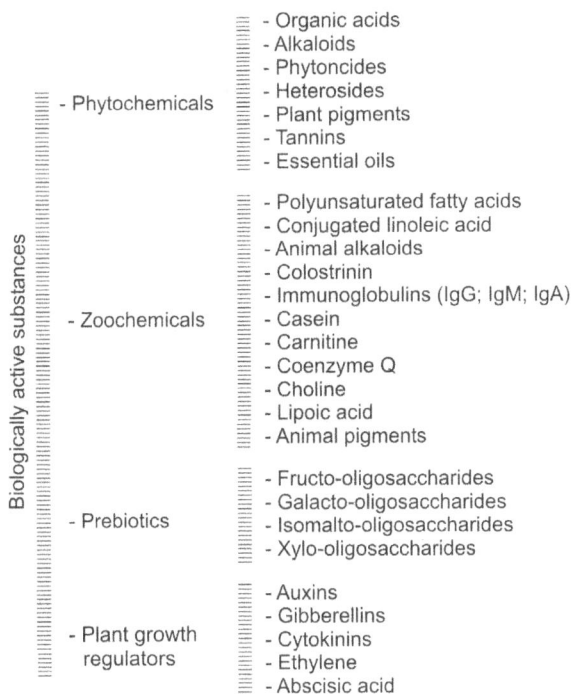

FIGURE 1.2   Biologically active substances (main classes of compounds).

The study of biologically active substances includes an extensive area of fundamental and applied scientific research, aimed at diversifying resources. It also aims to find new methods for chemical synthesis (in specialized laboratories) having as objective to expand the use of these compounds. Ab initio, it should be noted that the effects may be positive or negative, depending on the nature of the substance, the dose and the bioavailability.

At present, with the deepening of scientific research and modern technological/ biotechnological applications, it has become possible to circumscribe the problem of biochemical effectors and, obviously, of biologically active substances. This approach was gradually imposed in "food science" (today a distinct field even with the accreditation of this name) in pharmacology, pathobiochemistry, etc.

Biologically active substances are of major interest in terms of application in plant biology, pharmacology, agrobiology, food science, nanobiosciences, etc. (Arteca, 1996; Luch, 2009; Guaadaoui et al., 2014; Gârban, 2018; Ritter et al., 2019).

1.2.2.2   Nutrients

Following the physicochemical transformations that take place in the living organism, their own bioconstituents with distinct structural characteristics and functional attributes are formed from the nutrients. In this way, the intake of nutrients and the synthesis of bioconstituents ensure a perpetual morphological renewal of cells and tissues. Overall, the morphogenesis processes are ensured.

The supply of nutrients provides, in eodem tempore, the necessary energy for vital processes involving the main functions (e.g., circulation, respiration, digestion, etc.) even in resting state as well as functions related to physical activity. In the field of nutrition, especially in humans, the basic energy needs and the related energy needs are discussed. In a general context, the processes of *energogenesis* are of interest.

Nutrients are the basic chemical components of foods. Depending on their physiological and biochemical roles, the food nutrients are classified as: *macronutrients,* i.e., carbohydrates, lipids and proteins; *micronutrients,* i.e., vitamins and biominerals and *other nutrients,* i.e., water, fiber, biologically active substances (Figure 1.3).

Nutrients vary in terms of composition and quantum from one food to another depending on the food source (agri-foods, aquatic-foods); the methods are used for processing and further storage, conservation, etc.

For the functioning of the organism as a whole, food nutrients possess well-defined roles such as:

morphogenetic role – providing precursors for the biosynthesis of self-constituents as well as their continuous replacement. During morphogenetic processes, "bioconstituents" concur to the formation of molecular cell structures. These processes integratively influence tissues, organs, apparatuses and systems and eventually the entire organism.

energogenetic role – consisting of the capacity to provide the energy required for various vital processes. During biodegradation processes, the energy needed for vital processes is released. In such processes, carbohydrates, lipids and compounds with macroergic bonds are involved (e.g., ATP, creatine phosphate a.o.);

effector role – characteristic for various biochemical compounds that influence biological processes (as activators/inhibitors) and control the specific interactions of various biochemical pathways.

informational role – found in protein macromolecules, such as nucleic acids: deoxyribonucleic acid (DNA) and ribonucleic acid (RNA). Macromolecules can store and transmit the information in their constitutive sequence of nucleoside monophosphates (nucleotides). The genetic information (contained in genes) is transmitted within the organism during cell division – in gene and cell filiation, respectively (Gârban, 2009);

physicochemical role – characterized by the fact that organic biomolecules, together with inorganic ones (i.e., bioinorganic molecules), contribute to acid-base, osmotic and colloidal osmotic (oncotic) balances, respectively, as well as to transmembrane flow processes.

Biologically active substances enter the body together with different nutrients (e.g., macronutrients and micronutrients) in various ways, the main one being the gastrointestinal (digestive) way. To have an image of the biochemical aspects related to the composition of living matter, the biococonstituents of the organism with the nutrients in food are compared (Figure 1.4).

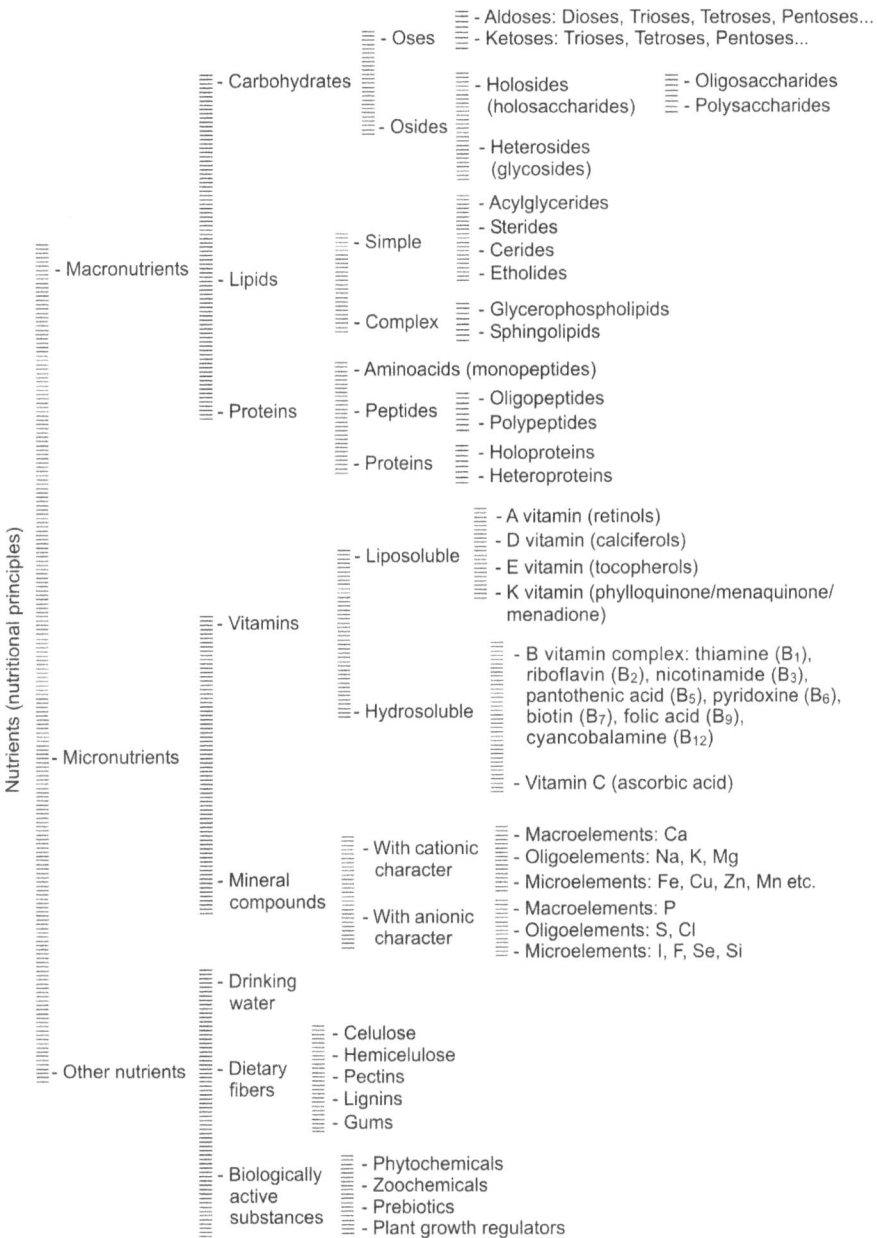

FIGURE 1.3 The chemical constituents of nutrients from food.

The development of normal physiological processes is ensured by the metabolization of *nutrients*. In metabolism, natural compounds in food are concerned, ab initio, which is the "raw material" of these processes.

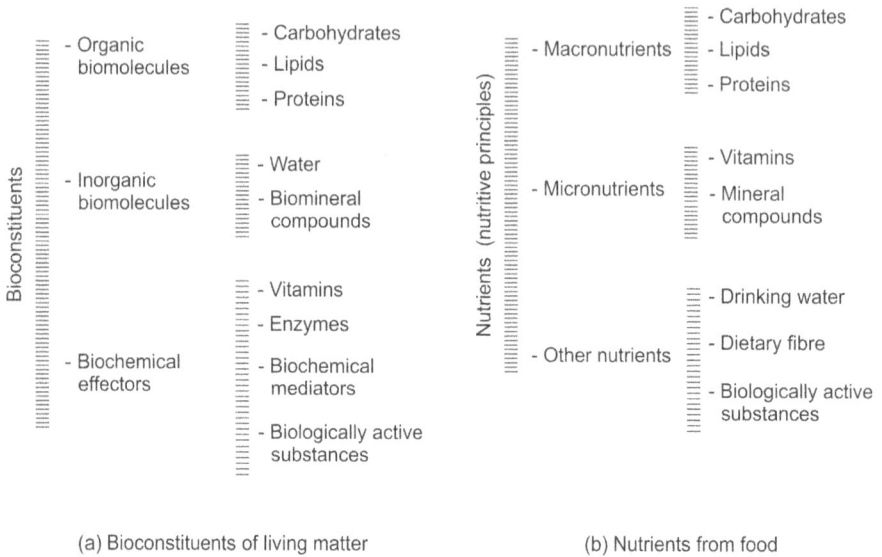

- Organic biomolecules
  - Carbohydrates
  - Lipids
  - Proteins
- Inorganic biomolecules
  - Water
  - Biomineral compounds
- Biochemical effectors
  - Vitamins
  - Enzymes
  - Biochemical mediators
  - Biologically active substances

Bioconstituents

(a) Bioconstituents of living matter

- Macronutrients
  - Carbohydrates
  - Lipids
  - Proteins
- Micronutrients
  - Vitamins
  - Mineral compounds
- Other nutrients
  - Drinking water
  - Dietary fibre
  - Biologically active substances

Nutrients (nutritive principles)

(b) Nutrients from food

FIGURE 1.4 The chemical composition of bioconstituents in the body and nutrients from food.

Data compiled from the literature (Devlin, 1992; Mincu, 1993; Hasler and Bloomberg, 1999; Ward and Bruce, 2003; Belitz et al., 2009, a.o.) presented in Table 1.1 draw attention to the nutritional intake by comparing the composition of plant-derived foods vs. animal-derived foods.

1.2.2.3 Chemical Xenobiotics

Along with nutrients, certain chemical xenobiotics can also access into the organism (food and/or pharmaceutical interest) which in turn will be subjected to biodegrading processes. The main processes of biodegradation and biosynthesis take place, in fact, during the *metabolizations of nutrients* and the *biotransformation of xenobiotics* – the fundamental dynamic features of biological processes.

The study of biotransformations is of interest in the case of chemical xenobiotics and is approached in a correlative manner, involving biodegradation (in this case xenobiodegradation) and biosynthesis (respectively xenobiosynthesis).

Chemical xenobiotics as extra-nutritional compounds detected in food can be of deliberate origin (food additives) and/or of accidental or even illicit origin (food pollutants) – Figure 1.5. Chemical contamination of food can occur directly or indirectly (Altug, 2002; Deshpande, 2002; D'Mello, 2003; Gârban, 2018).

The chemical xenobiotics present in the environment belong to different classes of substances, coming from industrial residues, phytosanitary actions, the use of medicines and disinfectants for medical use (veterinary and human), etc. Examples of xenobiotics of various origins are: (i) organic: pesticides (e.g., organochlorine derivatives, organophosphorus derivatives); polycyclic aromatic hydrocarbons; mycotoxins, etc.; (ii) inorganic: nitrates, nitrites, some heavy metals (Pb, Cd, Hg and Sn), etc.

TABLE 1.1
General Nutrient Intake: Foods of Plant Origin vs. Foods of Animal Origin

Nutrients		Distribution in Food of Plant Origin	Distribution in Food of Animal Origin
Macronutrients	Proteins	Cereals. Vegetables. Contains incomplete protein with lower nutritional value (does not contain all essential amino acids)	Meat. Dairy products. They contain complete proteins, with higher nutritional value (with essential amino acids)
	Lipids	Oleaginous plants. They contain, predominantly, fatty acids. They have a high content of unsaturated acids and less saturated fatty acids	Fish. Dairy products. Fatty acids are dominant, but they also contain monounsaturated fatty acids. Fish oil is an important source of EPA/DHA
	Carbohydrates	Cereals. Some root plants	Bee products. Milk and dairy products
Micronutrients	Vitamins	Fruits. Vegetables. Sources of hydro- and liposoluble vitamins such as vitamins C, K E and B (including folic acid-vitamin B9)	Meat. Dairy products. Vitamin sources represented by B complex group (especially vitamin B_{12})
	Minerals	Vegetables. Fruits. Plant foods contain significant amounts of K and Mg	Meat. Fish. Dairy products. The dominant minerals are: Fe, Zn and Ca (especially in dairy products)
Other nutrients	Water	75% in plant tissues. Predominant in fruits and vegetables.	60% in animal tissues. Predominant in milk and meat.
	Dietary fibers	Edible vegetal tissues especially cereals and vegetables are an important source of fiber	Edible animal tissues are low in fiber
	Biologically active compounds	Plants of spontaneous flora. Medicinal herbs. Flavoring plants	Products from aquatic animals. Milk and dairy products. Meat. Bee products

Among the pharmaceutical products can be listed: (i) antibacterial sulfamides; (ii) antibiotics; (iii) anti-tuberculosis and anti-leprosy drugs; (iv) antiprotozoal drugs; (v) anti-syphilis drugs; (vi) antifungals; (vii) antihelmintics; (viii) antiviral and (ix) cytostatic. For details, a vast literature can be consulted (Trevor et al., 2015; Ritter et al., 2019, a.o.).

Chemical xenobiotics with attributes of chemical contaminants/pollutants are among the "multiple stressors" that can affect terrestrial and aquatic organisms. Taken from the environment, chemical xenobiotics are "translocated" into plant and animal tissues, from where they eventually end up in food (Gil and Pla, 2001; Belitz et al., 2009; Baveye et al., 2011; Schröder and Collins, 2011).

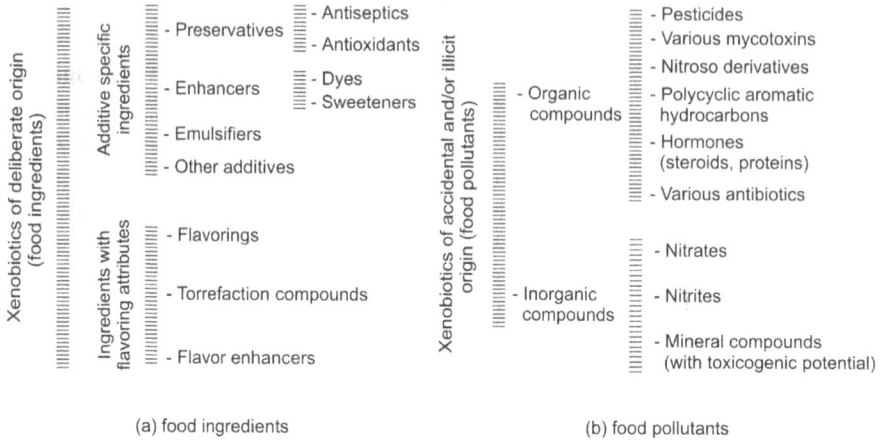

(a) food ingredients

Xenobiotics of deliberate origin (food ingredients)

Additive specific ingredients
- Preservatives
  - Antiseptics
  - Antioxidants
- Enhancers
  - Dyes
  - Sweeteners
- Emulsifiers
- Other additives

Ingredients with flavoring attributes
- Flavorings
- Torrefaction compounds
- Flavor enhancers

(b) food pollutants

Xenobiotics of accidental and/or illicit origin (food pollutants)

- Organic compounds
  - Pesticides
  - Various mycotoxins
  - Nitroso derivatives
  - Polycyclic aromatic hydrocarbons
  - Hormones (steroids, proteins)
  - Various antibiotics

- Inorganic compounds
  - Nitrates
  - Nitrites
  - Mineral compounds (with toxicogenic potential)

FIGURE 1.5 Chemical xenobiotics of food interest.

1.3 METABOLIZATION OF NUTRIENTS

1.3.1 BASIC CONCEPTS

The morphological, physiological and biochemical peculiarities of living matter are realized within transformations based on interactions that follow *native biochemical pathways* characteristic of the processes that ensure the metabolization of nutrients.

Metabolism – is defined by the complex set of biochemical interactions that underlie the physicochemical transformations of compounds of exogenous origin – nutrients in food and, to a lesser extent, of compounds of endogenous origin – bioconstituents in living matter (which are periodically renewed by the resulted food metabolites).

In nutrition, the fundamental problem lies in meeting the needs of nutrients with trophic and energy intake. Only in this way, the processes of morphogenesis and energogenesis can be carried out in optimal conditions – without affecting the health state. At the origin of these processes are the transformations that affect the *metabolization of nutrients* (catabolism – anabolism), ensuring the continuous maintenance of biochemical homeostasis and – in a broader context – the morphophysiological status of the organism.

1.3.2 PHASES AND PATHWAYS OF METABOLIZATION

1.3.2.1 Metabolization Phases

Metabolic processes can take place in the sense of biodegradation or biosynthesis of chemical compounds that reach/exist in the body, with two distinct phases: *catabolism* and *anabolism*.

The essential feature of metabolism lies in the development of biodegradation and biosynthesis reactions. In these reactions – carried out under the action of enzymes – many stage compounds with the name of intermediate metabolites are formed.

Catabolism – It is represented by the totality of biodegradation processes that concern the compounds of exogenous (nutrients from food) and endogenous origin (bioconstituents and metabolites from the internal environment).

The phase of catabolism concerns primarily compounds with complex structures and high molecular weight of *nutrients*, which decompose into simple compounds. Thus, molecules of protein, carbohydrate, or lipid nature of exogenous or endogenous origin can be degraded to smaller molecules, such as amino acids, lactic acid, glycerol, urea, CO_2, NH_3, etc.

During the catabolic processes, the chemical energy is released from the biomolecules subjected to degradation, and this energy is conserved in the ATP molecules (generally in nucleoside triphosphates), which subsequently ensure the microtransport of metabolites.

Anabolism – It is characterized by the totality of biosynthesis processes during which the *bioconstituents* of its own organism and the "circulating metabolites" (detectable in biological fluids) are formed. In this phase, compounds such as nucleic acids, proteins, polysaccharides and lipids are synthesized, starting from simple, small molecule precursors. Such precursors can be nucleobases, amino acids, monosaccharides, fatty acids, glycerol, etc.

A general, schematic representation of the metabolization phases, marking catabolism (phase I) and anabolism (phase II), is shown in Figure 1.6. In the catabolism phase, conceptually one can start from a *compound A* (e.g., nutrient) and reach a *compound B* (intermediate metabolite).

The second phase (anabolism), conceptually starts from a "compound B" (symbolizing an intermediate metabolite) and finally reaches a "compound C" (bioconstituent/circulating metabolite). The investigation of metabolism, depending on the object of study, can address: material metabolism (protein, carbohydrate, lipid and hydro-electrolyte) and energy metabolism that accompanies biodegradation and/or biosynthesis. In biochemistry, more precisely in *dynamic biochemistry*, the metabolization of nutrients with the highlighting of the succession of distinct processes is discussed: *absorption-distribution-metabolization-elimination*. This sequence of processes is usually rendered by the acronym ADME.

1.3.2.2 Metabolization Pathways

In metabolism, for each phase (catabolism and anabolism), there are specific biochemical pathways for compounds belonging to various classes, e.g., carbohydrates, lipids and proteins. Metabolization pathways are characterized by a sequence of biochemical interactions.

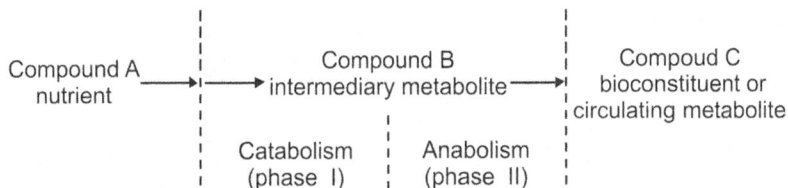

FIGURE 1.6 Characteristic phases for nutrients metabolization.

Metabolic processes, as shown, take place in *native biochemical pathways* with the participation of natural enzyme systems. These pathways are integrated into metabolization phases, which can be catabolic, anabolic and amphibolic.

1.4   BIOTRANSFORMATION OF CHEMICAL XENOBIOTICS

1.4.1   Basic Concepts

In general, in the body, various xenobiotics (gr. *xenos*-foreign; *bios*-life) – along with food or independently of it – can gain access. These are considered chemical, physical and biological agents that result in *food insalubrization* (contamination).

The group of chemical xenobiotics includes food contaminants, pharmaceuticals (various drugs obtained by extraction or synthesis), cosmetic ingredients and strictly toxic substances, i.e., biocidal products (Sipes and Gandolfi, 1991; Sandermann, 1992; Ronis and Cunny, 2008; Nassar et al., 2009; Gârban, 2018). Of particular interest are the chemical xenobiotics considered as food contaminants. These are non-nutritive substances grouped according to their biologically active specificity as *food additives* (contaminants of deliberate use) and *food pollutants* (contaminants of accidental or illicit origin).

Xenobiotics that enter into the body with foods are gradually subjected to biotransformation processes.

Biotransformation – is defined by the set of biochemical interactions at the origin of the physicochemical transformations of the chemical xenobiotics that enter into the body. These transformations of chemical xenobiotics are achieved by xenobiodegradation and xenobiosynthesis.

Compounds from the chemical xenobiotic group, during biotransformation, follow *specific xenobiochemical pathways* conditioned by the characteristics of the chemical structure. This approach can be a first step toward explaining the metabolic disorders that can be induced by the presence of xenobiotics along with nutrients.

Thus, it becomes possible – in the conditions of deepening the mechanisms of action – to understand the aspects related to the pathobiochemical effects and implicitly the ways of detoxification under normal conditions, e.g., xenobiotic conjugation reactions and adduct formation reactions.

1.4.2   Phases and Pathways of Biotransformation

1.4.2.1   Biotransformation Phases

Like in the case of nutrients metabolization, the biotransformation of the chemical xenobiotics occurs in two distinct phases.

> *Xenobiodegradation* – Represents the first phase of biotransformation in which a "compound X", biologically active or inactive, derived from exogenous intake, is (in general) subject to degradation in the body. In this phase, redox or hydrolysis reactions occur, which may result in a *biologically inactive compound Y* or a *biologically active compound Y'* with action on tissues and organs.

Xenobiosynthesis – It is the second phase in which the resulting compound, i.e. *compound Y* or *compound Y'* evolves into a final *compound Z*. In the case of the intermediate compound Y, there are synthetic reactions (e.g., conjugation, adducts formation) which lead to the final compound Z. The resulting new compound is generally biologically inactive. This compound can be stored or discarded. The general schematic representation of the biotransformation phases including xenobiodegradation (phase I) and xenobiosynthesis (phase II) is shown in Figure 1.7.

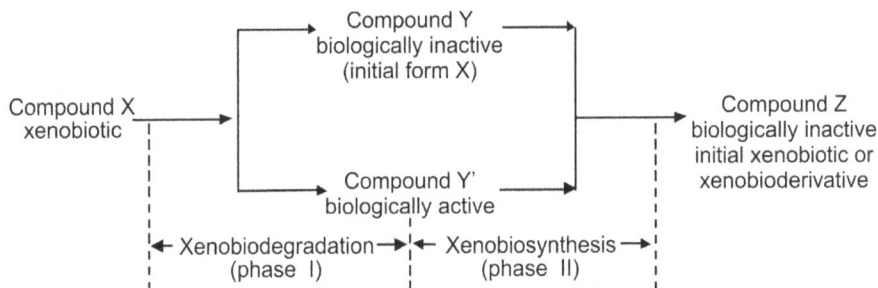

FIGURE 1.7   Characteristic phases for xenobiotic biotransformation.

In xenobiochemistry, the enzymes involved in the biotransformation of xenobiotics, which define a specificity conditioned by the chemical structure of the primary xenobiotic (in this case, "compound X"), are often discussed. Dynamic xenobiochemistry – through a conceptual similarity with nutrient metabolization – considers the sequence of distinct stages: *absorption-distribution-biotransformation-elimination*. It can usually be noted by the acronym ADBE.

1.4.2.2   Biotransformation Pathways

In biotransformation, as shown, the phases of xenobiodegradation and xenobiosynthesis are distinguished. In their case, there are distinct biotransformation pathways by the specificity of the sequence of chemical reactions.

Biotransformation is dependent on the structural characteristics and physical and chemical properties of the xenobiotic, highlighting *specific xenobiochemical pathways* usually called *biotransformation pathways*. The development of interactions is ensured by the presence of "specific enzymes".

The literature on biochemistry currently mentions the existence of *native enzymes* or *metabolic enzymes*, important in metabolic processes. In the field of xenobiochemistry, the notions of *substrate-looking enzymes* or *parametabolic enzymes* (Fishman, 1970; Vineis, 2002), etc. are used. These names – in fact inappropriate – refer to the enzymes involved in the biotransformation of xenobiotics. For these enzymes, the accreditation of the name *xenobioactive enzymes* can be agreed. Investigations in this area are not fully elucidated.

1.5   METABOLIZATION AND BIOTRANSFORMATION: SIMILITUDES AND DISCREPANCIES

As known, the physicochemical transformations in the body occur at the same time both for the nutrients (metabolization) – studied in biochemistry, and for the chemical xenobiotics (biotransformation) – studied in xenobiochemistry.

Studies regarding the processes of the chemical xenobiotics biotransformation led to the idea that they are also carried out in distinct phase, too. In this sense, a comparison between the chemical xenobiotics biotransformation and nutrients metabolization becomes possible (Testa, 1995; Cresteil, 1998; Ioannides, 2002, Wilson and Nicholson, 2003; Faber, 2011; Ritter et al., 2019). Their peculiarities can be discussed separately, interesting the similarities and discrepancies existing with biochemical and xenobiochemical specificity, respectively.

Similarities consist of number of phases – both metabolization of nutrients and biotransformation of chemical xenobiotics occurs in the organism in two phases, i.e. biodegradation and biosynthesis; involvement of enzymes in these phases – the reactions are carried out by biochemical pathways which involve the presence of enzymes, finally resulting the end products.

Discrepancies reside in the nature of compounds (nutrients or chemical xenobiotics); the difference between the biochemical pathways (native, specific); characteristics of the enzymes that generate "native biochemical pathways" and "specific biochemical pathways", the resulted end products (metabolites, respectively xenobioderivatives).

Metabolization of nutrients is studied by static (descriptive) biochemistry and dynamic biochemistry, i.e. material metabolisms and energy metabolism. In their case, the two generally accepted phases are: *catabolism* – which corresponds to the biodegradation phase and *anabolism* – characteristic for the biosynthesis phase.

In biochemistry, the interactions are carried out in *native biochemical pathways* – characteristic for metabolization of nutrients, with the participation of *natural enzymatic systems*. Compounds formed from nutrients as a result of physicochemical transformation processes evolve differently, depending on the composition (Guyton and Hall, 2006; Dekant, 2009; Gârban, 2018). Thus, in the case of nutrients, *intermediate metabolites* are formed. They can evolve into "bioconstituents" that are continuously reconfigured into various tissues or into *circulating metabolites* present in the blood, lymph, cerebrospinal fluid, etc.

Biotransformation of xenobiotics especially of the chemical ones is studied by static (descriptive) xenobiochemistry and dynamic xenobiochemistry (biotransformation). In the case of the *biotransformation of the chemical xenobiotics*, the name *xenobiodegradation* – define the phase of biodegradation of xenobiotics (with the formation of xenobioderivatives), respectively *xenobiosynthesis* – to characterize the biosynthetic phase of compounds derived from xenobiotics and to be eliminated.

Regarding the biotransformation, it can be discussed about *specific biochemical pathways* that ensure the xenobiodegradation/xenobiosynthesis of chemical xenobiotics. These processes involve the presence of *classical (natural) enzyme systems* – with excess synthesis (conditioned by the composition of the xenobiotic) or *atypical enzyme systems* (Wilson and Nicholson, 2003; Ronis and Cunny, 2008).

In the case of chemical xenobiotics, *biologically active* or *biologically inactive* intermediates are formed. In the case of biologically active intermediates, they integrate into the phases of xenobiodegradation and/or xenobiosynthesis, forming *xenobioderivatives*.

Given the confusion that can be made in the case of descriptions of metabolization and biotransformation, it is important to reiterate some terminological issues.

Thus, for example, it is correct to say "catabolism" in the case of nutrients – a situation in which the series of biochemical reactions are the subject of the study of biochemistry and physiology.

However, only the term "xenobiodegradation" is correct to be used in the case of xenobiotics of pharmaceutical interest (e.g., cytostatics, antimicrobial chemotherapeutics) or in the case of xenobiotics of food interest (e.g., organochlorine derivatives, heavy metals, etc.).

To understand the problems related to metabolization and biotransformation, a brief presentation of the access routes of nutrients and food chemical xenobiotics is required, using general notions of anatomy and physiology. For this purpose, "apparata and systems" are discussed, and some aspects regarding the access of biologically active substances in the body are detailed.

1.6　APPARATA AND SYSTEMS

Addressing issues related to the anatomy and physiology of the digestive tract involves a brief reconsideration of general knowledge about the apparata and systems that define the human body as a biological entity.

In this context, reference will be made to each apparatus (lat. Apparatus-device) and to each system (lat. System-system), noting that sometimes there is no strict discrimination of their name, e.g., the digestive system is called in latin anatomical terminology "apparatus digestorius" or "systema digestorium".

According to a general classification, which brings together the various apparata and systems in human anatomy, in histology and in the comparative anatomy of mammals, there is the possibility of grouping apparata and systems in relation to their functional integration. In this context, we can distinguish:

1. The support and mobility apparatus consist of:
   a. the bone system – which is the object of study of Osteology;
   b. the articular system – studied in Arthrology;
   c. the muscular system – which is the object of study of Myology.
2. The matter transport apparatus consists of:
   a. the *cardiovascular system* – at the level of which both the metabolites coming from the nutrients of exogenous origin are circulated – which gain access through food (enteral and/or parenteral), as well as compounds of endogenous origin;
      i. blood gases (e.g. oxygen, carbon dioxide);
      ii. residual metabolic products, e.g., urea, uric acid, creatinine, etc.
   b. *peripheral nervous system* – includes the "networks" of sensory and motor nerves. At this level, some of the neurotransmitters are released.

3. The matter import apparatus consists of:
   a. *the respiratory system* – which ensures, mainly, the supply of oxygen and the elimination of carbon dioxide;
   b. *digestive tract* – consisting of the digestive tract and its accessory glands – apparatus that ensures the exogenous supply of nutrients, subsequently converted, within the metabolic processes, into bioconstituents of the organism.
4. Apparatus for removing matter consists of:
   a. *reno-urinary tract* – consisting of the kidneys and the urinary tract, which in the terminal part have morphological differences depending on sex;
   b. *the genital tract* – which has sex-dependent morphophysiological differences (male gonad – testicle; female gonad – ovary), with specific features of the tract.
5. The body's correlation apparata (systems) is composed of:
   a. *the glandular system* – which, from a morphological point of view, can be divided into two "subsystems". These are:
      i. *the endocrine subsystem* – which targets all the glands with internal secretion at the level of which various *hormones* are produced – biochemical messengers, which ensure connections between various compartments of the body. Currently, endocrinology refers to a "glandular endocrine system" (distinct endocrine glands) and a "diffuse endocrine system" that refers to endocrine secretory cells disseminated in various organs (even at the level of organs of the digestive apparatus);
      ii. *the hematopoietic subsystem* – with an organization that has similarities with the glandular system – participates in the formation of different "blood elements" in morphological and physiological relation (i.e., red blood cells-erythrocytes; white blood cells-leukocytes; platelets-thrombocytes) shows localized areas for their biogenesis, e.g., hematopoiesis in the bone marrow and spleen.
   b. *central nervous system* – includes the brain and spinal cord. They ensure the functional integration of the human body through neurotransmitters formed at this level. Also, on specific (neuronal) pathways, it ensures a true correlation of the physiological activity of different apparatus and systems.

1.7 ACCESS OF BIOLOGICALLY ACTIVE SUBSTANCES INTO THE BODY

1.7.1 BASIC CONCEPTS

Biologically active substances access into the body similarly to nutrients, environmental xenobiotics, pharmaceuticals and cosmetics. To understand the biochemical mechanisms, generalities will be presented about the gastrointestinal, pulmonary and cutaneous routes.

The access of various substances in the body is conditioned by the level of exposure and the rate of their passage into the systemic blood circulation (Senna and Legrand, 2004; Kim and Nylander-French, 2009).

For this reason, it is important to evaluate the aspects of "biochemical kinetics", which involve knowing the details of the distribution, retention and elimination of various compounds that reach the tissues.

For the evaluation of the action of metabolites, respectively of residual xenobio-derivatives, the notions of "biochemical kinetics" and "xenobiochemical kinetics" are used in biochemistry and xenobiochemistry (subsequent field of the previous one). Specialized treatises that address the issue of xenobiotics of pharmaceutical interest refer to pharmacokinetics (pharmaceutical kinetics).

Similarly, in the case of chemical xenobiotics of strictly toxicological interest, generically referred to as "biocides" (e.g., biocides used as pesticides or as bactericides), the expression "toxicokinetic" is used. In their case, the pathobiochemical limits (which lead to lethal thresholds) are also of interest, which are discussed in *molecular toxicology*.

Accurate approach to biochemical kinetics for various compounds that enter the body and the study of their specific mechanisms can lead to risk prevention. Biochemical kinetics is concerned with the access of substances into the *systemic* bloodstream – a process controlled, in the first instance, by the epithelium lining the organs and which is an effective barrier.

In this situation, examples are: epithelial cells of the gastrointestinal tract (GIT) – important for food intake (exposure by ingestion) and epithelial cells of the skin – important for cosmetics (exposure by skin absorption). For xenobiotics present in the air, the transit takes place through the lungs during the gas exchange at the level of the alveoli (exposure by inhalation).

1.7.2 Gastrointestinal Route

1.7.2.1 General Data

Given the fact that various nutrients enter the body, mainly through the digestive tract (including most biologically active substances), aspects of anatomy and physiology related to the digestive tract are presented in more detail.

1.7.2.1.1 Anatomic Aspects

1. *The digestive tract* – It has the characteristics of a 10–12 m long duct, which – at the extremities – communicates with the external environment. This tract consists of various anatomical segments morphologically and physiologically differentiated.

   From an anatomical point of view, there are: the oral cavity, the pharynx, the esophagus, the stomach, the small intestine (duodenum, jejunum and ileum), the large intestine (colon, cecum, and rectum) and the anus.

   From a morphophysiological point of view, the first two segments (the oral cavity and the pharynx) play a mixed role, serving, at the same time, as a respiratory and digestive tract. The following segments are limited to alimentary channel quality (*canalis alimentarius*).

The wall of the digestive tract can be considered as a "barrier" between the external environment (in this case, the lumen of the tract) and the internal environment of the body in which nutrients and, in general, various chemicals access as a result of absorption processes.

From an *anatomo-physiological* perspective, from the point of view of digestion, the digestive tract highlights three parts (sequences), namely, (i) the *ingestive part*: the oral cavity, pharynx and esophagus – at which food is transported; (ii) the *digestive part:* stomach, small intestine – at which food is digested and absorbed to the greatest extent; (iii) *ejective part*: large intestine, anus – at the level of which the digestion residues accumulate and the elimination of unabsorbed residues takes place.

2. *The accessory glands of the digestive tract* – They are represented by organs arranged in various areas of the tract as glands, whose secretions contribute to the processes of nutrient metabolization and biotransformation of xenobiotics. The term *accessory glands* is often used in the literature. The *liver* and *pancreas* are studied as accessory glands. General anatomical and physiological aspects of these organs are mentioned.

The pancreas – It is located in the curved area of the duodenum. It is a mixed gland (amphicrine). *Exocrine secretion* – pancreatic juice discharged (in two channels) into the duodenum. *Endocrine secretion* – is represented by two hormones: insulin (produced in β cells) – with hypoglycemic action and glucagon (produced in α cells) – with hyperglycemic action.

The liver –It is the accessory gland at which "bile secretion" occurs. This secretion contains organic compounds (bile acids, cholesterol, phospholipids, etc.) and inorganic compounds (Na$^+$, K$^+$, Ca^{2+}, HCO$^-$, Cl$^-$ ions, etc.). All of these compounds are involved in digestion.

1.7.2.1.2 Physiological Aspects

There are three stages in the physiology of digestion: (i) the *oro-pharyngo-esophageal stage*; (ii) the *gastric stage* and (iii) the *intestinal stage*.

A diagrammatic representation of the stages of digestion and absorption is shown in Figure 1.8. The duration of the physiological processes in various stages/sub-stages is also presented separately.

1. *The oro-pharyngo-esophageal* stage – The first act of interest in this stage of digestion is the prehension of food by the lips. Prehension is a voluntary act, controlled by the central nervous system (CNS) and coordinated by the movements of the upper jaw (maxilla) and lower jaw (mandible).

Descriptively, digestion in the oral cavity is limited to three physiologically distinct processes: (i) *salivation*; (ii) *mastication* and (iii) *swallowing*). At this stage, the "food bowl" is formed. In the first stage, there are two stages in physiology: *buccal* with the duration of approx. 1 minute; *esophageal* lasting 1–5 minutes.

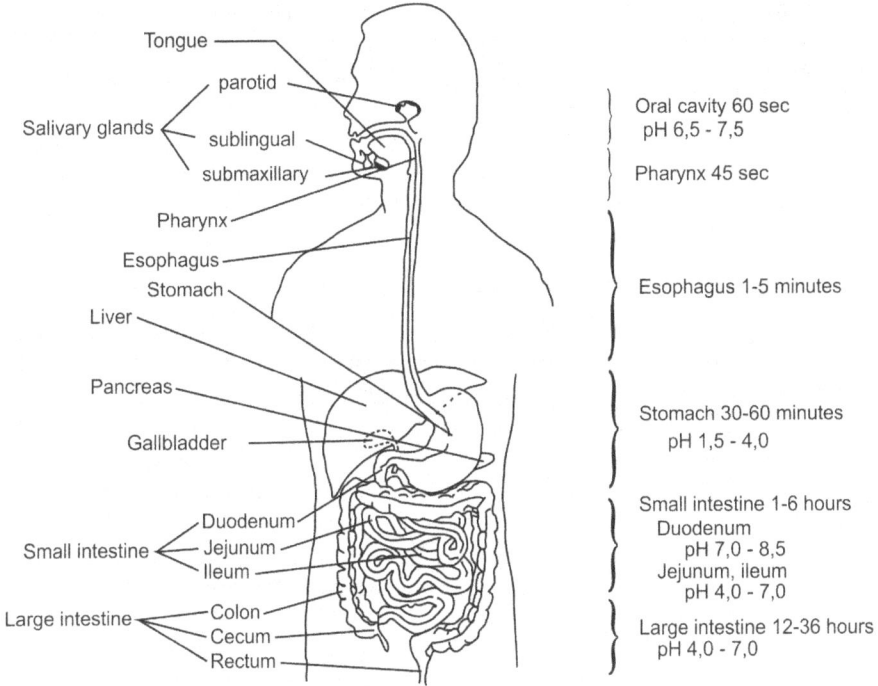

FIGURE 1.8 Duration of food transit through the digestive tract and specific pH level (diagrammatic representation).

2. *Gastric stage* – The food bowl that reaches the stomach is subjected to transformation processes that are performed in physical and physicochemical terms. Physicochemical changes are the result of the secretory and motor activity of the stomach (see Figure 1.8).

   Two aspects are discussed: (i) *the action of gastric secretion* and (ii) *the effect of gastric motility*. Reaching the stomach by swallowing, the "food bowl" is impregnated with gastric juice and gradually disintegrated. This is done through a prolonged brewing, which forms a semi-liquid mass, generically called "gastric chyme". The second (gastric) stage of digestion lasts 30–60 minutes.

   At the level of the stomach, there is also the gastric mucus which has the role of protecting the mucosa from the harmful effects that can be manifested by autodigestion in which pepsin and hydrochloric acid can participate.

3. *The intestinal stage* – Digestion is completed, in part, in the small intestine. From the stomach, the gastric chyme leaves with an acidic pH. Neutralization is achieved in the small intestine by the action of secretions: (i) *pancreatic*; (ii) *biliary* and (iii) *intestinal*.

   At this stage, the digestion is continued and the absorption of small molecules resulting from the nutrients present in the food is achieved – to a large extent. Physiologically, it is known that there are distinct digestion times: for the small intestine it is 1–6 hours; for the large intestine it is 12–36 hours. The duration of these periods can be correlated with the nutrients present in the diet (e.g., the presence of dietary fiber decreases digestion time, especially in the large intestine).

With regard to the action of the intestinal juice, it can be concluded that it ensures the continuity of the digestion of proteins, carbohydrates and lipids and the maintenance of the intraluminal osmolarity in the conditions of the absorption-desorption relation.

1.7.2.2 Peculiarities of the Transit

Nutrients and chemical xenobiotics in food enter the GIT through food intake. Obviously, the biologically active substances are in the same situation. Most ingested compounds are absorbed through the lining of the stomach and small intestine (West, 1991; Staats, 1991; Devlin, 1992; Johnson, 2006).

There are four layers in the stomach, namely, mucosa, submucosa, muscle and serosa. The first functional barrier in the stomach is the mucosa. At the level of the small intestine, a similar general structure is observed; however, there are differences, highlighted histologically and specific to the functionality of these segments of the GIT. The small intestine is approx. 5 m and has a large surface through which nutrients and chemical xenobiotics access.

At the gastrointestinal level, non-ionized substances (with lipophilic properties) are absorbed through the gastric lining and ionized substances (hydrophilic) are absorbed predominantly through the intestinal mucosa.

A model that illustrates the peculiarities of the absorption mechanism – valid for nutrients (implicitly biologically active substances) and chemical xenobiotics – considers the transit through the gastrointestinal tissue. The description of the exchange rate of the level of substances in the gastrointestinal tissue – based on mathematical relations – takes into account the volume of the substance in the gastrointestinal tissue – denoted V_g and the concentration of the substance in the gastrointestinal tissue – denoted C_g, depending on time (t). Thus, it can be written:

$$V_g \frac{dC_g}{dt} = Q_g \left(C_a - \frac{C_g}{P} \right) + kA_s \qquad (1.1)$$

where
  Q_g - rate of infusion (of the substance) in gastrointestinal tissue
  P - partition coefficient between gastrointestinal tissue and blood
  C_a - the concentration of the substance in the arterial blood
  k - the general absorption constant
  A_s - the level of the substance in the gastrointestinal tissue

The issue of nutrients and chemical xenobiotics (of food, pharmaceutical or agrobiological interest) uptake is discussed in connection with the subsequent stages. In the case of nutrients, the stages are: absorption-distribution-metabolization-elimination (ADME), corresponding to biochemical kinetics. For chemical xenobiotics, the stages are: absorption-distribution-biotransformation-elimination (ADBE). These last ones are specific for xenobiochemical kinetics, i.e. chemical xenobiotics as food ingredients, others used as medicines. In the specialized literature, these problems have been deepened especially in the case of drugs by pharmacokinetic studies.

Regarding this transport, it is known that for various chemicals, transmembrane trafficking is easier in the case of low molecular weight and lipophilic substances (Kim and Nylander-French, 2009). Passive transmembrane transport is performed on such substances.

In contrast, there are substances in which active transmembrane transport with energy intake is performed. In the case of the GIT, there are specialized transport systems in which carrier peptides, organic anions, organic cations occur and assure the transmembrane transport.

1.7.3   PULMONARY ROUTE

1.7.3.1   General Data

The lungs, as organs of the respiratory system, provide a large area of contact with the air and, in general, with the gases in the environment. Overall, the role of the lungs is reduced to: (i) providing the necessary oxygen from the inspired air (potentially accompanied by gaseous xenobiotics); (ii) elimination of volatile residues and toxins (e.g., carbon dioxide); (iii) defense against harmful substances (e.g., dust) and microorganisms (e.g., viruses and bacteria).

Of the total volume of the lung approx. 10% is the lung tissue itself, the rest being air and blood. From a functional point of view in the lungs the following can be distinguished: airways (gas conduction) and specific regions of gas exchange.

The gas exchange takes place at the alveolar level. The alveoli extend over an extremely large area (about 80 m²). The extremely thin alveolar wall allows quick and efficient gas exchange. Pulmonary activity is performed in two stages: *inspiration* – a process during which oxygen in the air (which can sometimes contain volatile xenobiotics) diffuses through the alveolar wall into the interstitial space and from there in the blood; *expiration* – the process by which carbon dioxide and other gaseous waste products diffuse in the opposite direction (from the blood into the interstitial space, alveoli and then the airways). The gas exchange (in-expiration) takes about 0.25 second.

1.7.3.2   Peculiarities of the Transit

Inhalation of certain pharmaceutical products (e.g., tobramycin antibiotic; long-acting bronchodilators for chronic obstructive pulmonary diseases), cosmetics (e.g., essential oils from perfumes, creams, etc.), various chemical xenobiotics (e.g., organochlorine compounds in occupational accidents) may occur during inhalation-expiration processes.

To evaluate inhalation exposure to xenobiotics (including pharmaceuticals), Ramsey and Andersen (1984) designed an experimental animal model for styrene in rats (similar to humans). This model is suitable for the assessment of inhalation exposure to *volatile organic compounds* (VOCs).

According to this model, inhalation, after the alveolar-blood transit, aims for the diffusion in tissues, e.g., liver, muscle, adipose tissue and easily infused tissues. Access according to the Ramsey-Andersen model takes place via the lungs, and the diffused volatile organic compounds are dependent on tissue volume (V_t) and VOC concentration in tissue (C_t), achieved over time (t). The definition of the *input-output* rate (i.e., input-output) in a certain tissue compartment is defined by the relation (Eq. 1.2):

$$V_t \frac{dC_t}{dt} = Q_t(C_a - C_v) \qquad (1.2)$$

where

    Q_t - tissue blood perfusion rate
    C_a- VOC concentration in arterial blood
    C_v - VOC concentration in venous blood.

According to this model, a difference can be made between the concentration of the xenobiotic in the arterial blood (afferent circulation) and the venous blood (efferent circulation), coming from various tissues and organs.

This observation is important because it leaves open the question of the difference between the rate of transit and, indirectly, of metabolization/biotransformation in various tissues and organs. This rate is dependent on the nature of the inhaled substance.

The approach to the problem of respiratory transit in this volume is of particular interest, given that there are volatile chemical compounds (odorants) in the air that enter with the air in the lungs. Such compounds may interfere with nutrient metabolization.

Knowing these problems and initiating preventive measures are of importance for human health and the protection of the environment, in general.

1.7.4   CUTANEOUS ROUTE

1.7.4.1   General Data

The skin is considered a complex organ, with a specific membranous covering of the body. In physiology, the skin has multiple roles, e.g., protection, thermoregulation, excretion and reception (tactile, thermal, and algetic).

As a whole, it is a barrier to the skin absorption of various substances (natural and chemical xenobiotics). It is also considered that the skin is involved in the biotransformation of xenobiotics as well as in the dynamic response of the immune system (Kim and Nylander-French, 2009).

Numerous chemicals, represented by certain pharmaceuticals and compounds of agrobiological interest, can partially or completely transit the skin and can interact with dermal macromolecules or components of the immune system. The skin presents itself as a target organ for chemicals from the environment and occupational exposures (usually called "occupational"). Dermal exposure to chemical xenobiotics and their dermal absorption are, more than once, the cause of associated diseases.

1.7.4.2   Peculiarities of the Transit

In this book, the issue of skin transit of biologically active substances and, in general, of various chemical compounds (including chemical xenobiotics) is related to their retention in plant and animal tissues. Such compounds affect the composition of food and subsequently accompanying the nutrients.

Also, in relation to the presence of xenobiotics as well as chemicals encountered in occupational exposures, references lead to the careful handling of products for phytosanitary, sanitary (veterinary and human), forestry use.

Looking at the aspects of the action of biologically active substances and/or xenobiotics present in cosmetics, it can be noted that they can be absorbed into the skin producing "local effects" or can transit the skin, producing "systemic effects" (Potts et al., 1992).

Regarding skin transit, it is known that the epidermis functions more as a barrier; at the level of the dermis, there are already blood capillaries, nerve endings (free or encapsulated), hair follicles, glands (sebaceous and sweat), and at the level of hypodermis, vascularization and innervation are more intense.

In the case of cutaneous absorption of various biologically active substances from pharmaceuticals, cosmetic products and other chemicals, the problem of transit and material flow (J) can be calculated according to the law given by Adolf Fich, which states the relationship (Eq. 1.3):

$$J = -D \frac{dC}{dx} \tag{1.3}$$

where

D - diffusion coefficient (area/time)

dC - concentration gradient when crossing the membrane (mass/volume)

dx - membrane thickness

A model limited to dermal absorption was designed for dermal absorption (McCarley and Bunge, 2001). According to this mathematical model of absorption, it is considered that the epidermis has a *horny* part – in which there are corneocytes (loaded with keratin) and a *viable* part – corresponding to the basal layer (germinative), in which primary keratinocytes including melanocytes, immunocompetent cells (Langerhans) are found, etc.

Starting from this mathematical model of the transit of a substance, aspects of skin absorption can be calculated.

For this purpose, it is considered the volume of the skin tissue (V_p), the concentration of the chemical compound in the skin tissue (C_p), evaluated according to time (t). The expression is made by relation (Eq. 1.4):

$$V_p = \frac{dC_p}{dt} = Q_p \left(-\frac{C_p}{P_{ps}} \right) + K_p A_e \left(C_d - \frac{C_p}{P_{pa}} \right) \tag{1.4}$$

where

Q_p - the rate of infusion of the studied substance

P_{ps} - partition coefficient of the substance in the skin/blood ratio

K_p - permeability coefficient

A_e - skin exposure area

C_d - dermal concentration

P_{pa} - the partition coefficient of the substance in the exposed skin/area ratio

Data on the access of biologically active substances of food, pharmaceutical and agrobiological interest (i.e., plant growth bioregulators) and, in general, on the access of chemical xenobiotics into the human body is a precursor stage to their integration into the "internal environment".

After access to the internal environment, biologically active substances participate in specific metabolization/biotransformation reactions. To facilitate the understanding of the problems related to the transit of biologically active substances (various substances), which produce beneficial effects or side effects, the issue of biochemical barriers is briefly discussed.

1.8   BARRIERS IN BIOCHEMISTRY

In the general context, the problem of the biochemical barriers in the body is also presented. All of this facilitates the understanding of the sequence of ADME processes – in the case of nutrients, compared to the ADBE processes – in the case of chemical xenobiotics.

The concept of "barrier" is used in biochemistry, physiology, nutrition, pharmacology, xenobiochemistry, pathology, etc. This concept refers to the impact between the products resulting from the metabolization of nutrients (implicitly biologically active substances) and the biotransformation of xenobiotics. This concept also takes into account the fact that there is an impact of chemical xenobiotics present in food, chemical xenobiotics as pharmaceuticals as well as strictly toxic chemical xenobiotics on tissues and cells.

At present, there is a certain terminological extension regarding the concept of "barrier". In this sense, the following terms are accredited: *biochemical barrier* – defined by the specifics of metabolic processes (catabolic-anabolic), stability in the biological environment (conditioned by pH, ionic strength, solubility, temperature, etc.), competitive interactions with other products of exogenous and/or endogenous origin, etc.; *physiological barrier* – characterized by data on digestion, absorption, distribution (transport, storage), elimination; *psychological barrier* – signaled (in nutrition and xenobiochemistry) by the existence of organoleptic changes, which reduce or cancel the edibility, e.g., change in color, taste, smell of food, etc.; *economic barrier* – understood as the high-cost price of a particular food or drug, often caused by expensive technologies for processing.

From a metabolic point of view, the biochemical barriers of the organism are of particular interest. In this sense, their biochemical particularities are approached, for which it is necessary to define the chemical compounds in food; the morphophysiological particularities for which it is necessary to know (even briefly) the morphological structures and the related physiological processes; varied biochemical interactions and, inherently, pathobiochemical and pathophysiological consequences (Richterich, 1971; Bell et al., 1976; Edelman, 1988; Kjeldsberg and Knight, 1993; Greenwood, 1995; Gârban, 1999; Altug, 2002; Kaplan and Pesce, 2010; Voet and Voet, 2011).

Natural biochemical barriers are influenced by: the enzymatic systems that participate in the specific reactions of certain biochemical pathways (anabolic and/or catabolic); pH of the environment; bioavailability; chronobiochemical specificity, etc.

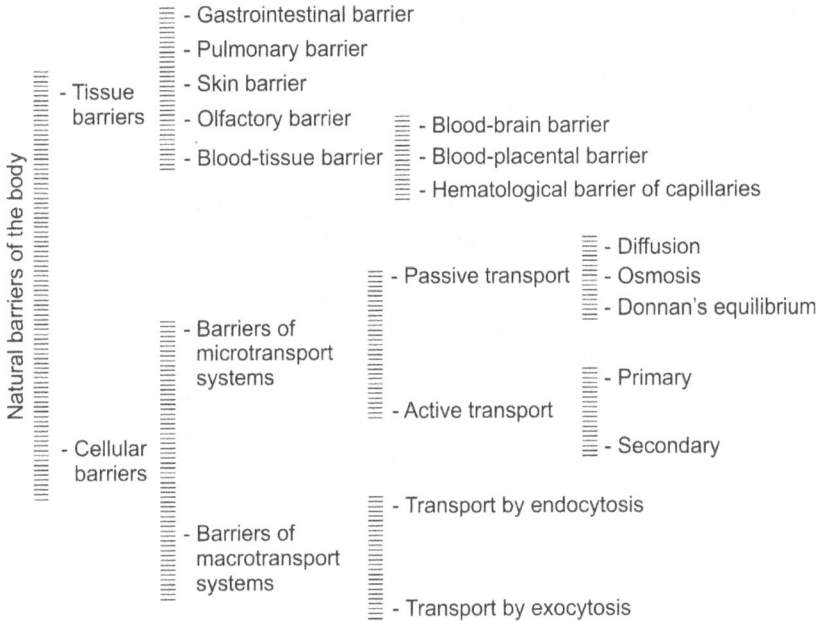

FIGURE 1.9 Barriers of the body (Gârban, 2017).

The properties of these barriers are validated in relation to: nutrients, chemical xenobiotics of food interest, pharmaceuticals, cosmetics and strictly toxic compounds. With reference to barriers, the following will be discussed in more detail: (i) the *tissue barriers* of the organism and (ii) the *membrane barriers* of the cells.

In Figure 1.9, various types of important barriers in biochemistry/pathobiochemistry, respectively, in physiology/pathophysiology are presented.

So, in the body, there are *tissue barriers* (in the digestive tract, respiratory, integumentary and circulatory systems) and *membrane barriers* (in cells and even in cellular organs).

1.8.1 Tissue Barriers of the Body

This chapter discusses (i) the gastrointestinal barrier represented by the wall of the GIT (mainly the intestinal wall), the blood capillaries in the intestinal villi; (ii) pulmonary barrier; (iii) skin barrier; (iv) olfactory barrier and (v) blood-tissue barriers (i.e., blood-brain barrier, blood-placental barrier and hematological barrier of capillaries).

1.8.1.1 Gastrointestinal Barrier

Ingested food and drugs (some with xenobiotic specificity) administered orally are subject to transit through this barrier. During transit, biodegradation takes place for small molecule compounds, termed generically "metabolites" – in the case of nutrient metabolization, respectively "residual xenobioderivatives" – in the case of chemical xenobiotic biotransformation. Their absorption continues, a process that consists of the transit of the wall of the digestive tract or – more rigorously expressed – of the gastrointestinal barrier.

Transit through the GIT is preferred in the small intestine and is followed by entry into the venous blood system, then the liver, heart and finally the arterial circulation in all organs and tissues of the body.

Absorption in the small intestine, especially in the proximal area (i.e., duodenum), is more intense targeting small molecules of carbohydrates, lipids, proteins, mineral compounds, vitamins, etc. At the intestine level, the transit is made through the intestinal villi, which give a large surface area for absorption.

The properties of the gastrointestinal barrier reveal changes in pathological conditions. Absorption is deficient in pathological conditions caused by enzymatic deficiencies, mucosal lesions, changes in the intestinal flora, the presence of xenobiotics with toxicogen potential, etc., (Sittig, 1985; Anderson, 1986; Fennema, 1996; Fraser et al., 2006) and even some xenobiotics of food interest (e.g., food additives) and/or of pharmaceutical interest (e.g., chemotherapeutic) – see Luck (1980), Salas et al. (1990, a.o.).

1.8.1.2 Pulmonary Barrier

As a whole, the pulmonary barrier consists of: (i) the upper respiratory tract – represented by the nasal cavities, pharynx and trachea; (ii) lower respiratory tract – represented by the bronchi and lungs. At the pulmonary level, the bronchi become arboreal forming secondary bronchioles whose development and branching lead to the lung lobes. At the terminal level, their surface is covered with an extremely thin sheet that constitutes the wall of the pulmonary alveolus. In physiology, the exchange of substances takes place in the alveolar endothelium. For example, the retention of atmospheric oxygen later dissolved in the blood and the elimination of carbon dioxide are of interest. The first stage is consumed in inspiration, and the second stage is consumed in expiration. Also, the lung barrier affects the access of volatile biologically active substances (of pharmacologic and cosmetic interest).

A more detailed look at the mechanisms involved in the presence of volatile substances at the alveolar level and, obviously, their transit, also requires a biochemical and pathobiochemical approach. The problem of the access of xenobiotics through the lung barrier concerns not only odoriferous compounds from cosmetic products but also xenobiotics of pharmaceutical interest in the case of the use of chemotherapeutics administered in the form of aerosols.

This approach is also important in the case of environmental chemical xenobiotics as products existing in nature or resulting from the purge of residual toxic substances from industry.

1.8.1.3 Olfactory Barrier

The olfactory barrier can be considered as a "tissue barrier" in which the olfactory mucosa participates as an essential formation (surface approx. 250 mm^2). The olfactory barrier is located in the upper segment of the nasal fossae and, to a lesser degree, on the middle segment in the zone of the cribriform plate of the ethmoid bone.

The olfactory analyzer as a whole is represented by three distinct sequences:

a. *olfactory mucosa* – which is the peripheral part of the olfactory analyzer and located in the upper region of the nostrils (consisting of epithelium and chorion);
b. *conduction path* – formed by two groups of neurons that ensure the transmission of olfactory excitation (information) from olfactory cells to the brain;
c. *cortical projection* – transmitted through nerve fibers to the thalamus and then to the cerebral cortex.
d. Obviously, the olfactory barrier targets odorous substances – certain pharmaceuticals and cosmetic ingredients.

1.8.1.4   Cutaneous Barrier

The transit of various chemicals through the "cutaneous barrier" is largely conditioned by the stratum corneum of the epidermis and to a lesser extent by the dermal and hypodermic layers. For these reasons, the transit from the outside to the blood-irrigated layers of the hypodermis is a "resorption process" (Heymann, 2003).

The resorption process as a whole includes three distinct stages:

a. *penetration* – involves the entry of the substance into the stratum corneum of the epidermis;
b. *permeabilization* – involves the transit from the stratum corneum to the underlying layers;
c. *actual resorption* – resides in the passage of the substance from the skin level in the vascular system (lymphatic and/or blood) from where it can exert action on other tissue compartments.

The physicochemical resorption process is a process of *passive diffusion* – which occurs transepidermally and transfollicularly or of *active diffusion* – which occurs through ions and with the help of promoters. The substances in which the *cosmetic ingredients* are dissolved influence the resorption differently. In the case of vegetable oils, they have shown to increase resorption. Alcohol/water cosmetic solutions also facilitate resorption.

1.8.1.5   Blood-Tissue Barriers

From the bloodstream, nutrients of exogenous origin, that reached the body through enteral and/or parenteral nutrition, as well as some chemical xenobiotics (e.g., drugs, food additives, contaminants, etc.) transit, in part, to the tissues by overcoming the blood-tissue barrier (Forth et al., 1977; Wurtman and Wurtman, 1983; Khera, 1987; Guyton and Hall, 2006).

In fact, one can discuss a "barrier group" with distinct biochemical characteristics. Thus, in terms of morphophysiology, biochemistry, biophysics and pharmacology, one can distinguish: (i) *blood-brain barrier*; (ii) *blood-placental barrier* and (iii) *hematological barrier of capillaries*. Particular aspects of them are discussed in the following sections.

1.8.1.5.1 Blood-Brain Barrier

The blood-brain barrier stands between the blood and the nerve tissue in the brain, in fact, with the entire CNS. Morphophysiologically, this barrier reveals the existence of specialized microstructures that limit the access of metabolites (Gruetter et al., 1992; Kjeldsberg and Knight, 1993) and other substances (e.g., xenobiotics and/or xenobiotic derivatives resulting from biotransformation), from blood plasma to neurons in the brain (Wurtman and Wurtman, 1983).

The access of metabolites and, in general, of various substances to the CNS (brain and spinal cord) is achieved in the cerebrospinal fluid (CSF) or directly in the brain.

1.8.1.5.2 Blood-Placental Barrier

At the level of the blood-placental barrier, the transit of various metabolites (exogenous and/or endogenous origin) and (in some cases) of drugs takes place. The transit is made from the maternal organism to the conception product (embryo in the early stage, respectively fetus in the advanced stage of pregnancy). It is self-evident that the nutrients that cross the blood-placental barrier as various metabolites contribute to the morphogenesis of conception product until parturition.

No less important is the study of this barrier in the presence of chemical xenobiotics of food, pharmaceutical or toxicological interest. It is known that in such situations there is an imminent risk for the following effects: *thanatogenic* (gr. Thanatos-God of death) characterized by embryonic resorption (in the early stg of gestation) and fetal mortality (in the late stage);

1.8.1.5.3 Hematological Barrier of Capillaries

The transit of substances from the blood capillaries to the tissues has certain characteristics. Thus, it can be stated that there are capillary hematological barriers at the tissue level, including various tissues, e.g., liver, muscle, etc. At the level of the capillaries, there are intracellular spaces with diameters of 4–6 nm crossed by small molecule metabolites from various carbohydrates, lipids, proteins, minerals, vitamins, biologically active substances, etc.

Chemical xenobiotics such as pharmaceuticals as well as compounds with toxicogene potential resulted from the biodegradation of various xenobiotics considered as environmental contaminants present in food, water, air can also transit (Léger and Béréziat, 1989; Rhoades and Pflanzer, 1992).

Some pharmacologicals – important fact to know in therapeutics – may alter the capillary transit. Endogenous substances, like estrogenic steroids, may cause a similar situation.

1.8.2 MEMBRANE BARRIERS OF CELLS

Transport through cell membranes is performed, depending on the nature and size of the compounds. There are two types of barriers, named after transportation systems. These have been accredited under the names of: (i) *barriers of microtransport systems* and (ii) *barriers of macrotransport systems*. General information on them is given below.

In biochemistry, molecular biology and physiology, the following types of biological membranes are discussed: (i) membranes of cellular organelles; (ii) typical cell membranes; (iii) tissue membranes specific to epithelial tissue, e.g., alveolar endothelium, renal epithelium, digestive mucosa, etc. and (iv) specialized tissue membranes, e.g., synaptic membranes, myelin membranes, etc. The integration of the micro- and macrotransport systems in the morphophysiological ensemble of the cell ensures the maintenance of the biochemical homeostasis of the intracellular environment.

1.8.2.1   Barriers of Microtransport Systems

Microtransport systems ensure the transmembrane transit of small molecules, e.g., glucose, urea, etc. and ions, e.g., Na^+, K^+, Ca^{2+}, Mg^{2+}, etc. by two types of transport: (i) *passive transport* and (ii) *active transport.*

1.8.2.1.1   Passive Transport

Passive transmembrane transport is performed in the direction of the concentration gradient (also called "electrochemical gradient") from higher concentration to lower concentration. This type of transport is of interest to ions and small molecules and is based on distinct physicochemical phenomena with specific mechanisms: (i) *diffusion (free/mediated)*; (ii) *osmosis* and (iii) *Donnan equilibrium.*

Changes in the gradients of molecular and/or ionic concentration are at the origin of these transport mechanisms. Experiments performed with semipermeable membranes (e.g., cellophane sheets) in order to separate the protein-NaCl mixture allowed the understanding and explanation of the dialysis process.

The membrane used, which is permeable to small molecules (e.g., NaCl) and impermeable to proteins, allowing the explanation of the physical phenomenon known as "passive transport". This type of transport, characteristic of certain substances present in cells, is studied accurately in biochemistry and physiology.

1.8.2.1.1.1   Transport by Diffusion   The diffusion mechanism – as a physicochemical phenomenon – lies in the motion, in solution, of the particles represented by ions and molecules.

In the case of diffusion of particles through a porous membrane (*in vitro*), as in the case of diffusion through the cell membrane (*in vivo*), it is possible to apply Fick's law. This law shows that the number of moles (n) of the substance diffusing in the unit of time (t) is dependent on the diffusion coefficient (D), the unit area (S) and the concentration gradient (negative – the concentration is decreasing). Fick's law can be reproduced by the relation (Eq. 1. 5):

$$\frac{dn}{dt} = D - S\left(-\frac{dc}{dx}\right) \qquad (1.5)$$

where
   D - diffusion coefficient
   S - the surface (section) through which the diffusion takes place
   dc/dx - concentration gradient

The concentration gradient (dc/dx) is made explicit – as shown – by changing the relative concentration with the change of distance. The negative sign signifies the decrease of the concentration with the transition from a higher concentration to a lower concentration through the transmembrane transit. Diffusion – as a passive microtransport system – is achieved through two mechanisms: (i) *free (simple) diffusion*; (ii) *mediated (facilitated) diffusion*.

i. *Free diffusion* – The mechanism of this physicochemical process involves the transmembrane transport of water, small molecule metabolites and ion transport. Oxygen (O_2) is also transported to the cell.

   In the free diffusion through the "double lipid layer" of the cell membrane, the ions are transported from a higher concentration to a lower concentration. The transport velocity (v) is dependent on the difference in concentration (Δc), which is expressed as a function, by the relation (Eq. 1. 6):

$$v = f(\Delta c) \qquad (1.6)$$

Free (simple) diffusion has some important general characteristics in membrane transit:
   - absence of saturation – condition in which the diffusion is dependent on the difference in concentration (concentration gradient or electrochemical gradient of ions), on the size of the molecules and on the temperature;
   - the slowness of the transit, because the dissolved molecules require time for integration, displacement and release from the phospholipid double layer before passing to the other side of the cell;
   - the existence of hydrophobic (nonpolar) molecules or, if they are hydrophilic (polar), the overall charge must be zero considering that the double phospholipid layer is "impermeable" for the simple diffusion of charged molecules, due to the high degree of hydration. Transmembrane transit by simple diffusion is possible for gases (e.g., atmospheric oxygen), volatile biologically active substances (of pharmaceutic, cosmetic and food interest), water, urea, ethanol, etc.
ii. *Mediated diffusion* – In the case of this mechanism, there are some differences: – transport through "ion channels"; – transport by means of "transport proteins" (also called "carrier proteins").

Transport through mediated diffusion is also done through some *molecular canalicular formations* consisting of protein molecules with a transmembrane arrangement.

   The discovery of *ion channels* and *water channels* is of medical importance. Many diseases are attributed to the difficulty of operating these channels. Consequently, it also retains attention in terms of therapeutic means (dietary and medicinal).

   Transport via *ion channels*. In the case of cations and anions, there is a transport mechanism based on the existence of specific proteins – integrated into biomembranes – that generate the so-called "ion channels".

   Ion channels appear as membrane formations consisting of "channel proteins" – specific proteins that intervene in the transit by diffusion for various ions (e.g., Na^+,

K⁺, Ca²⁺, Mg²⁺, Cl⁻, etc.). Regarding ion channels, it has been admitted since 1970, that they have the ability to select ions. An "ion filter" was presumed to exist. Research has made difficult progress.

Conclusive results were obtained by X-ray crystallography, which allowed the visualization of the structure for the ion channel.

There are specific operators (chemical or electrical mediators) for the transit of ion channels based on the potential difference. They can alter membrane permeability and selectivity (Gârban, 1999).

Transport by *carrier proteins*. In the case of this transport mechanism, a certain substance intervenes, which transits the membrane in the presence and under the action of a "transport system", which is represented by "transport proteins" located at the biomembrane level.

Transport takes place in a concentration gradient, straight from a higher concentration to a lower concentration in the presence of "carrier proteins". In older textbooks, they were called "permeates".

The carrier proteins bind to the substance to be transported, taking it from one side of the membrane and releasing it from the other side. In this way, for example, glucose is transported.

1.8.2.1.1.2 Transport by Osmosis   Osmosis is characterized by the phenomenon of transmembrane diffusion of water molecules. In this sense, osmosis can be considered as a "water-dependent diffusion". The passage of water through the membrane barrier is ensured by the difference in concentration of the water-soluble compounds existing between the intracellular environment and the extracellular environment.

Water channels – They are membrane formations made up of *canalicular protein entities* that ensure the passive transport of water and small organic molecules (e.g., urea). Research on water channels, known generically as *porins* (water-selective pores), has expanded knowledge about porins and the transit of water through biological membranes. The study of osmosis defined the notion of osmotic concentration expressed by specific units called osmole (abbreviated Osm).

The physicochemical properties are also evaluated by osmotic pressure in biological systems. In a given environment, in vitro or in vivo, the osmotic pressure is defined by the hydrolytic peculiarities of the solvents being directly proportional to their concentration.

The value of the *osmotic pressure* (π) is given by a relatively simple relation, which takes into account the concentration (c) of the dissolved substances and appeals to the van't Hoff law. The usual expression is (Eq. 1.7):

$$\pi = R.T.c \tag{1.7}$$

where
  R - universal ideal gas constant (R = 8310 J/kmol K)
  T - absolute temperature (degrees K)
  c - concentration (in M/L or mM/L)

If T is constant – so there is no change in temperature – osmotic pressure can be directly related to osmolarity. The osmolarity of biological solutions and/or liquids can be determined directly by the use of a freezing point osmometer.

Most osmometers are calibrated so that they can return the value in Osm or mOsm on reading. The evaluation of the osmolarity in relation to the isosmotic (normal) states is made in the situations in which the osmolarity increases – hyperosmolarity, or decreases – hypoosmolarity.

In biochemistry and physiology, the osmotic pressure of human blood plasma is explored, which is altered by nutrients within physiological limits. Its expression is done in mOsm/L. Thus, it was found that the normal osmotic pressure of human plasma is 310 mOsm/L, being evaluated in relation to the cryoscopic point of −0.56°C.

Substances present in the blood develop a partial (independent) osmotic pressure, which explains the high value of pressure in plasma and, in general, in biological fluids.

The evaluation of osmotic pressure takes into account the fact that part of it is due to proteins, representing the so-called *colloid osmotic pressure* or *oncotic pressure*.

1.8.2.1.1.3 *Transport Conditioned by the Donnan Equilibrium* The Donnan equilibrium is a phenomenon of conditioned transmembrane diffusion. This occurs when there is a non-diffusible ion (often called a "non-permeable ion") on one side of the membrane relative to the membrane.

Through the diffusion phenomenon, it is possible to achieve the electric balancing of the particles on either side of a selectively permeable membrane (semipermeable) by evaluating the diffusion forces and the electrostatic forces, which contribute to the establishment of the final equilibrium.

This equilibrium was studied by chemist Frederick G. Donnan (1911). The characteristic and defining mechanism for the Donnan equilibrium has made it possible to explain the electrochemical phenomena that occur at the ionic level in the case of the membrane with selective permeability.

1.8.2.1.2 *Active Transport*

Active transmembrane transport occurs in the *opposite direction* to the concentration gradient, so from lower concentration to higher concentration. For this reason, it is also called transport in *counter-gradient of concentration*. Characteristic for active transport through biomembranes is the fact that it requires an *energy input*. So, it involves prior energy activation.

The evaluation of the type of active transmembrane transport takes into account several main features: ensuring the supply of energy, provided mainly by adenosine triphosphate – ATP (A - Pi ~ Pi ~ Pi), which transforms into adenosine diphosphate – ADP and energy – releasing phosphate – Pi molecule; the dependence of the transit speed on the concentration of the substance, e.g., the growth of the concentration of the substance present outside the cell, causes an increase in penetration (this up to a certain maximum speed limit, which cannot be exceeded); the existence of transmembrane "penetration zones" with protein structure, at the level of which the functionality in "pump" mode is highlighted.

There are known two types of active transport:

a. primary active transport by ion pump mechanism (explained by the specifics of the Na^+/K^+ pump and the proton pump);
b. secondary active transport by coupled mechanism.

The particularities of these mechanisms, which are important for active transmembrane transport, are discussed below.

1.8.2.1.2.1 Primary Active Transport by Ion Pump Mechanism   In the case of the ion pump, the "transport system" of a protein or lipoprotein nature – connects phosphate-type energizing components with macroergic binding (~ Pi), having as energy source adenosine triphosphate (ATP). To explain the specifics of this transport, the functional particularities for: *Na^+/K^+ pump* and *proton pump* are further discussed.

> *Na^+/K^+ Pump* – It is better known for the cells of the animal/human body. Under these conditions, a concentration-dependent transition of Na^+ and K^+ ions occurs. Adenosine triphosphate – ATP is involved in the transition, which is transformed into adenosine diphosphate – ADP and a P_i residue.
>
> *Proton Pumps* – In the case of this pump, "channel proteins" intervene which ensure the transport of H^+ from inside the cell to its outside. The pump is running in two stages:
> - inward opening – during which an ATP molecule is hydrolyzed, the channel protein has a specific conformation. In this first stage, the H^+ ions enter the intracellular level;
> - outward opening – during which the channel protein changes its conformation. It is the second stage in which ions are removed from the extracellular level.

By resuming the above-mentioned steps, the functional cycle of the proton pump is ensured.

1.8.2.1.2.2 Secondary Active Transport by Cotransport or Coupled Mechanism
The "coupled" transport mechanism has the characteristics of indirect transmembrane transport. In the case of this transport type, the electrochemical potential difference created by the ions pumping inside/outside of the cell is of importance.

In the case of this transport system, there are known two modes of transit the biomembranes: *uniporter* – where only one type of molecule transits through the membrane, e.g. Ca^{2+} transport across the muscle cell membrane (sarcoplasmic membrane); *co-transporter* – in which two different molecules are transported and can be *symporter* – when the molecules or ions are transported in the same direction, e.g., Na^+ and glucose transport; Na^+ and amino acids and *antiporter* – when two substances (especially ions) are transported in different directions, e.g., various Na^+/K^+ pumps; Na^+/Ca^{2+}; Cl^-/HCO^-; Na^+/H^+.

It has been experimentally found that biological membranes, such as cell and mitochondrial membranes, can be penetrated by ions and small molecules through an active transport process.

1.8.2.2 Barriers of Macrotransport Systems

Within the specific barriers of macrotransport systems, we may include *endocytosis* and *exocytosis*. These barriers have characteristic mechanisms by which the macro-molecules in solution are transported. However, macrotransport systems are also of interest in the incorporation of microorganisms.

1.8.2.2.1 Transport by Endocytosis

Endocytosis is a form of transmembrane transport specific to the barriers of macro-transport systems that ensure the *influx* of substances into the cellular environment. By endocytosis, large molecules enter the cell, e.g., proteins, lipids, etc. In the process of endocytosis, there is a first stage in which the internalization (invagination) of the cell membrane takes place.

After this, in the second stage, the phagosome is formed which transports the substances taken from the extracellular environment during the internalization stage. Then, through the dehiscence of the phagosome, its contents are taken up by a primary lysosome – in which there are only digestive enzymes.

During the takeover, a so-called phagosome-lysosome *fusion* takes place, during which the substances (e.g., nutrient molecules, drug molecules) are transported to be digested. The described steps reveal the influx of substances at the cellular level. The Golgi complex in the cytosol located in the vicinity of lysosomes participates in this process.

The endocytosis process is characteristic for cells. The endocytosis category actually includes two distinct processes:

    i. *phagocytosis* – specific, in general, for solid particles (taking into account proteins, cell fragments, bacteria, etc.). Leukocytes and macrophages are involved in phagocytosis. This process is important in terms of immunochemistry.
    ii. *pinocytosis* – specific for liquid particles (e.g., lipids). Through this process, macromolecular compounds are taken up from the external environment in solution, by emitting an ectoplasmic membrane – resulting from the plasma membrane or by constituting a tubular invagination to take over the products (e.g., liquid droplets).

Endocytosis is achieved by "enclaving" the products taken from the external environment of the cell. The gluco-, lipo- and proteolytic enzymes released, after the dislocation of the lysosomal membrane, act on them in the internal (cytoplasmic) environment.

1.8.2.2.2 Transport by Exocytosis

Exocytosis is a characteristic form of transmembrane transport for the barriers of macrotransport systems. Functionally, exocytosis, in physiology, is characterized as a "reverse process" of endocytosis. The process of exocytosis contributes to the *efflux* of substances from the cell into the extracellular environment.

In the case of exocytosis, the initiation of the process occurs at the level of the secondary lysosome. Such a lysosome is formed after the "fusion" of phagosome-lysosome. In the secondary lysosome enzymes and molecules taken over during the internalization and formation of the phagosome are found. Such molecules – known generically as "substrates" – interact with enzyme molecules. Following the substrate-enzyme interaction, the digestive vesicle is formed, which is oriented toward the plasma membrane of the cell. Part of the contents of the digestive vesicle is absorbed into the cytosol.

In the final stage, the contents of the digestive vesicle are eliminated in the extracellular fluid. Along with this, products made by the endothelial reticulum and the Golgi complex, which participate in the formation of this vesicle, are also eliminated. The membrane of the vesicle, after removing its contents, at the end of cytosis, integrates into the plasma membrane.

Thus, the notion of "membrane recycling" for lysosomal membranes has been accredited in cell physiology. In the literature, the term "hemocytosis" is sometimes used to define the release of cell synthesis products into the extracellular environment. In this way, the release of glandular excretion products in the case of hormones is commonly discussed.

1.8.2.2.3 Specificity of Macrotransport Systems

Macrotransport systems provide transit for proteins, hormones, growth factors, plant growth bioregulators, enzymes (e.g., transfers involved in iron transport) and lipoproteins (some involved in cholesterol transport).

The various proteins that enter the cell are "ligands". They are retained on the cell membrane where the "receptors" are located in certain cell regions. From the invagination of the cell membrane, the "internalization" of the compounds that enter the cell occurs.

Thus, protein macromolecules (ligands) bind to specific receptors to form the "ligand-receptor" assembly. After the invagination of the "transport system", the formation of a vesicle takes place. Next, the proteins are separated from the receptors by splitting the vesicle.

The first part of the vesicle, which contains protein ligands, is directed to the lysosomes, where the proteins will be destroyed by the release of a small molecule or substances with which they are combined (e.g., cholesterol, vitamins, iron, etc.).

A second part of the vesicle, containing the molecular receptors, is directed to the cell membrane, where the receptors are reactivated (in a recycling process).

However, it should be noted that biochemical barriers occur in the case of compounds of food interest (macronutrients, micronutrients and other nutrients – including biologically active substances) and compounds of pharmaceutical interest. They are also involved in the transit of compounds of agrobiological interest, such as plant growth bioregulators, important in plant physiology through their stimulating or retardant effect.

1.9 BIOMARKERS IN THE INVESTIGATION OF BIOLOGICALLY ACTIVE SUBSTANCES

The use of *biological markers* (biomarkers) in the investigation of biologically active substances has proven to be an area of excellence in assessing their presence in foods and pharmaceuticals.

For this purpose, studies based on in vitro tests (molecular and cellular), laboratory animal experiments, epidemiological research and interventional research (meaning studies on volunteers) were undertaken – see Fung et al. (2000), Gârban et al. (2006), Biesalski et al. (2009), Milencovic et al. (2017).

There are various definitions in the literature for biological markers. Thus, Silbergeld and Davis (1994) considered that "biological markers are physiological signals that reflect exposure, early cellular response with inherent or acquired susceptibilities that require a new strategy for solving toxicological problems".

Currently, the term "biomarker" is widely used to include almost any measurement that reflects an interaction between a biological system and an environmental agent, which can be chemical, physical and biological (http://www.inchem.org/documents/ehc/ehc/ehc155.htm).

1.9.1 Types of Biomarkers

The types of biomarkers differ depending on the areas of interest (Schulte, 1989; Grandjean et al., 1994; Mendelson et al., 1998). They are grouped mainly into three types:

a. *Exposure biomarkers* – are exogenous substances (natural or xenobiotic), but can also be endogenous substances (metabolites or residual xenobioderivatives) resulting from metabolic and biotransformation reactions. These biomarkers can be taken from a compartment (tissue/biological fluid) of the body and quantified analytically.
b. *Response* or *effect biomarkers* – originate in biochemical, physiological, behavioral or other changes in an organism. Depending on the magnitude of the value of a biomarker, the effects can be directly recognized or can be associated with a found or possible change in health or disease.
c. *Susceptibility biomarkers* – are indicators of an innate or acquired ability of an organism to respond to the challenge of exposure to a particular substance (natural or xenobiotic compound), in this case, a biologically active substance.

1.9.2 Application of Biomarkers

1.9.2.1 Evaluation of Biologically Active Substances

In the case of biologically active substances, especially in the category of those that are physiologically active and pharmacologically active, it is possible to evaluate their effect by using biomarkers. It is reiterated that the biomarker is considered, in the general sense, a *chemical test* or a *biological test* that analyzes a biological material providing data on biological exposure, effect, or susceptibility.

Biomarkers are used in medicine, biology, food science, pharmacology, ecology and environmental chemistry because they can provide data on the organism, environment, or ecosystem from where the samples were taken. The development of scientific and technological knowledge led to the diversification of the possibilities for modifying traditional foods and the discovery of new food sources.

Using scientific discoveries based on chemistry, molecular biology and genetics, it has become possible to obtain foods with certain special characteristics. In such situations, it has been found that some ingredients introduced in food (mainly from the group of biologically active substances) have beneficial effects on health.

Thus, by evaluating the action of some compounds or mixtures of compounds from the group of phytochemicals and zoochemicals, beneficial effects on health were found and validated (in vitro and in vivo). For this purpose, the bioavailability, the dose-effect level and – very importantly – the safety of the action on the target tissues, etc. were studied. Biomarkers were used for chemical extracts from plant tissues (phytochemicals), e.g., polyphenols, caffeine, plant sterols and from animal tissues (zoochemicals), e.g. omega-3 fatty acids, etc. In evaluating the role of various biologically active compounds (of food and pharmaceutical interest), biomarkers are also followed under related clinical aspects.

Thus, in the case of various diseases, a certain investigated *biomarker* was correlated with the so-called *clinical endpoint* that expresses the evolution of the disease. This mode of evaluation is explained in Table 1.2 (according to Biesalski et al., 2009).

In the case of biomarkers used for evaluation, it is important to note that representative values are measured by *biochemical parameters* – specific to clinical chemistry or by *physiological parameters* – specific to functional explorations and even *genetic tests* based on the investigation of chromosomal characteristics. Among the blood biochemical parameters detected, the following compounds can be found: total cholesterol (CT), HDL-cholesterol (HDL-C), LDL-cholesterol (LDL-C), LDL-oxidized (oxy-LDL), etc. (Rodriguez-Mateos et al., 2015).

TABLE 1.2
Examples of Biomarkers and "Clinical Endpoint"

	Cardiovascular Disease
Biomarkers	Serum cholesterol, blood pressure, etc.
Clinical endpoint	Heart attack, vascular attack
	Osteoporosis
Biomarkers	Markers of bone formation, bone resorption and bone density
Clinical endpoint	Bone fractures
	Cancer, e.g., Prostate Cancer
Biomarkers	Prostate specific antigen (PSA)
Clinical endpoint	Prostate cancer or metastasis

Source: According to Biesalski et al. (2009).

Among the investigated *physiological parameters* are systolic and diastolic blood pressure; pulse etc. (Hollman et al., 2011; Thijssen et al., 2011).

In recent research – based on molecular and cellular biology – some *genetic tests* were performed with the help of biomarkers. In this way, aspects related to "genetic polymorphism" were highlighted, which are at the origin of some enzymatic changes, e.g., catechol-O-methyl transferase (COMT) genes; 7α-hydroxylase cholesterol (CYP7A1).

The effect of natural biologically active compounds must be known for their preventive use in chronic degenerative diseases. To this end, the discovery of diseases must be made early. In such situations, they can be introduced into the diet, after noticing some symptoms that can be related to cardiovascular diseases, osteoporosis, and various forms of cancer. It is known that diseases can occur two to three decades after the detection of the pathogenic risk.

1.9.2.2   Specificity of the Action of Biologically Active Substances

In the literature, the issue of defining biologically active substances has also received attention in the sense of applying measurements based on mathematical equations. This has been done analogously with the study of thermodynamic activity.

A correlation was tried between the biological activity A (actually a numerical coefficient) and the concentration c of the substance under study.

To ensure the quantification of the activity, the use of some units of size (therefore, a numerical coefficient) was followed. An example is a situation in which the biologically active substance has catalytic activity (similar to enzymes). In such cases, the quantification of the catalytic activity is done in units of arbitrary measurement (Katal). Later, measurements specific to the international system (IS) were used.

It is considered that the concentration measurement is insufficient to define the role of a certain biologically active substance. For this reason, the procedure is the same as for enzymes, in which, as shown, the catalytic activity is determined. Two types of "unit systems" were used in the measurements: arbitrary units – known since the beginning of the studies in the respective field and fundamental units – based on the IS.

For biologically active substances that act at low concentrations, the working values are expressed in μkat/L or nkat/L (Jackson et al., 2007; Gârban, 2018).

Since 1999, instead of the classical arbitrary system (katal), the IS has been used, in which the expression is made in relation to time and concentration: katal (kat) has the dimensions $[s^{-1} \cdot mol]$. A special problem in the applications of biomarkers in the evaluation of biologically active substances lies in the specificity of the body's reaction to such substances.

It has been found, for example, that there is interindividual variability in relation to action on the body. In this regard, the effects of biologically active substances of food interest from plants and the so-called "cardiometabolic status" in relation to health were studied. The study by Milenkovic et al. (2017) followed various aspects of the interindividual variability of biomarkers on cardiometabolic status (Figure 1.10). It has been noted that there is an impact related to various factors: age, sex, genetic polymorphism, pathophysiological condition, etc.

From this schematic representation, it can be observed that biologically active substances from the following classes were studied: polyphenols (flavan-3-oils,

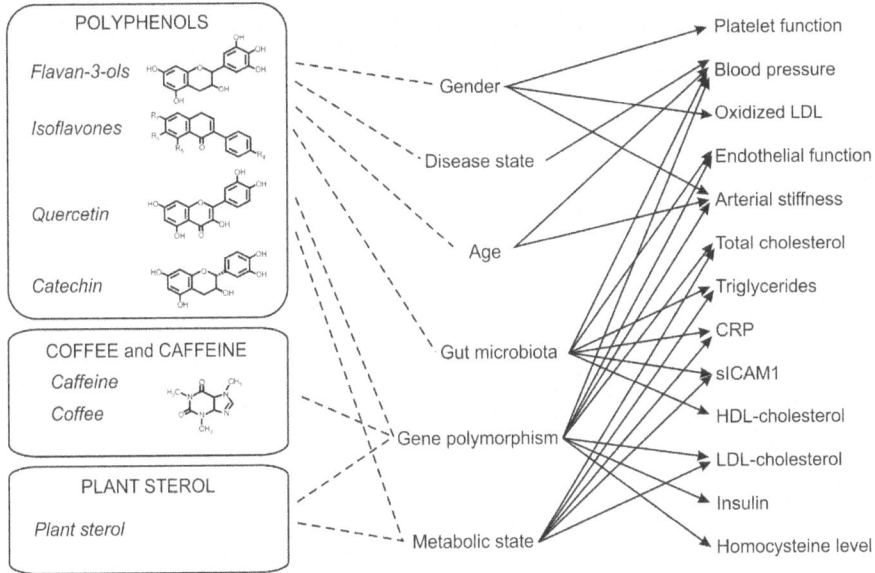

FIGURE 1.10 Schematic representation of factors involved in interindividual variability for biomarkers in relation to cardiometabolic status (after Milenkovic et al., 2017).

isoflavones, quercetin and catechin); coffee and caffeine; plant sterols (phytosterols). Verification of the action of biologically active substances is carried out by testing them on cell culture or experimental animal models.

For this reason, the notion of "bioassay" (biological assay) is used in the literature. As a benchmark, standard preparations are used in relation to which the activity of biologically active substances is studied.

In general, over time, research has been done for various biologically active substances of plant origin (phytochemicals), biologically active substances of animal origin (zoochemicals); prebiotics and plant growth regulators.

1.10 BIOLOGICALLY ACTIVE SUBSTANCES IN ETHNOBIOLOGY – OVERVIEW

In a volume on biologically active substances, the contextual approach to ethnobiology issues is required. This field has experienced a more accelerated development in the last decades.

According to the International Society of Ethnobiology, "Ethnobiology is the scientific study of dynamic relationships among peoples, biota and environments".

Motivation, in relation to humans, starts with the consideration that the information provided by the knowledge and practices of local people (especially ethnic groups) has come to the attention of the international bodies such as World Health Organization and Food and Agricultural Organization. Data on "indigenous knowledge and practices" have been shown to be important for biodiversity conservation, environmental management, and so on (Pieroni et al., 2005). In a broader context, ethnobiology issues make numerous references to ethnomedicine, ethno-agriculture, etc. From the plethora of fields derived from them, ethnopharmacology and ethnonutrition hold attention. There are currently specialized journals in the field of "ethnonutrition" (e.g., *Journal of Ethnic Foods*) and "ethnopharmacology" (e.g., *Journal of Ethnobiology and Ethnomedicine*).

Ethnopharmacology is developed from the field of "traditional medicine" based on the use of biologically active substances of plants and animals, practiced by different ethnic groups (e.g., indigenous peoples). The notion of ethnomedicine is considered to be synonymous with that of "traditional medicine" (i.e., folk medicine).

Traditional medicine maintained its information mainly through "oral tradition". In ethnopharmacology, the aim was to identify biologically active substances of prophylactic and therapeutic interest (Brusotti et al., 2014; Bergonzi et al., 2022). Thus, new drugs were initially obtained in the form of natural extracts.

Subsequently, the composition and structure of the active ingredient from the natural products were discovered. Finally, the synthesis of the organic compounds was performed and included as ingredients in pharmaceutical products.

Ethnonutrition – based on ethno-agriculture – involves an approach to issues related to biologically active substances that may be of food interest. To this end, existing information highlights data on the composition of selectively used ingredients in the diet of certain ethnic groups.

The problems of ethnonutrition can also be studied in correlation with agricultural biodiversity (agrobiodiversity) and, in general, lead to the characterization of food diversity. There is even the notion of "ethnic food" which defines food from the cultural heritage of an ethnic group (Kwon, 2015). For this purpose, knowledge of local ingredients from plant and/or animal sources is used.

REFERENCES

Altug T. - *Introduction to Toxicology and Food: Toxin Science, Food Toxicants, Chemoprevention*, CRC Press, New York, 2002.

Anderson J.W. - *Dietary Fiber in Nutrition, Management of Diabets. Dietary Fiber - Basic and Clinical Aspects*, Plenum Press, New York, 1986.

Arteca R. - *Plant Growth Substances: Principles and Applications*, Chapman and Hall, New York, 1996.

Barnes J. - Quality, efficacy and safety of complementary medicines: fashions, facts and the future. Part II: efficacy and safety, *Br. J. Clin. Pharmacol.*, 2003, 55, 331–340.

Baveye P., Block J.C., Goncharuk V.V. - *Bioavailability of Organic Xenobiotics in the Environment: Practical Consequences for the Environment*, Springer, London, 2011.

Belitz H.-D., Grosch W., Schieberle P. - *Food Chemistry*, 4th edition, Springer Verlag, Berlin-Heilderberg, 2009.

Bell G.H., Smith E.D., Paterson C.R. - *Textbook of Physiology and Biochemistry*, 9th edition, Churchill Livingstone, Edinburgh-London-New York, 1976.

Bergonzi M.C., Heard C.M., Garcia-Pardo J. - Bioactive molecules from plants: discovery and pharmaceutical applications. *Pharmaceutics*, 2022, 14(10), 2116. doi: https://doi.org/10.3390/pharmaceutics14102116.

Biesalski H.-K., Dragsted O.L., Elmadfa I., Grossklaus R., Michael Müller M., Schrenk D., Walter P., Weber P. - Bioactive compunds: definition and assessment of activity, *Nutrition*, 2009, 25, 1202–1205.

Bruneton J. - *Pharmacognosie - Phytochimie, Plantes Médicinales, 4-ème édition, revue et augmentée*, Tec & Doc - Éditions médicales internationales, Paris, 2009.

Brusotti G., Cesari I., Dentamaro A., Caccialanza G., Massolini G. - Isolation and characterization of bioactive compounds from plant resources: the role of analysis in the ethnopharmacological approach. *J. Pharm. Biomed. Anal.*, 2014, 87, 218–228. https://doi.org/10.1016/j.jpba.2013.03.007.

Cresteil T. - Onset of xenobiotic metabolism in children: toxicological implications, *Food Addit. Contam.*, 1998, 15(Suppl), 45–51.

Crews H.M., Hanley A.B. (Eds.) - *Biomarkers in Food Chemical Risk Assessment*, Royal Society of Chemistry, London, 1995.

Dekant W. - The role of biotransformation and bioactivation in toxicity, pp. 57–86, in *Molecular, Clinical and Environmental Toxicology, Vol. 1: Molecular Toxicology* (Luch A., Ed.), Birkhäuser Verlag, Basel-Boston-Berlin, 2009.

Deshpande S.S. - *Handbook of Food Toxicology*, Marcel Dekker Inc., New York, 2002.

Devlin T.M. (Ed.) - *Textbook of Biochemistry with Clinical Correlations*, 3rd edition, Wiley and Sons Inc., New York-Chichester-Brisbane-Toronto-Singapore, 1992.

D'Mello J.P.F. - *Food Safety: Contaminants and Toxins*, CABI Publishing, Cambridge, U.K., 2003.

Donnan F.G. - Theorie der Membrangleichgewichte und Membranpotentiale bei Vor-handensein von nicht dialysierenden Elektrolyten. Ein Beitrag zur physikalisch-chemischen Physiologie, *Z. Elektrochem. Angew. Phys. Chem.*, 1911, 17(10), 572–581.

Edelman G.M. - *Topobiology: An Introduction to Molecular Embriology*, Basic Book Inc., New York, 1988.

Faber K. - *Biotransformations in Organic Chemistry. A Textbook*, 6th edition, Springer Verlag, Berlin-Heidelberg- New York, 2011.

Fennema O.R. (Ed.) - *Food Chemistry*, 3rd edition, Marcel Dekker Inc., New York, 1996.

Fishman W.H. (Ed.) - *Metabolic Conjugation and Metabolic Hydrolysis, Vol. I*, Academic Press, New York, 1970.

Forth W., Henschler D., Rummel W. (Eds.) - *Allgemeine und Spezielle pharmakologie und Toxicologie*, 2.Auflage, Bibliographisches Institut, Wissenschafts - Verlag, Mannheim-Wien-Zürich, 1977.

Fraser L.R., Beyret E., Milligan S.R., Adeoya-Osiguwa S.A. - Effects of estrogenic xenobiotics on human and mouse spermatozoa, *Human Reprod.*, 2006, 21(5), 1184–1193.

Fung E.T., Wright G.L. Jr., Dalmasso E.A. - Proteomic strategies for biomarker identification: progress and challenges, *Curr. Opin. Mol. Ther.*, 2000, 2, 643–650.

Gârban Z. - *Biochemstry: Comprehensive Treatise, Vol. I. Basics of biochemistry* (in Romanian), 4th edition, Editura Didactică şi Pedagogică, Bucureşti, 1999.

Gârban Z., Gârban G., Ghibu G.-D. - Biomarkers: theoretical aspects and applicative peculiarities. Note II. Nutritional biomarkers, *J. Agroaliment. Proc. Technol.*, Timişoara, 2006, XII(2), 349–356.

Gârban Z. - *Molecular Biology: Concepts, Methods, Appplications* (in Romanian), 6th edition, Editura Solness, Timişoara, 2009.

Gârban Z. - *Biotransformation of Xenobiotics of Food and Pharmaceutic Interest: Specific Interactions* (in Romanian), Editura Academiei Române, Bucureşti, 2017.

Gârban Z. - *Quo Vadis Food Xenobiochemistry*, 3rd edition, Publishing House of the Romanian Academy, Bucharest, 2018.

Gil F., Pla A. - Biomarkers as biological indicators of xenobiotic exposure, *J. Appl. Toxicol.*, 2001, 21, 245–255.

Godlewska-Żyłkiewicz B., Świsłocka R., Kalinowska M., Golonko A., Świderski G., Arciszewska Ż., Nalewajko-Sieliwoniuk E., Naumowicz M., Lewandowski W. - Biologically active compounds of plants: structure-related antioxidant, microbiological and cytotoxic activity of selected carboxylic acids. *Materials (Basel)*, 2020, 13(19), 4454. https://doi.org/10.3390/ma13194454.

Goyal A., Sharma A., Kaur J., Kumari S., Garg M., Sindhu R.K., Rahman M.H., Akhtar M.F., Tagde P., Najda A., Banach-Albińska B., Masternak K., Alanazi I.S., Mohamed H.R.H., El-Kott A.F., Shah M., Germoush M.O., Al-Malky H.S., Abukhuwayjah S.H., Altyar A.E., Bungau S.G., Abdel-Daim M.M. - Bioactive-based cosmeceuticals: an update on emerging trends. *Molecules*, 2022, 27(3), 828. https://doi.org/10.3390/molecules27030828.

Grandjean P., Brown S.S., Reavey P., Young D.S. - Biomarkers of chemical exposure: state of the art, *Clin. Chem.*, 1994, 40, 1360–1362.

Greenwood J. - *New Concepts of a Blood-Brain Barrier*, Plenum Press, New York-London, 1995.

Gruetter R., Novotny E.J., Boulware S.D., Rothman D.L., Mason G.F., Shulman G.I., Shulman R.G., Tamborlane W.V. - Direct measurement of brain glucose concentrations in human by 13C NMR spectroscopy. *Proc. Nat. Acad. Sci.* 1992, 89, 1109–1112.

Guaadaoui A., Benaicha S., Elmajdoub N., Bellaoui M., Hamal A. - What is a bioactive compound? A combined definition for a preliminary consensus, *Int. J. Nutr. Food Sci.*, 2014, 3(3), 174–179.

Guyton A.C., Hall J.E. - *Textbook of Medical Physiology*, 11th edition, Elsevier-Saunders, Philadelphia, 2006.

Hasler C., Blumberg J. - Phytochemicals: biochemistry and physiology, *J. Nutr.*, 1999, 129(3), 756S–757S.

Heller R., Esnault R., Lance C. - *Physiologie végétale. Tome 2, Développement*, 6-ème édition, Dunod, Paris, 2004.

Hernandez D.F., Cervantes E.L., Luna-Vital D.A., Mojica L. - Food-derived bioactive compounds with anti-aging potential for nutricosmetic and cosmeceutical products. *Crit. Rev. Food Sci. Nutr.* 2021, 61(22), 3740–3755. https://doi.org/10.1080/10408398.2020.1805407.

Heymann E. - *Haut, Haar und Kosmetik, Eine chemische Wechselwirkung - Handbuch für Korperpflegeberufe, Apotheker und Dermatologen*, 2 Auflage, Verlag Hans Huber, Bern-Göttingen-Toronto - Seattle, 2003.

Hollman P.C., Cassidy A., Comte B., Heinonen M., Richelle M., Richling E., Serafini M., Scalbert A., Sies H., Vidry S. - The biological relevance of direct antioxidant effects of polyphenols for cardiovascular health in humans is not established., *J. Nutr.*, 2011, 141, 989S–1009S.

Ioannides C. - *Enzyme Systems that Metabolize Drugs and Other Xenobiotics*, John Wiley and Sons Ltd., Chichester, 2002.

Jackson M.C., Esnouf P.M., Winzor J.D., Duewer L.D. - Defining and measuring biological activity: applying the principles of metrology, *Accredit. Qual. Assur.*, 2007, 12, 283–294.

Johnson L.R. - *Physiology of the Gastrointestinal Tract*, Elsevier Academic Press, Burlington, 2006.

Kaplan L.A., Pesce A.J. - *Clinical Chemistry: Theory, Analysis, Correlation*, 5th edition, Elsevier/Mosby, St. Louis, 2010.

Katz H.-S., Weawer W.W. (Eds.) - Bioactive food components, pp. 201–204, in *Encyclopedia of Food and Culture, Vol. 1, Acceptance to Food Politics*, Charles Scribner's Sons, imprint of The Gale Group, Inc. - a division of Thomson Learning, Inc., New York, 2003.

Khera K.S. - Maternal toxicity in humans and animals: effects on foetal development and criteria for their detection, *Teratog. Carcinog. Mutagen.*, 1987, 7, 287–295.

Kim D., Nylander-French A.L. - Physiologically based toxicokinetic models and their application in human exposure and internal dose assessment, pp. 37–55, in *Molecular, Clinical and Environmental Toxicology, Vol. 1: Molecular Toxicology* (Luch A., Ed.), Birkhäuser Verlag, Switzerland, 2009.

Kjeldsberg C.R., Knight J.A. - *Body Fluids: Laboratory Examination of Amniotic, Cerebrospinal, Seminal, Serous [and] Synovial Fluids*, 3rd edition, Publisher Amer. Soc. Clin. Pathol. Press, Chicago, 1993.

Kwon D.Y. - What is ethnic food? *J. Ethn.Foods*, 2015, 2, 1.
Léger C.L., Béréziat G. - *Biomembranes et Nutrition*, Colloques International, 12–14 Juin, Paris 1989, Editions INSERM, Paris, 1989.
Liu R.H. - Health promoting components of fruits and vegetables in the diet, Supplement to *Adv. Nutr.*, 2013, 4, 384S–392S, https://doi.org/10.3945/an.112.003517.
Luch A. (Ed.) - *Molecular, Clinical and Environmental Toxicology, Vol. 1: Molecular Toxicology*, Birkhäuser Verlag, Switzerland, 2009.
Luck H. - *Antimicrobial Food Additives*, Springer Verlag, Berlin-New York, 1980.
McCarley K.D., Bunge A.L. - Pharmacokinetic models of dermal absorption, *J. Pharm. Sci.*, 2001, 90, 1699–1719.
Mendelson L., Mohr L.C., Peeters J.P. (Eds.) - *Biomarkers: Medical and Workplace Applications*, Joseph Henry Press, Washington DC, 1998.
Milenkovic D., Morand C., Cassidy A., Konic-Ristic A., Tomás-Barberán F., Ordovas M.J., Kroon P., De Caterina R., Rodriguez-Mateos A. - Interindividual variability in biomarkers of cardiometabolic health after consumption of major plant-food bioactive compounds and the determinants involved, *Adv. Nutr.*, 2017, 8(4), 558–570.
Mincu I. - *Human-Food Impact* (in Romanian), Editura Medicală, Bucureşti, 1993.
Nassar F.A., Hollenberg F.P., Scatina A.J. (Eds.) - *Drug metabolism Handbook: Concepts and Application*, John Wiley and Sons Inc., London, 2009.
Pieroni A., Leimar Price L., Vanderbroek I. - Welcome to journal of ethnobiology and ethnomedicine. *J. Ethnobiol. Ethnomed.*, 2005, 1, 1. https://doi.org/10.1186/1746-4269-1-1.
Potts R.O., Bommannan D.B., Guy R.H. - Percutaneous absorption, pp.13–28, in *Pharmacology of the Skin* (Mukhtar H., Ed.), CRC Press, Boca Raton, 1992.
Ramsey J.C., Andersen M.E. - A physiologically based description of the inhalation pharamacokinetics of styrene in rats and human, *Toxicol. Appl. Pharmacol.*, 1984, 73, 150–175.
Rhoades R., Pflanzer R. - *Human Physiology*, 2nd edition, Saunders College Publishing, Fort Wort- Philadelphia-San Diego-New York, 1992.
Richterich R. - *Klinische Chemie*, 3. Aufl., Karger, Basel, 1971.
Ritter J.M., Flower R., Henderson G., Loke Y.K., MacEwan D., Rang H.P. - *Rang & Dale's Pharmacology*, 9th edition, Elsevier, Edinburgh, 2019.
Rodriguez-Mateos A., Cifuentes-Gomez T., Gonzalez-Salvador I., Ottaviani J.I., Schroeter H., Kelm M., Heiss C., Spencer J.P. - Influence of age on the absorption, metabolism, and excretion of cocoa flavanols in healthy subjects, *Mol. Nutr. Food Res.*, 2015, 59, 1504–1512.
Ronis J.J.M., Cunny H.C. - Developmental effects on xenobiotic metabolism, pp. 257–272, in *Molecular and Chemical Toxicology*, 4th edition (Smart R.C., Hodgson E., Eds.), John Wiley and Sons Inc., Hoboken, USA, 2008.
Salas J., Pollastrini M.T., Barea M., Fernandez P.M. - Seguridad en el uso de aditivas - productos carnicos, *Aliment. Equipos y Tecnol.*, 1990, 2, 169–178.
Sandermann H. - Plant metabolism and xenobiotics, *Trends Biochem. Sci.*, 1992, 17, 82–84.
Schröder P., Collins C.D. (Eds.) - *Organic Xenobiotics and Plants:From Mode of Action to Ecophysiology, Vol. 8*, Springer Verlag, Dordrecht-Heidelberg-London-NewYork, 2011.
Schulte P.A. - A conceptual framework for the validation and use of biomarkers, *Environ. Res.*, 1989, 48, 129–144.
Senna M., Legrand A.-P. - Inorganic/organic composites for biological, pharmaceutical and medical applications, *Ann. Chem. Sci. Mater.*, 2004, 29(1), 1–5.
Silbergeld E.K., Davis D.L. - Role of biomarkers in identifying and understanding environmentally induced disease, *Clin. Chem.*, 1994, 40(7 Pt 2),1363–1367.
Sipes I.G., Gandolfi A.J. - Biotransformation of toxicants, pp. 88–126, in *Casarett and Doull's Toxicology: The Basic Science of Poison* (Amdur M.O., Doull J., Klaassen C.D., Eds.), 4th edition, Pergamon Press, New York, 1991.
Sittig M. - *Handbook of Toxic and Hazardous Chemicals and Carcinogens*, 2nd edition, Noyes Publ., New York, 1985.

Staats D.A., Fisher J.W., Connolly R.B. - Gastrointestinal absorption of xenobiotics in physiologically based pharmacokinetic model. A two compartment description, *Drug Metab. Dispos.*, 1991, 19, 144–148.

Teodoro A.J. - Bioactive compounds of food: their role in the prevention and treatment of diseases, *Hindawi - Oxid. Med. Cell. Longevity,* 2019, Article ID 3765986, 4 pages, https://doi.org/10.1155/2019/3765986.

Testa B. - *Metabolism of Drugs and Other Xenobiotics*, Academic Press, New York, 1995.

Thijssen D.H., Black M.A., Pyke K.E., Padilla J., Atkinson G., Harris R.A., Parker B., Widlansky M.E., Tschakovsky M.E., Green D.J. - Assessment of flow-mediated dilation in humans: a methodological and physiological guideline, *Am. J. Physiol. Heart Circ. Physiol.*, 2011, 300(1), H2–H12.

Trevor A., Katzung B., Knuidering-Hall M. - *Katzung & Trevor's Pharmacology Examination and Board Review*, 11th edition, McGraw-Hill Education/Medical, New York-Chicago-Athens-London-Madrid, 2015.

Vineis P. - The relationship between polymorphisms of xenobiobiotic metabolizing enzymes and susceptibility to cancer, *Toxicology*, 2002, 181–182, 457–462.

Voet D., Voet J.G. - *Biochemistry*, 4th edition, John Wiley and Sons Inc., New York, 2011.

Ward R.E., Bruce J. - Zoonutrients and health, *Food Technol.*, 2003, 57, 33–39.

West J.B. (Ed.) - *Best and Taylor's Physiological Basis of Medical Practice*. 12th edition, Wiliams and Wilkins, Baltimore, 1991.

Wilson I.D., Nicholson J.K. - Topics in xenobiochemistry: do metabolic pathways exist for xenobiotics? The micrometabolism hypothesis, *Xenobiotica*, 2003, 33(9), 887–901.

Wurtman J.R., Wurtman J. - *Nutrition and the Brain, Vol. 6, Physiological and Behavioral Effects of Food Constituents*, Raven Press, New York, 1983.

2 Natural Compounds of Plant Origin

2.1 INTRODUCTION

Numerous *biologically active substances*, called by the generic term *bioactive substances*, are present in very small quantities (along with various nutrients) in natural foods and/or processed foods. They can also be found in phytopharmaceuticals/pharmaceuticals for internal or external use.

Various biologically active substances of plant origin – phytochemicals (e.g., natural pigments, essential oils, etc.) – highlight certain specific effects. Some biologically active substances are known in *ethnonutrition* – being used over time as ingredients in food preparation. These are of interest to gastronomy and food engineering.

Other biologically active substances are known in ethnopharmacology and are used in traditional medicine in the form of extracts (solutions and ointments), capsules (with powder from leaves, stems, roots, etc.) and teas. The initiation of the knowledge of the usefulness of biologically active substances was done in ethnonutrition, but the deepening of knowledge was done mainly in ethnopharmacology (Süntar, 2020; Jacob et al., 2021).

The name ethnonutrition also refers to some food ingredients from plant products used by certain ethnic groups (existing populations in geographically distinct areas). Observations accumulated throughout time in the case of the natural compounds of vegetal origin led to the detection of ingredients having spice attributes and being used in the diet of different ethnic groups. Among the widely used spices, the following can be exemplified: *hot spices* (e.g., pepper, mustard and paprika); *spicy spices* (cinnamon, vanilla and bay leaf); *alliaceous spices* (onion and garlic).

In the field of ethnopharmacology, the European thematic colloquia were representative. The first colloquium took place (in 1990) at Metz (Fleurentin et al., 1990), the second in 1993 at Heidelberg (Schröder et al., 1996) and continued (as triennial scientific events). During these scientific events, data were published on various sources of biologically active plant substances studied in different geographical areas of the Earth.

Such information was at the origin of obtaining extracts containing well-characterized chemical compounds in terms of chemical composition and structure. At a later stage, data on chemical composition and structure proved useful for "laboratory organic synthesis", and the substances obtained were then used in the pharmaceutical and even nutritional fields (e.g., food additives and fortified foods).

Advances in scientific research, fundamental and applied, of biologically active substances, have eloquently demonstrated that many problems of *nutrition, pharmacology, cosmetics* and agrobiology are related to the problems of biochemistry/pathobiochemistry and physiology/pathophysiology.

DOI: 10.1201/9781032702520-2

This observation showed the interactive nature of the study of biologically active substances – a favorite of natural compounds of plant origin – along with highlighting the complexity of data on chemical structure, biological activity, structure-activity relationship (SAR) and involvement in material metabolisms (carbohydrate, lipid, protein and hydro-electrolytic) and in energy metabolism. Isolation of various phytochemicals was done from plants existing in the spontaneous flora and/or from cultivated plants (e.g., medicinal plants and aromatic plants).

Various analytical (sometimes bioanalytical) methods for identification (e.g., chromatography) and structural characterization (e.g., spectrometry) were used for this purpose. The exceptional importance of biologically active substances, mainly represented by phytochemicals, aroused special interest in the development of knowledge on (i) *nutrition and diet therapy* – constituted in an independent scientific field, which pursued the physiological properties of active nutrients; (ii) *pharmacology and pharmacognosy* – increasingly confronted with the use of pharmacologically active natural products and (iii) *cosmetology* – based on the knowledge of the chemical composition of the ingredients in cosmetics, often remarking the concomitant approach to the problems of pharmacology and cosmetology (e.g., dermato-cosmetology).

To understand the involvement of modern biochemistry of biologically active substances in cosmetology issues, for example, it is mentioned that clearance determinations are currently being made for various cosmetic products in relation to their action on the skin.

This example is not uncommon or circumstantial in the scientific research undertaken on cosmetics in relation to the biochemical action of the ingredients in their composition.

Over time, studies have been carried out on the preferred extension and deepening of studies on natural compounds of plant origin, following the chemical structure, the physiological/pharmacological/toxicological role and, specifically, the relationship between the chemical structure and biological activity.

2.2   CHEMICAL COMPOSITION AND NATURAL DISTRIBUTION

Biologically active substances of plant origin in terms of chemical composition and structure belong to various classes of compounds. Compounds with a linear, aromatic and heterocyclic structure can be highlighted. One, two or more (often mixed) functional groups are grafted onto such compounds. For these reasons, a general assessment and a structural description of the biologically active substances cannot be made.

A closer look at the natural compounds of plant origin in the body attests to the fact that they are substances of organic nature, i.e. carbohydrates, lipids, proteins and inorganic nature, i.e. water and mineral bioelectrolytes. With regard to biologically active substances, it is known that their natural distribution is lower and much more diversified by their chemical composition. Among the specialized works published in the international literature, some informative treatises or papers with current issues in the fields of *human nutrition* (Guthrie, 1975; Hudson, 1990; Alais and Linden, 1993; Hasler and Blumberg, 1999; Gârban, 2000); *pharmacology* and *pharmacognosy* (Guenther, 1950; Britton, 1983; Chadwich and Marsh, 1990; Kritchevsky et al., 1990; Braquet et al., 1991; Charlwood and Banthorpe, 1991; Sandermann, 1992;

Lawson, 1993; Waterman, 1993); *cosmetology* (Raab, 1985; Pugliese, 1987; Müller, 1989; Heymann, 1994; Ziolkowsky, 1995) are mentioned for information purposes.

Obviously, there are currently numerous encyclopedias and treatises that address restrictively topics related to biologically active substances whose consultation can be an adjunct to readers interested in aspects of contemporary research of theoretical interest (chemical structure) and applied (biological activity). In Romania, various studies of general interest or regarding phytochemicals in botanical aspect (morphology and taxonomy), nutritional, phytotherapeutic, agrobiological, etc. have been carried out (Grinţescu, 1945; Borza, 1968; Gârban, 2009). Related studies on the theoretical and applied aspects of nutrition have often been published (Neamţu, 1983; Gârban, 2000), phytochemistry, pharmacology and pharmacognosy (Radu et al., 1981; Ciulei et al., 1993) and cosmetics.

Extension of research on bioconstituents and biochemical effectors has sometimes been done with the ignorance of initiating coherent studies on biologically active substances. Existing data in the literature based on the deepening of research by modern methods, e.g., spectrophotometric, chromatographic, magnetochemical, isotopic, etc., have brought and bring exceptional scientific information and under the applicative report – it was possible to proceed to the evaluation of the biological specificity – active in nutrition, pharmacology, cosmetology, agrobiology.

2.3 CLASSIFICATION AND NOMENCLATURE

On biologically active substances, assumed their great diversity of composition and the multitude of biologically active effects (only partially known) – as shown – various ways of classification have been ascribed.

The classification may take into account biological-taxonomic criteria (classes of microorganisms, plant species, and animal species); chemical criteria (structure, properties, and reactivity); biochemical criteria (participation in metabolic interactions, the relationship between chemical structure and biological activity); physiological criteria (effects induced on various organs and systems, e.g., digestive tract, cardiovascular system, etc.). In matters of nomenclature, concerning the various classes of biologically active substances, it can be stated that the origin of trivial (usual) names is often taken from botany and zoology, from plant and animal physiology, from organic chemistry, even from certain technologies in which such biologically active substances are used.

Chemical names follow the rules established for organic chemistry and obviously, in this situation, take into account the criteria developed by the International Union of Pure and Applied Chemistry (IUPAC) and IUBMB. A classification of phytochemicals – with more extensive accreditation – anticipates their more detailed presentation. In this sense, seven classes of biologically active substances from the group of phytochemicals are mentioned: (i) *organic acids*; (ii) *alkaloids*; (iii) *phytoncides*; (iv) *heterosides*; (v) *natural pigments*; (vi) *tannins* and (vii) *essential oils*.

With reference to the biologically active substances present in the plant kingdom, numerous fundamental and applied research have been undertaken. This context mentions the existence of information on the applicability in pharmacology (e.g., glycosides, alkaloids) – being pharmacologically active; nutrition (organic acids, natural pigments, etc.) – being physiologically active; cosmetology and cosmetic dermatology (e.g., essential oils, glycosides) – being physiologically active; toxicology (e.g., alkaloids) – being toxicologically active. The conceptualization of the notion of "biologically active substances" based on the evaluation of chemical structure and biological activity in biochemistry interests various fields, e.g., plant biochemistry, food, pharmaceutical, medical and, last but not least, cosmetology. This observation reveals the circumscription of a modern, interactive and related approach from basic research to applied research.

2.4 PHYTOCHEMICAL CLASSES

2.4.1 INTRODUCTORY EXPOSURE

The presentation of the various natural compounds of plant origin, with the specification of the main classes of compounds, the mention of some representatives and the highlighting of the natural distribution is given in Table 2.1. The study of these compounds is relevant in nutrition, pharmacology and toxicology. Classes of compounds are presented below.

> *Organic acids* – Organic acids with an aliphatic or aromatic structure are included in the class of biologically active substances of nutritional and/ or pharmaceutical interest. There are also compounds with the structure of aldehyde acids (e.g. pyruvic acid), ketone acids (e.g. acetylacetic acid), hydroxylated dicarboxylic acids (e.g. tartaric acid), hydroxylated tricarboxylic acids (e.g. citric acid), etc. Thus, a wide variety of compounds may be included which are characterized by the presence of the carboxyl group.

TABLE 2.1
Compounds of Plant Origin: Representatives and Natural Distribution

No.	Specification	Representatives (General Examples)	Natural Distribution (in Tissues and Food)
1	Organic acids	Oxalic, malic, lactic, mandelic, quinic, tartaric, citric, etc.	Vegetables and fruits (confer a sour taste)
2	Alkaloids	Caffeine, theophylline, theobromine, papaverine, nicotine, etc.	Coffee beans, cocoa tea leaves, tobacco, poppy fruits and seeds
3	Phytoncides	Allicin, alliin, anthocyanin derivatives, sativin,	Vegetables (garlic and onion); fruits (blueberries, blackberries)
4	Heterosides (glycosides)	Digitoxigenin, gitoxigenin, digoxigenin, strophanthidin, sapogenin, solagenin	Leaves, bulbs, fruits, seeds, roots
5	Natural pigments	Carotenoids, flavonoids, indoles, tetrapyrroles, etc.	Flowers, fruits, tree stalks, roots
6	Tannins	Gallotannins, catechin tannins	Leaves, fruits, seeds, tree stalks, (bark of certain trees), roots
7	Essential oils	Myrcene, linalool, limonen, borneol, camphor, etc.	Flowers, fruits, tree stalks

Alkaloids – Biologically active substances in the class of alkaloids are heterocyclic organic compounds with nitrogen whose main physicochemical characteristic is basicity. In the structure of alkaloids are found various nitrogen heterocycles such as pyrrolidine, piperidine, pyridine, quinoline, indole, purine, etc.

Phytoncides – Natural products of this class are obtained by extraction from plants. These products are characterized by antibiotic action, which is why they have been called "plant antibiotics", compared to the actual antibiotics from microorganisms.

Heterosides (Glycosides) – They are biologically active chemicals consisting of two components: the carbohydrate component – oligosaccharide, and the non-carbohydrate component – represented by an aglycone. Depending on the nature of the aglycone, various types of glycosides can be formed: O-heterosides (oxygen glycosides); S-heterosides (sulfur glycosides) and N-heterosides (nitrogen glycosides). Note that the generic name for glycosides is heteroside.

Natural pigments – They are chemical compounds – preferably of a plant nature – that give various colors to the food in which they are or are introduced during their processing. The variety of colors is given by the so-called "chromophore groups". The types of biologically active natural pigments differ a lot, distinguishing mainly two subgroups, namely: *non-nitrogenous pigments*, e.g., carotenoid pigments, flavonoid pigments, quinone pigments; *nitrogenous pigments*, e.g., tetrapyrrole pigments, indole pigments, etc.

Tannins – Biologically active substances in the tannin class are organic compounds with a heterogeneous structure of polyphenolic type, distinguishing two subgroups: *gallotannins* – natural esters of glucose with gallic acid, or derivatives thereof; *catechin tannins* – macromolecular compounds resulting from the polycondensation of several catechin molecules or catechin derivatives.

Essential oils – They are biologically active substances of plant origin whose biogenesis takes place in: flowers, fruits, leaves, seeds, etc. However, essential oils of animal origin (e.g. musk) are also known. From a chemical point of view, essential oils - also called "volatile oils" - contain: saturated or unsaturated acyclic hydrocarbons, aromatic hydrocarbons, lower and upper alcohols, aldehydes, ketones, organic acids, heterocyclic compounds, ethers, esters, etc. This group of substances includes terpenoid compounds – isoprene derivatives; tropolone compounds – unsaturated homocyclic compounds derived from the tropone; various compounds – very different structure, but with specific biological activity, which reveals some similarities. This general classification, as found, has a broader conceptual accreditation. It can be noted that at the origin of the classification are mainly considerations of fundamental (theoretical), but also of applicative (industrial) interest with references to natural sources, processing paths, fields of use, etc. Within each class of compounds, data are presented regarding the chemical structure, biological activity, structure-activity relationship, applicative aspects, etc.

It is necessary to mention that certain biologically active substances for the human body are chemical xenobiotics originating from plants (more numerous examples are among alkaloid compounds). In this regard, in Chapter 1, references were also made to xenobiotics.

2.4.2 ORGANIC ACIDS

2.4.2.1 Synoptical Data

This class of compounds includes aldehyde acids (aldo-acids), ketone acids (keto-acids), hydroxyl acids (hydroxy-acids), etc. They are found more frequently in various products of plant origin, but also in products of animal origin resulting from metabolic processes. Some of the compounds in this class are found as "storage substances" in tissues, others are intermediate (unstable) metabolites in metabolic processes.

Organic acids have recently been found in natural products, which – from the point of view of nutrition and pharmacology – can be included in the group of biologically active substances. Organic acids are free or in the form of derivatives, e.g., esters – in the case of lipids; peptides – in the case of proteins; salified compounds. They are also present in the ionic state – as metabolites in plant and animal tissues (Bodea et al., 1964–1966; Neamțu, 1983; Charlwood and Banthorpe, 1991; Simmonds, 1992; Ensminger et al., 1995; Heller, 1997; Gârban, 2018; Shi et al., 2022).

2.4.2.2 Representative Compounds

The presentation of the structural formulas of the main organic acids, considered as biologically active substances, is done by grouping them in organic acids with aliphatic chains (mono- and polyhydroxy) and organic acids that have cyclic chain fragments (aromatic or saturated).

Among the organic acids with aliphatic chains are mentioned monocarboxylic acids: *glyoxylic acid* (I); *glycolic acid* (II); *lactic acid* (III); *pyruvic acid* (IV); *β-hydroxybutyric acid* (V); *glyceric acid* (VI) and dicarboxylic acids: *acetoacetic acid* (VII); *α-keto-glutaric acid* (VIII); *malic acid* (IX); *maleic acid* (X); *fumaric acid* (XI); *tartaric acid* (XII) and *citric acid* (XIII). The molecular structure of these compounds is presented in Figure 2.1.

> *Glyoxylic acid* – is present in various plant products such as gooseberries, apples, grapes and in general, in unripe fruits. This compound is not found in ripe fruits.
>
> *Glycolic acid* – is among the compounds present in the intermediate metabolism.
>
> *Lactic acid* (α-*hydroxy propionic*) – is in the racemic form (±). It was isolated by Scheele in 1780 from sour milk and by Berzelius in 1808 from aqueous meat extract (as evidenced by the presence of lactic acid in the muscles of superior animals). Lactic acid also results from various carbohydrates such as glucose, sucrose, lactosethrough fermentation processes under the action of enzymes produced by lactobacilli (e.g. *Lactobacillus casei*, *Lactobacillus delbrueckii* ssp. *bulgaricus* a.o.). In industry, it is obtained by fermenting molasses during specific processing in the sugar industry.

HOOC–CHO   HOOC–CH$_2$–OH   HOOC–CH(OH)–CH$_3$ HOOC–CO–CH$_3$
glyoxylic acid   glycolic acid   lactic acid   pyruvic acid
(I)   (II)   (III)   (IV)

HOOC–CH$_2$–CH(OH)–CH$_3$ HOOC–CH(OH)–CH$_2$OH
β - hydroxybutyric acid   glyceric acid
(V)   (VI)

HOOC–CH$_2$–CO–COOH   HOOC–CH$_2$–CH$_2$–CO–COOH   HOOC–CH(OH)
|
HOOC–CH$_2$
acetoacetic acid   α - ketoglutaric acid   malic acid
(VII)   (VIII)   (IX)

HOOC–CH   HOOC–CH   HOOC–CH–OH   CH$_2$–COOH
‖   ‖ | |
HOOC–CH   CH–COOH   HOOC–CH–OH   HO–C–COOH
|
CH$_2$–COOH
maleic acid   fumaric acid   tartaric acid   citric acid
(X)   (XI)   (XII)   (XIII)

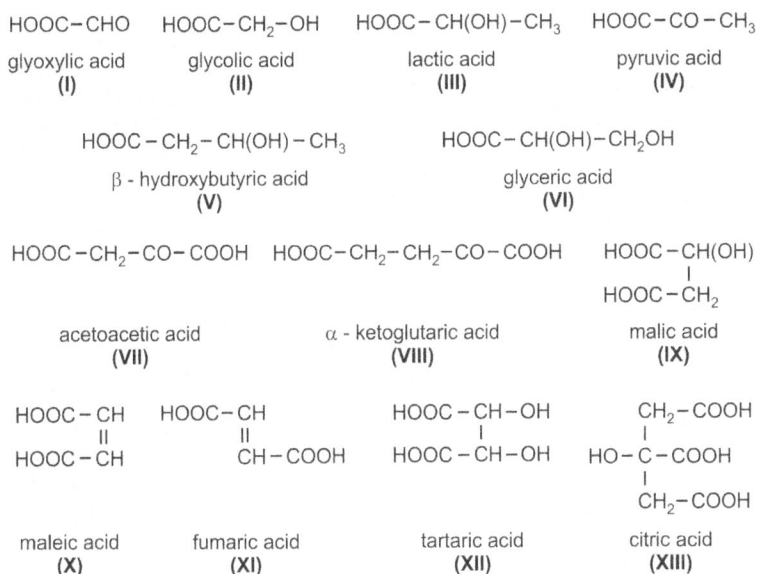

FIGURE 2.1   Mono- and polyhydroxy aliphatic carbonyl organic acids.

Pyruvic acid – is formed by decarboxylation of hydroxymaleic acid which, in turn, can be obtained from tartaric acid (Berzelius, 1835). Alcohol can be obtained from pyruvic acid by alcoholic fermentation, under the action of the enzyme decarboxylase (obtained from brewer's yeast). The reaction takes place in a neutral medium. In the field of biochemistry, pyruvic acid is known as a metabolite that is found in specific interactions of carbohydrate metabolism.

β-Hydroxybutyric acid – is found in plant tissues as a metabolite resulting from the oxidation of fats. It is found in larger quantities in the blood of diabetics, being eliminated in the urine from which it can be determined.

Glyceric acid – belongs to the group of aldonic acids (formed by the oxidation of aldoses). It is a metabolite present in plant tissues, but is also found in animal tissues.

Acetyl-acetic acid – is found in small amounts in plant and animal products, being an intermediate in the oxidative processes of lipids. In the human body, it occurs in high concentrations in the case of diabetes, a circumstance in which it can be determined from the urine of patients.

α-Ketoglutaric acid – is an intermediate metabolite in plant and animal organisms.

Malic acid (*hydroxysuccinic acid*) – is found in sour fruits, e.g., unripe apples, gooseberries and various plants (mountain ash, rhubar). Malic acid was detected in small quantities in wine and its calcium salt in the tobacco leaves.

Maleic acid and fumaric acid – are present mainly in the form of isomers: trans- (maleic), respectively cis- (fumaric) which occur in living organisms, characterizing some chemical transformations specific to carbohydrate metabolism. They are dicarboxylic acids with a chain of four carbon atoms.

Tartaric acid (*dihydroxysuccinic acid*) – is in both racemic and enantiomeric forms. It is isolated from potassium tartrate known as wine stone. It is found in larger quantities in wine yeast at the end of fermentation. Tartaric acid is used in the food industry in the preparation of lemonade, candy, obtaining effervescent powders.

Citric acid (*hydroxytricarboxylic acid*) – is widespread in nature. In the vegetable kingdom, it was isolated from various fruits (e.g., lemon, currant, raspberry, etc.), from vegetables, and even from sugar beet. From a biochemical point of view, citric acid is important because it intervenes in the carbohydrate metabolism present in anaerobic processes, being found in the cycle of tricarboxylic acids. Citric acid and tartaric acid are also used in the food industry. For this purpose, they are obtained by fermentative processes, using enzymes produced by various mycelium.

Among the organic acids with cyclic chains (aromatic or saturated) are: *mandelic acid* (XIV), *tropic acid* (XV) and *quinic acid* (XVI) - Figure 2.2.

Mandelic acid (*α-hydroxy-phenylacetic acid*) – is in the amygdalin glycoside - present in the seeds of bitter almonds, being released by hydrolysis.

Quinic acid (*1,3,4,5-tetra-hydroxycyclohexanecarboxylic acid*) – was found in the bark of the quinine tree (hence the name), in sugar beet, blueberries, coffee beans, and other plants (isolated even from hay).

Tropic acid (*α-phenyl-β-hydroxypropionic acid*) – was detected in the structure of some esters from the class of alkaloids, i.e. atropine and hyoscyamine.

2.4.2.3   Organic Acids in Biochemistry

Organic acids are commonly found as metabolic products in plant tissues, animal and human tissues.

There are plants with a high content of organic acids, often called "acidic plants". This group of plants include: sorrel (*Rumex acetosella*), rhubarb (*Rheum officinale*), begonia (*Begonia* sp.). Some of organic acids are used as food additive, more exactly as preservatives due to their antiseptic effects. Among the currently used preservatives from the group of organic acids are: acetic acid E 260, lactic acid E 270, propionic acid E 280. The use of antiseptic substances as food preservatives - by inhibiting the growth of bacteria (e.g. *Clostridium*, *Salmonella*, etc.) and fungi (molds) can ensure food safety.

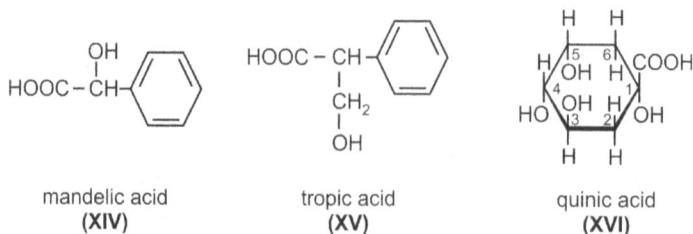

mandelic acid
(XIV)

tropic acid
(XV)

quinic acid
(XVI)

FIGURE 2.2   Organic acids with cyclic chains.

The role of *antioxidants* and their beneficial effects are currently being discussed in biochemistry.

In food biochemistry and in correlation with it, in the chemistry and technology of food processing, the problem of the use of antioxidants as additives is discussed. Some of the additives used as antioxidants are organic acids, e.g., citric acid E330, tartaric acid E334, ascorbic acid E300, and ascorbates.

An overview of the additives used as antioxidants reveals the complexity of this field. Some general information is provided below.

Antioxidants – are chemical compounds used to protect food from oxidation or self-oxidation (i.e. rancidity) and often to maintain the altered color by oxidation.

So, antioxidants play a role in prolonging shelf life. These groups include compounds with different structure and properties. Examples are:

a. *antioxidant agents*, e.g., ascorbic acid E 300 and Na or Ca ascorbates; natural extracts rich in tocopherol E 306 and α-synthetic tocopherol E 307; octyl gallate E 311 and dodecyl gallate E 312;

b. *substances with oxidizing action, but also with other functions*, e.g., sulfur dioxide E 220; sodium sulfate E 222, lecithin E 322, etc.;

c. *substances that potentiate the action of antioxidant agents*, e.g., citric acid E 330; tartaric acid E 334; orthophosphoric acid E 338 and its Na, K or Ca salts.

The interaction between food and antioxidant additives occurs in various ways. Mention is made on the ability of some antioxidants to retain oxygen, thus blocking oxidation. In other cases, the antioxidants disrupt the oxidation chain by interacting with the peroxide radical and deactivates it.

The problem of organic acids and their role as biologically active substances are confirmed by the fact that they are found in the form of specific metabolites in various biochemical pathways of biodegradation or biosynthesis of constituents in the human body (Rawn, 1991; Devlin, 1992).

2.4.3 ALKALOIDS

2.4.3.1 Synoptical Data

Alkaloids are nitrogenous organic compounds, mostly isolated from the plant kingdom, which are distinguished by the fact that they are "biologically active" substances having (conditioned by the chemical structure) physiological-active, pharmacological-active and toxicological- active specificity.

Alkaloids whose biological activity is limited by the physiological-active specificity highlighted in the case of food consumption were also isolated, e.g., caffeine, theophylline, etc.

This subgroup may also include alkaloids found in small amounts in various plant- or animal-derived foods. Structurally, alkaloids are mainly heterocyclic nitrogen compounds (Bodea et al., 1964–1966; Troll, 1973; Goodwin and Mercer, 1983; Brossi, 1989; Alais and Linden, 1993; Ciulei et al., 1993; Kutchan, 1995; Dewick, 1997; Bruneton, 2009; Gârban, 2018).

In plants, alkaloids are free or bound to organic acids (e.g., benzoic, citric, tartaric, oxalic, malic, succinic, meconic, etc.) or even inorganic (e.g., phosphoric and sulfuric) forming specific "salts". The extraction of alkaloids can be performed with organic solvents, preferably with alcohol, chloroform and benzene.

The generic name *alkaloids* (gr. *alkali* - alkaline; *eidos* - similar) was given to these compounds of plant origin because they are basic in nature and form salts with various acids. This name was introduced in 1819 by the pharmacist William Meissner (1792–1859) from Halle and has been retained, although it disagrees with the "theory of the structure of molecules". A similar situation can be exemplified to "vitamins".

It was originally believed that the formation of nitrogen compounds in the living world is a strict prerogative of plants. Subsequently, by studying the metabolism, other nitrogen compounds were detected, e.g., ammonia, biogenic amines, etc.

Ab initio a significant number of natural products, all with the common name of alkaloids, received names suffixed by "*ine*". Examples are nicotine, caffeine, atropine, morphine, etc.

Usually, in biochemistry, alkaloids are considered to be derived from amino acids. In plant tissues, they are in the form of complex mixtures, which are secondary metabolites, precursors of alkaloids.

During the extraction operations in liquid media of strong bases, the compounds were stabilized, being called "alkaloids". In the decades that followed (after 1819), these compounds were described as "Lewis bases" containing a nitrogen heterocycle or, improperly, considered an amine function. Owing to the non-nitrogenous electronic doublet, alkaloids were considered Lewis bases.

From a historical point of view, it is mentioned that since ancient times people have known the physiological, pharmacological and toxicological effects of certain plants. Thus, from antiquity, a "resinous extract" that was used as a sleeping pill (in the East) was obtained from poppy fruit. The stimulant effect of coca leaves, the antimalarial effect of cinchona bark (in the tropical regions of South America inhabited by the Incas) were also known.

The toxic effects of hemlock and night shade were also known (including in Europe). Below are general data on the chronological discovery of alkaloids and the contribution of various chemists, pharmacists, etc. It is mentioned in this sense the discovery of *morphine* (1803) – independent of Séguin, Courtois and Desorne; *noscapine* (1817) – discovered by Robiquet; *strychnine, quinine and brucine* (1818) by Pelletier and Caventou; *caffeine* (1820) by Runge and Robiquet; *coniine* (1927) by Gieseke; *nicotine* (1828) by Posselt and Reimann; *atropine* (1831) by Mein, Geiger, and Hesse; *codeine* (1832) by Robiquet; *theobromine* (1842) by Woskressenski; Merck's Poppy (1848); *cocaine* (1860) by Niemann; *apomorphine* (1870) by Matthiesen and Wright; *pilocarpine* (1875) by Gerard and Hardy; theophylline (1888) of Kassel; Schmidt's *scopolamine* (1892); *yohimbine* (1896–1897) by Spiegel and Thomas; *lobeline* (1921) by Wieland; *ergot alkaloids* (1918–1950) by Jacobs and Stoll; *reserpine* (1952–54) by Schlittler and Müller, etc.

The first alkaloid was extracted in 1803 independently of Séguin, Courtois and Derosne from opium. After crystallization, it was called "*magisterium opii*", noting its alkaline character, but attributing the alkalinity of the potassium hydroxyl used during extraction. In 1817, this alkaloid was purified by the pharmacist Sertürner from Hanover and was named "*morphine*" after the god Morpheus (the god of sleep and dreams).

The distribution of alkaloids in plants is a characteristic of specie, but it is also chronobiochemically conditioned depending on the stage of development of the plant (i.e. "phenophase") as well as soil (pedological) or climatic factors. Thus, it can be concluded that alkaloids are "nitrogenous bases" that, in the free state, are insoluble in water, and in the form of salts with organic or inorganic acids become soluble in water.

A characteristic of the biogenesis of alkaloids lies in the fact that their synthesis in green plants is done in some tissues and the storage in other tissues. For example, nicotine is synthesized in the roots of tobacco (*Nicotiana tabacum*) and accumulates in the leaves.

2.4.3.2 Heterocyclic Structure of Alkaloids

The study of alkaloid composition, chemical structure and activity, as well as their biogenesis has aroused continued interest in organic chemistry and biochemistry (Liener, 1969; Bonner and Varner, 1976; Ciulei et al., 1993; Bruneton, 2009).

The isolation of alkaloids was imposed for theoretical and applied reasons, given the stimulating action on material and energy metabolism, pharmacological effects and toxicological implications of these compounds.

Currently, the notion of *alkaloids* defines only nitrogen heterocyclic substances distinct by their basic character, specific physicochemical properties (colorless, optically active, bitter taste, precipitates with certain salts, e.g., KI, $PtCl_2$, etc. and with acids, e.g., phospho-wolframic, phospho-molybdenum, silico-tungsten, picric, etc.), biologically active properties on animal and human organisms.

The discussion of alkaloids is not the subject of detailed treatment in this volume, given the complexity of the issue in this area. However, general mention should be made of some alkaloids found in foods of plant origin, noting the possible toxicogenic risk. Details on alkaloids can be found in specialized treatises in the field of organic chemistry, plant biochemistry, pharmacology, toxicology, etc. (Brossi, 1989; Pelletier, 1998; Khan et al., 2013; Funayama and Cordell, 2015).

The production and localization of alkaloids in plants is explained by the fact that plant organisms do not have specific ways of eliminating nitrogenous products, similar to animals (e.g., renal elimination of urea, uric acid and other non-protein nitrogenous derivatives). Viewed in this context, alkaloids can be estimated as *storage substances* of nitrogen from heterocyclic compounds (Wiss, 1960; Neamțu, 1983; Dewick, 1997; Bhambhani et al., 2021).

Various amino acids are precursors for the N-heterocyclic structure of alkaloids. Thus, for example, pyrrolidine and piperidine heterocycles are formed from L-ornithine; from L-aspartic acid also results pyridine; L-tyrosine is synthesized from isoquinoline; indole is synthesized from L-tryptophan.

However, there are situations when heterocyclic compounds are synthesized from other precursors. And in this case, examples are: from xanthine results purine heterocycle; quinazoline and quinoline are formed from anthranilic acid. Further alkaloids are formed from these intermediate metabolites, represented by various heterocyclic compounds.

An overview of various alkaloids, considering the types of compounds, the molecular species of alkaloids and the main plant extraction sources are given in Table 2.2. It can be noted that in the biosynthesis of alkaloids, in many cases, amino acids or other metabolic intermediates are involved. The action of alkaloids on the human body as biologically active substances (with physiological, pharmacological or toxicological action) is exemplified by the effects induced by them.

TABLE 2.2
Alkaloids with N-Heterocyclic Structure

No.	Name	Structure	Name of Alkaloids (Examples)	Plant Source (Latin Name)	Primary Precursor (in Biosynthesis)
1	Acridine		Rutacridone	*Ruta graveolens, Ruta chalepensis*	
2	Quinazoline		Vasicine	*Adhatoda vasica*	Anthranilic acid
3	Quinoline		Quinine	*Cinchona officinalis*	L-tryptophan
			Dictamine	*Dictamnus albus*	Anthranilic acid
4	Quinolizidine		Lupinine	*Lupinus luteus*	L-lysine
			Lupanine	*Lupinus mutabilis*	
			Sparteine	*Sarothamnus scoparius*	
5	Indole		Yohimbine	*Corynanthe yohimbe*	
			Ajmalicine	*Rauvolfia* spp *Catharanthus roseus*	L-tryptophan
			Agroclavine	*Claviceps purpurea*	L-tryptophan
			Lysergic acid	*Claviceps purpurea*	L-tryptophan
6	Isoquinoline		Morphine	*Papaver somniferum*	L-tyrosine
			Papaverine	*Papaver somniferum*	L-tyrosine
			Narcotine (noscapine)	*Papaveraceae*	
			Berberine	*Berberis vulgaris*	
			Salsoline	*Salsola kali*	
			Lycorine	*Lycoris radiata*	L-tyrosine
			Codeine	*Papaver somniferum*	
			Thebaine	*Papaver bracteatum*	
7	Piperidine		Anabasine	*Anabasis aphylla*	L-lysine
			Sedamine	*Sedum sarmentosum*	L-lysine
			Lobeline	*Lobelia inflata*	L-lysine
			Coniine	*Conium maculatum*	L-lysine
			Conhydrine	*Conium maculatum*	L-lysine
8	Pyridine		Nicotine	*Nicotiana tabacum*	L-aspartic acid
			Nicotyrine	*Nicotiana tabacum*	
			Anabasine	*Anabasis aphylla*	
			Ricinine	*Ricinus communis*	
			Arecoline, arecaidine	*Areca catechu*	
			Tenellin	*Beauveria tenella*	L-phenylalanine

(Continued)

TABLE 2.2 (*Continued*)
Alkaloids with N-Heterocyclic Structure

No.	Types of Heterocyclic Compounds		Name of Alkaloids (Examples)	Plant Source (Latin Name)	Primary Precursor (in Biosynthesis)
	Name	Structure			
9	Pyrrolidine		Stachydrine	*Medicago sativa*	L-ornithine
			Betonicin	*Betonica officinalis*	
			Hordenine	*Hordeum vulgare*	
			Hygrine	*Erythroxylum coca*	L-ornithine
			Cuscohigrine	*Erythroxylum coca*	L-ornithine
10	Pyrolysidine		Isatidine, Retrorsine	*Senecio isatideus*	L-ornithine
11	Purine		Caffeine	*Coffea arabica*	Xanthine
			Theophylline	*Camellia sinensis*	
			Theobromine	*Theobroma cacao*	Xanthine
12	Tropane		Atropine	*Atropa belladonna*	L-ornithine
			Hyoscyamine	*Hyoscyamus niger*	L-ornithine
			Scopolamine	*Scopolia carniolica*	L-ornithine
			Cocaine	*Erythroxylum coca*	L-ornithine
			Ecgonine	*Erythroxylum coca*	L-ornithine

In this sense, it is mentioned that: *morphine* acts on the central nervous system (CNS) producing drowsiness; *papaverine* has a similar action; *cocaine* acts on the peripheral nervous system and is suitable for use as an anesthetic; *coniine and conhydrine* (present in hemlock) are poisons that act on the CNS with direct effects on the respiratory system; *strychnine* is also poison; *quinine* used as a medicine in case of malaria; *codeine* is used in antitussive drug therapy, etc. In general, alkaloids act on the nervous system and smooth muscles. The physiological mechanisms are largely known, noting that the biochemical (molecular) mechanisms are only partially known.

2.4.3.3 Representative Compounds

Given the specificity of this book, some references will be made to the natural alkaloids present in products of plant origin of food and/or pharmaceutical interest. The presentation of some examples of representative alkaloids, with the rendering of the structural formulas, is made taking into account the specificity of the heterocycles that define the various classes of compounds.

2.4.3.3.1 Alkaloids with Purine Heterocycle

The main compounds in this class (group) are: *caffeine* (1,3,7-trimethyl-xanthine) - I; *theophylline* (1,3-dimethyl-xanthine) - II; *theobromine* (3,7-dimethyl-xanthine) - III, present in the leaves and berries of various plants. Their structural formulas are shown in Figure 2.3. Alkaloids most commonly found in products used for human consumption are also mentioned.

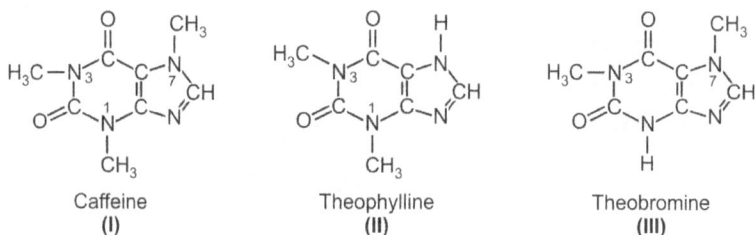

Caffeine
(I)

Theophylline
(II)

Theobromine
(III)

FIGURE 2.3   Alkaloids with purine heterocycle.

Caffeine – is present in coffee beans (*Coffea arabica*) in varying amounts (1.0%–2.8%). It is soluble in cold and hot water.

Theophylline – is found in tea leaves (*Camellia sinensis*) along with caffeine. It has a stimulating effect on the CNS and the heart. It also has a diuretic effect. It is soluble in hot water, which explains its presence in tea.

Theobromine – is present in the seeds produced by the cocoa tree (*Theobroma cacao*) in a proportion of 1.5%–3.0%, in coffee beans and tea leaves. It has a more stimulating action and more intense effects. The crystalline substance has a bitter taste. It is soluble in cold and hot water.

2.4.3.3.2   Alkaloids with Tropane Heterocycle

Tropane is a heterocycle that has a bicyclic tertiary amine structure. It is the source of more than 20 alkaloids that release an acid and an amino alcohol upon hydrolysis. The released amino alcohol is a derivative of tropane. Alkaloids with this composition have been isolated from various plants belonging to the families of solanceae, convolvulaceae, dioscoraceae, etc.

The main amino-alcohols derived from *tropane* (IV) are: *tropine* (V); *scopine* (VI); *ecgonine* (VII); *3,7-dihydroxytropane* (VIII) (Figure 2.4). The chemical names of the various compounds in the alkaloid group can be found on the IUPAC websites.

Atropine is formed from tropine and (+) tropic acid, *hyoscyamine* is formed with (−) tropic acid, and *convolamine* is formed with veratric acid. Similarly, *scopolamine* is formed from scopine and (−) tropic acid, *cocaine* is formed from ecgonine and benzoic acid with methanol, and from ecgonine, cyanic acid and methanol from *cinnamyl-cocaine*. Finally, *veleroidine* may result from 3,7-dihydroxytropane and isovaleric acid.

The main alkaloids with tropic heterocycle are: *atropine* (IXa; IXb), *hyoscyamine* (X), *scopolamine* (XI) and *cocaine* (XII) – briefly presented below (Figure 2.5)

Atropine – is the ester of tropine with (±) tropic acid. It was preferably isolated from nightshade (*Atropa belladonna*). This compound contains tropine and (±) tropic acid (hence the equimolecular mixture). (+)Atropine (IXa) was formed with (+) tropic acid, and (-) atropine (IXb) was formed with (-) tropic acid. The separation of these compounds was done by fractional crystallization. Atropine was isolated in 1831 by Main. Its correct structure was determined by Wilstätter (1898).

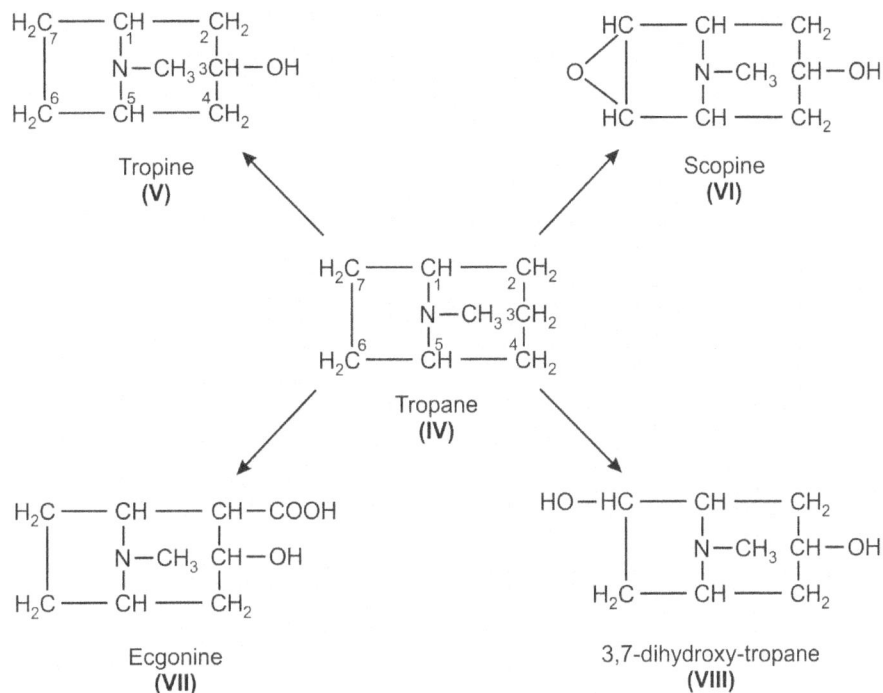

FIGURE 2.4 Tropane and its hydroxylated/oxygenated derivatives.

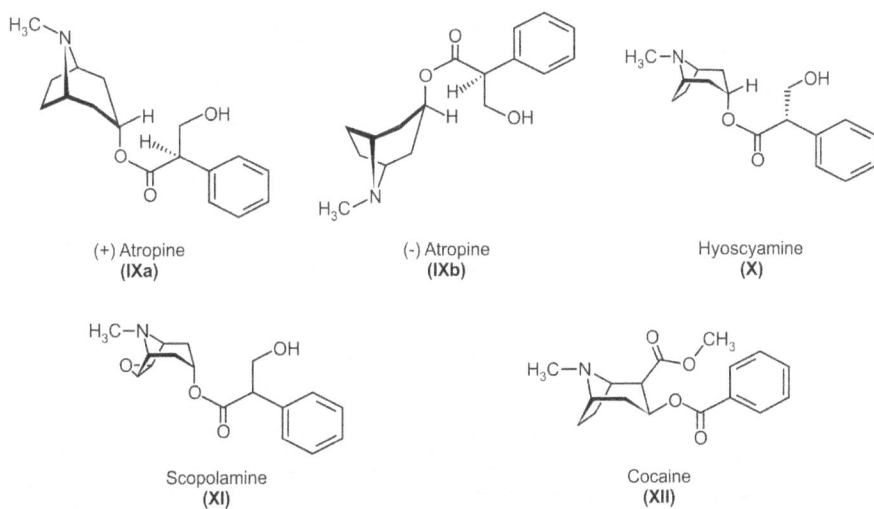

FIGURE 2.5 Molecular structure of major tropane alkaloids.

Atropine is used in medicine as inhibitor of cholinergic receptor. At low doses (about 0.2–0.3 mg), it causes bradycardia (reduced heart rate), and at high doses (about 0.5–0.75 mg), it has tachycardic effects (accelerated heart rate).

In ophthalmology, atropine is used for the effect of mydriasis (pupil dilation) to facilitate examination of the retina of the eye.

Hyosciamine – is the ester of tropine with (–) tropic acid. It is mainly isolated from *Hyosciamus niger* and *Datura stramonium*. It has many similarities in physicochemical and physiological properties with atropine. In plants hyosciamine is found in larger quantities. In terms of physiology (pharmacology) it is known for its sympathetic effects. These are manifested by tachycardia, mydriasis, decreased secretions (salivation and perspiration), as well as a decrease in intestinal transit.

Scopolamine – was formed as a *scopine* ester with (–) tropic acid. It is found in larger quantities in nightshades. Scopolamine is a levogir compound. In the case of mild hydrolysis with lipase (in the body) or in a neutral medium (in the laboratory) from scopolamine, inactive scopine and (–) tropic acid are formed.

Cocaine – is presented as a double ester of the hydroxy-amino acid (–) *ecgonine* with methyl alcohol and benzoic acid. The original heterocycle is the tropane. This alkaloid was extracted from the leaves of Erythroxylon coca, a shrub from South America. Coca leaves were used by South American Indian aborigines as a stimulant/exciting.

It has a narcotic effect, stimulating nervous central system (NCS). It is one of the most widespread drugs that cause the "addiction" of the consumer. The discovery of the primary extract has been disputed by various chemists in Germany and Italy. However, it is known that pure cocaine was isolated in 1923 by Richard Willstätter. The physiological/pharmacological action of cocaine is to inhibit the effects of dopamine, norepinephrine and serotonin. Cocaine intervenes at the level of neuronal synapses, raising the sensitivity threshold of the receptors. This creates a state of euphoria and addiction to cocaine.

2.4.3.3.3 Alkaloid with a Pyridine Heterocycle

There are alkaloids with a pyridine heterocycle (e.g., ricinine) and alkaloids with two heterocycles, namely, pyridine and pyrrolidine (e.g., nicotine and nicotyrine).

The alkaloid *ricinine* (XIII) with the unique pyridine heterocycle is shown separately (Figure 2.6).

Ricinine – is an alkaloid named (1-methyl-3-cyano-4-methoxy-2-pyridone) and was isolated from young castor plants (*Ricinus communis*). There is a methoxy group and a cyan group in the structure of ricinine, both of which can generate toxic compounds by biotransformation in the body. It is characterized by bitter taste and high toxicity (especially seeds).

Ricinine has also been isolated from *Piper nigrum, Discocleidion rufescens, Nicotiana tabacum,* etc. The compound has been shown to be toxic to humans and animals.

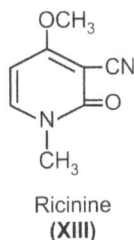

Ricinine
(XIII)

FIGURE 2.6 Pyridine alkaloid.

A toxic compound called *ricin* has also been found in castor (*Ricinus communis*) plants. This is a *glycoprotein* with a mass of 60,000 Da. It consists of two polypeptide chains. Chain A is responsible for its toxic properties by being involved in inhibiting protein synthesis by attacking RNA in ribosomes. Chain B is involved in fixing ricin on the cell wall.

2.4.3.3.4   Alkaloids with a Pyridine Heterocycle and a Pyrrolidine Heterocycle
This category includes *nicotine* (XIV), *nicotyrine* (XV) and *anabasine* (XVI) - Figure 2.7.

Nicotine (β-pyridine-N-methyl-pyrrolidine) – has been isolated from various tobacco species, e.g. *Nicotiana tabacum* (approx. 2%) and *Nicotiana rustica* (approx. 8%) - Solanaceae family. It is in the form of salts with citric and malic acid. Along with nicotine in *Nicotiana* sp. there are about 10 other alkaloids.

Tobacco use was historically attested, having its origins in Peru, and then in the Maya people of America. It was brought to Europe in 1518 by the Spanish people. From here it was taken to France by the French ambassador Jean Nicot in 1559. Then it spread throughout Europe.

Nicotine is extracted by steam entrainment. The extraction product is a colorless or light-yellow colored liquid with a characteristic odor. Small amounts of nicotine were also found in tomatoes (*Solanum lycopersicum*) and potatoes (*Solanum tuberosum*).

Nicotine is a psychoactive substance with anxiolytic and antidepressant character, exciting the nervous system (central and peripheral), with hypertensive effects. It also activates digestive secretions (salivary, gastric and intestinal). It enters the body by inhalation, ingestion, but also by contact. If inhaled, it is absorbed through the pulmonary blood capillaries.

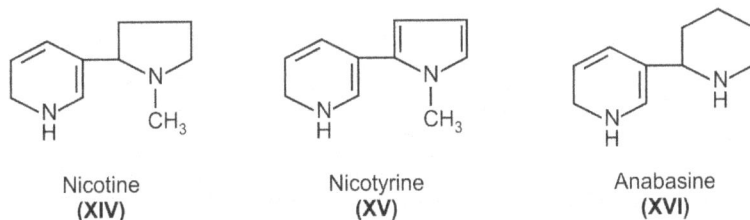

| Nicotine | Nicotyrine | Anabasine |
| **(XIV)** | **(XV)** | **(XVI)** |

FIGURE 2.7 Alkaloids with pyrimidine and pyrrolidine heterocycles.

After crossing the heart (left), it continues through the blood-brain barrier and reaches the brain in 10–20 seconds (note that it does not cross the hepatic system). Nicotine levels in blood plasma are 6–10 times higher than in venous plasma. There are more recent studies (Bullen et al., 2010), comparative, which show that the pharmacokinetic properties differ from conventional cigarettes compared to electronic cigarettes. For example, the maximum plasma peak for conventional cigarettes is after 14.3 minutes, and for e-cigarettes after 19.6 minutes (this time increases if an inhaler is used).

The physiological influence of nicotine lies in increasing blood pressure and heart rate. In this situation, the release of adrenaline also increases. Numerous metabolic changes occur.

There is also a transient effect ameliorating memory, ability to concentrate. Smoking also induces a mental dependence due to the presence of nicotine. Withdrawal from smoking can cause undesirable phenomena, e.g. headache. In extreme situations, depression occurs, which can last 3–4 days or even a few weeks after quitting. A special effect of nicotine lies in the abnormal production of dopamine, which creates a state of euphoria.

When the dopamine level decreases, the smoker is in a position to resume smoking. Nicotine overdose and intoxication occur when active smoking persists. In such situations, characteristic symptoms appear; daytime hyperactivity, nausea, vomiting, pallor, palpitations/tachycardia, headache, insomnia, lipothymia, vertigo, abdominal pain, general weakness, etc.

Nicotine addiction, although with minor consequences, has some similarities to heroin and cocaine addiction. Chronic use of cigarettes has been shown to be harmful both due to nicotine and especially to the combustion products of cigarette paper (generally polycyclic aromatic hydrocarbons). The effects, in the long run, lie in severe respiratory disorders ending in lung cancer, etc.

Nicotyrine – has a nicotine-like structure, being a *bisdehydronicotine* derivative. It is formed as an intermediate compound during nicotine biosynthesis. It looks like a colorless liquid.

Anabasine – is an alkaloid present in higher quantities in the plant *Anabasis aphylla* from the *Amaranthaceae* family and in smaller quantities in *Nicotiana* sp. It is a levogir compound, isomer with nicotine.

2.4.3.3.5 Quinoline Heterocycle Alkaloids

The quinoline heterocycle (2,3-benzopyridine) was originally isolated from coal tar (Runge, 1834). Figure 2.8 shows the structure of *quinoline* (XVII). These tars also contain numerous methylated derivatives. However, quinoline heterocycle alkaloids have also been isolated from the bark of *Cincona* and *Remijia* trees in the Andes Mountains.

The bark of these trees was used by the natives in the treatment of malaria. This treatment was introduced in Europe in the 17th century. Subsequently, these trees have been cultivated in Indonesia and India.

The main alkaloids isolated from the bark of "quinine trees" were *quinine* (XVIII) and *cinchonine* (XIX). Structurally, quinine differs from cinchonine in that it possesses a methoxy group at the C6 position of the quinoline nucleus. The structural formulas of their alkaloids are given in Figure 2.9.

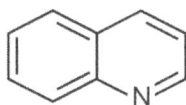

FIGURE 2.8 Quinoline (XVII).

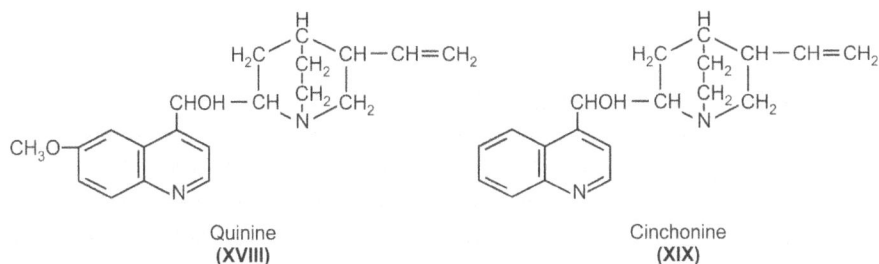

Quinine
(XVIII)

Cinchonine
(XIX)

FIGURE 2.9 Alkaloids with quinoline heterocycle.

In the plant, the alkaloids are bound to specific acids: quinic acid and quinovic acid. The isolation of alkaloids was done by Pelletier and Caventou (1820). The structural specificity of the alkaloids was confirmed by oxidation reactions. Quinidine and cinchonidine – their stereoisomers – have also been isolated from the above-mentioned alkaloids.

> *Quinine* – may be anhydrous and crystallized. It is slightly soluble in alcohol and ether and hardly soluble in water and benzene. In clinical pharmacology, it is known to be used in the treatment of malaria. Subsequently, a general antipyretic effect has been reported, correlated with the fact that it acts on the nerve center of the body's thermoregulation.
> *Cinchonine* – is soluble in acids and alcohol and insoluble in water. Numerous derivatives are formed from quinone during biogenesis in the quinine shell. Subsequently, quinoline derivatives used as antimalarial drugs were also obtained, using laboratory synthesis.

2.4.3.3.6 Alkaloids with Isoquinoline Heterocycle

In the constitution of these alkaloids, there is an aromatic heterocycle called *isoquinoline* (XX) or benzopyridine (Figure 2.10). This is a structural isomer of quinoline.

The term isoquinoline (considered classical) is also used in the case where it refers to isoquinoline derivatives. Thus, for example, the 1-benzyl-isoquinoline derivative is the structural backbone of alkaloids in the group which includes papaverine.

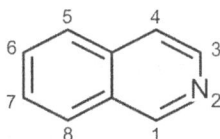

FIGURE 2.10 Isoquinoline (XX).

2.4.3.3.6.1 With Classical Isoquinoline Heterocycle In this group are: *morphine* (XXI); *codeine* (XXII); and *thebaine* (XXIII) – Figure 2.11.

> **Morphine** – also known as (5, 6) -7,8-didehydro-4,5-epoxy-17-methylmorfi-
> nan-3,6-diol (cf. IUPAC) – is an alkaloid found in poppy latex with opium
> content. Its amount in opium is 8%–14%. The latex is extracted from the
> poppy (*Papaver somniferum*) capsules in the wet state. Morphine has many
> chemical derivatives, among which the most important are: heroin, hydro-
> morphone, naloxone, ethylmorphine, tramadol, etc.
>
> It is estimated that the poppy has been cultivated since the year 7000
> (BC) in Mediterranean Europe, as well as in the valleys of the Rhine,
> Rhone, Po and Danube (Danube) rivers. In the year 3400 (BC), it was
> known in Mesopotamia. Later, Alexander the Great brought it to Persia
> and India, where it was cultivated. In the year 400 (AD), it was also cul-
> tivated in Egypt.
>
> In the Middle Ages, however, it was prohibited in Europe. In 1522,
> Paracelsus created "laudano" - an opium-based alcoholic beverage with
> moderate analgesic use. In the US Civil War, it is estimated that there were
> about 400,000 victims of morphine addiction, known and used to give
> "strength to soldiers."
>
> Starting with morphine, in 1874, heroin (diacethylmorphine) was syn-
> thesized. The synthesis of heroin from morphine was done by acetylating
> the two hydroxyl radicals of the molecule. It was later marketed in 1898 by
> German Bayer Laboratories.
>
> Morphine is a compound used as a drug that has an opioid agonist effect.
> It acts preferentially on the NSC. Its effects are manifested by supraspi-
> nal analgesia, euphoria, respiratory depression, miosis (shrinking of the
> pupil of the eye), inhibition of digestive motility (followed by constipation).
> Action of morphine on cognitive ability is negative targeting anterograde
> and retrograde memory. Repeated use gives analgesic effects, but also pro-
> duces addiction.
>
> Biotransformation (in metabolic processes) is accomplished by con-
> jugation reactions, e.g. conjugation with *glucuronic acid* (resulting in 2-,
> 3-, 6-glucuronide-morphine). Biotransformation can also be done by

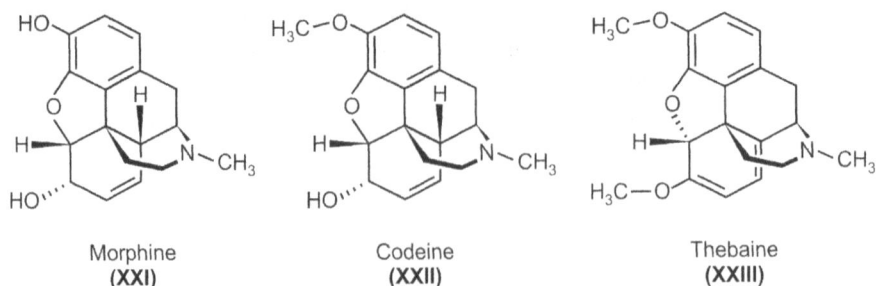

| Morphine | Codeine | Thebaine |
| (XXI) | (XXII) | (XXIII) |

FIGURE 2.11 Alkaloids with classical isoquinoline heterocycle.

N-demethylation (in a small proportion). Excretion is performed at the renal (90%) and biliary (10%) levels. Therapeutic administration (in the form of a medicinal product) of morphine is done orally or by subcutaneous injection, intravenously, even inhalation (using a nebulizer).

Codeine – is also an alkaloid obtained from opium in which the extract is 0.7%–2.5%. Codeine is also known as *methylmorphine*. It is also an analgesic, but weaker than morphine. It is used in antitussive and antidiarrheal drugs.

Biotransformation occurs in the liver, being inactivated by conjugation with glucuronic acid, resulting in *codeine-6-glucuronide* (inactivation in a proportion of about 80% by glucuronide-conjugation). It was also obtained by laboratory synthesis (later and industrial) by O-methylation of morphine.

Thebaine – is also known as *codeine-methyl-enol-ester*. The name thebaine comes from the ancient city of Thebes in Egypt. This alkaloid has structural similarities to morphine and codeine. It is obtained from the extract of *Papaver bracteatum* (Iranian poppy). Other compounds with alkaloid attributes can be obtained industrially from the original extract, such as *oxycodone, oxymorphone, nalbuphine, naloxone, etorphine*, etc.

2.4.3.3.6.2 With 1-Benzylisoquinoline Heterocycle From this group, there are several representative compounds: *papaverine* (XXIV); *narcotine* (XXV) also called *noscapine*; *laudanosine* (XXVI) – Figure 2.12.

Papaverine – chemical name 1- (3,4-dimethoxybenzyl) -6,7-dimethylquinoline, is an isoquinoline alkaloid isolated from poppy latex (*Papaver somniferum*).

Structurally papaverine has a 1-benzylisoquinoline nucleus, which are grafted to methoxy ($-O-CH_3$) groups. Papaverine was discovered in 1848 by Georg Merck (1825–1873) who was a student of the German chemists Justus von Liebig and August von Hofmann. The discoverer of papaverine was the son of Emanuel Merck (1794–1855) - who was the founder of the "Merck Corporation" - a German chemical and pharmaceutical company.

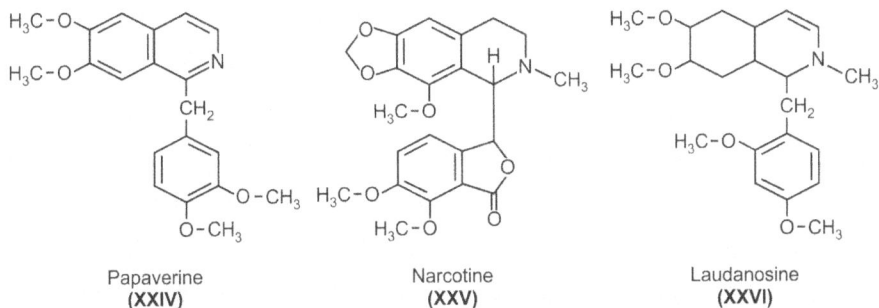

Papaverine
(XXIV)

Narcotine
(XXV)

Laudanosine
(XXVI)

FIGURE 2.12   1-benzyl-isoquinoline heterocycle alkaloids.

It is mainly used in analgesic treatments for visceral spasms of the gastrointestinal tract, biliary and ureteral crises. It is also used as a vasodilator in the brain and coronary diseases, as well as in subarachnoid hemorrhages.

It is also used in erectile dysfunction (ensuring blood flow to the corpora cavernosa). The mechanism of action *in vivo* is not fully known, but a selective inhibition of the enzyme phosphodiesterase has been observed, which causes an increase in cyclic adenosine monophosphate (cyclic APM) levels.

The less common side effects of papaverine are polymorphic ventricular tachycardia and increased enzymatic markers (transaminases and alkaline phosphatases). Drowsiness, dizziness, etc. were also observed.

Following chronic administration, motor and cognitive impairment and increased anxiety were observed in mice.

Narcotine – a benzoquinoline alkaloid obtained from poppy latex, but is also present in other plants of the *Papaveraceous* family. It has an analgesic effect. It was first used as an antitussive drug. It also has anti-proliferative properties.

Laudanosine – is a benzyl-tetrahydroisoquinoline alkaloid, also known as *N-methyl-tetrahydropapaverine*. It is found in small amounts in poppy latex (0.1%) from where it was first isolated in 1871. As expected, knowing the structure, by partial dehydrogenation laudanosine is converted to papaverine. Laudanosine interacts with GABA receptors and glycine receptors, opioid and nicotinic receptors.

2.4.3.3.7 Alkaloids with Steroid Nucleus

The best-known steroid alkaloids are *solanidine* (XXVII) and *tomatidine* (XXVIII). The chemical structure of these substances is shown in Figure 2.13. Steroid alkaloids contain 27 carbon atoms, one nitrogen atom and have a –OH group at the C_3 position of the cyclopentane perhydro-phenanthrenic nucleus. These alkaloids are present as aglycones in certain heterosides (glycosides). In plants, these alkaloids are bound to carbohydrate fragments to form glycosidic (holoside) compounds. For this reason, in the literature, it also appears with the name "*glycoalkaloids*".

Solanidine – isolated from potatoes (*Solanum tuberosum*) – where it is found in larger quantities in the peel of tubers from younger plants.

The properties of solanidine are similar to those of steroids. Solanidine is the solanine aglycone - a heteroside that also contains a triglyceride (Glc-Gal-Rha).

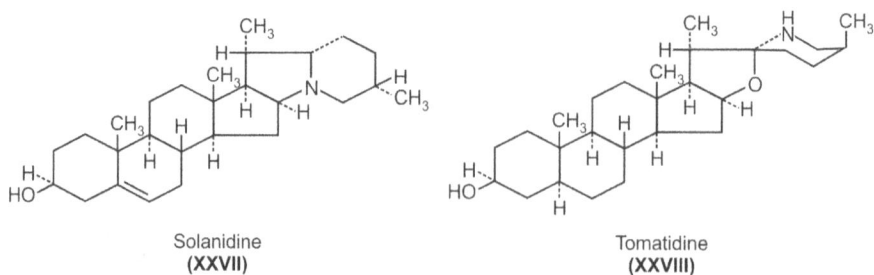

Solanidine
(**XXVII**)

Tomatidine
(**XXVIII**)

FIGURE 2.13 Alkaloids with steroid nucleus.

Tomatidine has been isolated from tomato leaves (*Solanum lycopersicum*) and other plants in the *Solanaceae* family.

Tomatidine is the aglycone of tomatine – a heteroside that also contains tetrasaccharides (Glc-Glc-Gal-Xyl). The effects of solanidine and tomatidine glycoalkaloids have also been studied in relation to embryology in female mice.

It was found to cause a decrease in the weight of the products of conception (more accentuated in solanidine −36.1 compared to tomatidine −11.9), a decrease in the number of products of conception (in solanidine −27.0, and in tomatidine −15.5).

Other alkaloids with a steroid core are: *veratridine* - extracted from the *Veratrum album* belonging to the *Melanthiaceae* family genus *Schoenocaulon*; *conessine* - obtained from the bark and seeds of the Indian shrub (*Holarrhena antidysenterica* belonging to *Apocynaceae* family); *muldamine* - isolated from *Veratrum californicum* with teratogenic effects; *zygacine* - isolated in 1913 from plants belonging to the genera *Toxicoscordion* and *Zigadenus*, etc.

2.4.3.3.8 Other Alkaloids

Piperine – is an alkaloid with a piperidine heterocycle present in peppercorns. It is insoluble in water. Piperine gives a specific taste to peppercorns.

Hordein – is present in small quantities in barley grains and in larger quantities in germinated barley. Stimulates heart activity and increases blood pressure. It also activates diuresis (elimination of urine).

Without being alkaloids (by definition, we retain the N-heterocyclic character of alkaloids), compounds with homocyclic structure, basic character and biologically active properties are also known. One such non-alkaloid compound is *capsaicin* which is an aromatic amide.

It was isolated from the fruits of the pepper (*Capsicum annuum*). The presence of capsaicin gives the spicy taste of these vegetables.

2.4.3.4 Alkaloids in Biochemistry

The presence of various alkaloids with attributes of biologically active substances in plant products (and sometimes in animal products) is of interest for the effects induced on the body when accessing food, but especially when used in pharmacology. Although they are low in natural products, alkaloids are important for their physiological/pharmacological role and, in extreme cases, for their toxicological effects.

These considerations were the basis for the inclusion of alkaloids in the group of biologically active substances, being considered physiologically active when accompanying nutrients from food or pharmacologically active, when used therapeutically, but also toxicologically active in accidental circumstances.

It is generally known that in initially isolated forms the alkaloid molecules exhibit high toxicity. At low doses, however, it has physiological, pharmacological (acute or chronic) activity with some delayed effects, or even toxicological (with some delayed effects). Thus, the use of small amounts of caffeine in coffee, cocaine in coca leaves, and nicotine in tobacco, was accepted in various human cultures, often very old. It was difficult to assess the reason for the various effects, e.g. stimulants, tonics, psychotropics, psychoactive, vomiting, sedatives, analgesics, etc.

It is now known that pure alkaloid molecules are often very toxic: e.g. strychnine, atropine, aconitine, cocaine, etc. In certain controlled doses, it is suitable for use in therapy, for example: morphine and codeine are used in therapeutic protocols for anesthetic purposes; quinine is considered an antimalarial agent; vinblastine and vincristine are used as anticancer agents.

Investigations into chemical compounds present in plants, in general, and in medicinal plants in particular, have led to the isolation of a significant number of nitrogenous compounds, including proteins, amines, amides, some vitamins and antibiotics, and numerous alkaloids. From a biochemical point of view, the alkaloids, as presented, are organic substances with nitrogen content in the molecule, mostly of vegetable origin but also of animal origin. Their structure being extremely varied, a classification based on the structure has a conventional character, distinguishing:

a. *true alkaloids* – this subgroup includes natural compounds of plant origin in which nitrogen is in the structure of heterocyclic nuclei. Such compounds may be derived from amino acids such as L-lysine, L-ornithine, L-tryptophan, etc.;

b. *pseudoalkaloids* – are substances that contain nitrogen integrated into a heterocycle, but which do not come from amino acids. This subgroup may include terpene alkaloids and sterol alkaloids;

c. *protoalkaloids* – compounds of vegetal origin in which nitrogen has an extra-cyclic arrangement. This subgroup generally includes biogenic amines which come from the decarboxylation of amino acids, e.g., ephedrine, mescaline, etc.

To have an image of the relationship between the original compound and the product resulting from tissue biogenesis Funayama and Cordell (2015) designed a scheme of the interrelation initial compound – final metabolite (Table 2.3). This table does not show the specific N-heterocycles that actually are at the origin of alkaloids.

It is generally estimated that alkaloids, as biologically active substances, are products of the secondary metabolism of superior plants. Their amount is up to 8% in the plant kingdom. The spread in the plant kingdom is very varied. For example, it is found in monocotyledonous plants belonging to the families *Liliaceae* and *Amaryllidaceae*. More commonly, however, it is found in dicotyledonous plants belonging to the families *Solanaceae, Papaveraceae, Rutaceae, Rubiaceae, Leguminosae, Ranunculaceae, Compositae,* etc.

The distribution of alkaloids in plant organs differs greatly in terms of quantitative ratio. For example, it predominates in the roots or rhizomes of *Atropa belladonna, Rauvolfia* sp., *Veratrum album*; in leaves of *Erythroxylon coca, Datura stramonium, Hyoscyamus niger, Nicotiana tabacum,* etc.; in the shell of *Berberis vulgaris, Cinchona* sp., etc.; in fruits at *Papaver somniferum*; in seeds at *Colchicum autumnale, Coffea arabica, Theobroma cacao,* etc.

It is noted that the number of alkaloids differs greatly from one plant to another. Sometimes their number exceeds 40 molecular species. Usually however, there is a primary alkaloid and a number of related structural secondary alkaloids. For example, in opium, there is morphine, thebaine, codeine, as substances related to the morphine nucleus.

TABLE 2.3
Source of Alkaloids

No.	The Original Compound	The Main Derived Alkaloids
1	Tryptophan	6,6′-Dibromoindigo, auxine, C-curarine, camptothecin, evodiamine, harmine, ibogaine, indigo, LSD, nigakinone; physostigmine, psilocin, psilocybin, pyrolnitrin, quinine, rezerpine, rincofiline; rutecarpine, serotonin, rosporin; streptonigrin, strychnine, trichotomine, VCR, VLB; yohimbine
2	Lysine	Anabasin, lobeline, matrine, pelletierin, piperine, sparteine
3	Proline	Prodigiosin, pyrrole-2-carboxylic acid, stachydrine
4	Glutamic acid	Acromelic acid, GABA, ibotenic acid, kainic acid
5	Histidine	Cimetidine, histamine, pilocarpine
6	Phenylalanine and tyrosine	Aristolohic acid, berberine, chelidonine, coclaurine, colchicine, d-tubocurarine, dopamine, emetine, galantamine, DOPA, licorine, morphine, phenylethylamine, thyroxine
7	Ornithine and arginine	Agrobatine, atropine, cocaine, ephedrine, kukoamine, nicotine, putrescine, spermine, tetrodotoxin
8	2,3-Diaminopropionic acid	Albizziine, quisqualic acid
9	Anthranilic acid	Acridone, acronicine, echinopsine, febrifugine, iodinin, kokusagine, melochinone, damascenin, phenazinomycin, pyocyanin, rutacridone, vasicine
10	Nicotinic acid	Arecoline, nicotinamide, nicotinic acid, ricin, vitamin B6
11	Purine compounds	5-Fluorouracil, caffeine, folic acid, pteridine, vitamin B1
12	Porphin compounds	Chlorophyll and its derivatives
13	Terpenoid derivatives	L-Deoxy-xylulose, aconitine, actinidine, atisine, batrachotoxin, chaconine, conessine, delphinine, gentianine, homobatracotoxin, ignavine, jesaconitine, loganine, mesaconitine, mevalonic acid, napelline, nupharidine, peiminoside, secologanine, solanine, sweroside, swertiamarin, taxol, verticine, zygacine
14	Polyketide compounds	Coniine, nigrifactin, piericidin A1, pinidine

Source: after Funayama Shinji and Cordell (2015) – modified.

A similar example is presented by *Secale cornutum* alkaloids having similar structures originating lysergic nucleus. In other cases, there is a circumscription of the distribution of alkaloids, which they have encountered favorite (almost exclusively) in certain plants, e.g., morphine in *Papaver somniferum*, vincristine in *Catharanthus roseus*, quinine in *Cinchona* sp.

There is also a distribution of certain alkaloids in plant species belonging to the same family, e.g., hyoscyamine and scopolamine are found in plants of the family *Solanaceae*. Often, however, a certain alkaloid is found in many species of plants, e.g., nicotine is found in various plants that taxonomically belong to 13 botanical families.

Alkaloid biogenesis has been and continues to be a vast field of research. It has been found that, in general, biosynthetic processes take place in young tissues of plants with more intense metabolic activity.

Also, in the research of alkaloids, the aim was to obtain, by processes of semi-synthesis/laboratory synthesis of compounds with similar properties and effects natural products. In the case of opium alkaloids, for example, they are thought to be:

- *natural compounds* – morphine, codeine, thebaine, papaverine, narcotics, etc.;
- *semi-synthetic compounds* – hydromorphone, hydrocodone, and heroin;
- *synthetic compounds* – fentanyl, pethidine, methadone, and propoxyphene.

Owing to the compounds that are at the origin of the alkaloids, alkaloids are distinguished whose biogenesis starts from various amino acids. Hereinafter, for information, here are some examples.

Lysine – is the precursor amino acid from which N-heterocyclic compounds mixed like pyridine, piperidine and quinoline (benzene pyridine), and then alkaloids are formed.

Ornithine – is the precursor of N-heterocyclic derivatives: pyrrolidine, tropane, pyrrolysine. Specific alkaloids are then formed.

Tyrosine – is at the origin of alkaloids which have N-heterocycles as intermediate precursors *tetrahydroisoquinoline, phenylethylisoquinoline* (isoquinoline being an isomer of quinoline).

Tryptophan – in turn can generate complex indole N-heterocycles with ergoline, yohimbine, reserpine, ajmaline, quinolinic structure – from which finally derives typical alkaloids.

Histidine – generates an imidazole nucleus from which, during the metabolic reactions, form specific alkaloids.

Also, there are alkaloids whose biogenesis starts from *other groups of compounds.* Examples are:

Purine – with specific mixed nucleus (pyrimidine and imidazole) generates straight xanthine precursor resulting in: caffeine, theobromine and theophylline.

Compounds resulting from amination reactions – can generate intermediates, which evolves into terpenoid or steroid core alkaloids.

Anthranilic acid – in turn, can generate intermediates with a structure of quinoline N-heterocycle or acridine N-heterocycle. Finally, they evolve to specific alkaloids.

Alkaloids are known to have been found in many cases use in phytotherapy. Gradually, with the progress made in pharmaceutical biochemistry and pharmacognosy, various extraction compounds of interest have been obtained.

2.4.4 PHYTONCIDES

2.4.4.1 Synoptical Data

Phytoncides have specific to biologically active natural compounds isolated from plants and considered "plant antibiotics".

The name phytoncides was introduced by Russian biologist Boris Tokin from University of St. Petersburg in 1928 and derives from the effects of extracts against insects, animals, microbes synthetized by shrubs and trees in the forest. Later they were also isolated from plants from agricultural crops (Chadwick and Marsh, 1990; Ahmed and Wang, 2021).

Phytoncide compounds – the name given to highlight the origin plant - have bacteriostatic and bactericidal properties (obviously lower comparatively with antibiotics from microorganisms). The effects of garlic are well known since antiquity, being used as a remedy in certain ailments. Certain phytoncides are used in the nutrition and pharmaceutic fields (Cavallito and Bailey, 1944; Lawson, 1993; Shang et al., 2019). Also, phytoncides have been found to have fungicidal, insecticidal, or even rodenticide properties.

Sometimes phytoncides are found together with other biologically active substances, e.g., essential oils (Thangaleela et al., 2022).

In the field of plant protection, specialists draw attention to avoid confusion terminology of the term "phytoncides" with substances used as "herbicides" sometimes referred to as "phytocides".

2.4.4.2 Representative Compounds

Biologically active substances in the phytoncide class have been isolated from various vegetables, e.g., allicin, alliin, ajoene – isolated from garlic (*Allium sativum*); dihydroalliin – extracted from onions (*Allium cepa*). Also, from the plants were more obtained *caffeic acid, ferulic acid, chlorogenic acid, benzoic acid, p-hydroxybenzoic acid* – isolated from carrot (*Daucus carota*). There were various phytoncides extracted from other vegetables: horseradish, parsley, tomatoes, peppers, etc.

Figure 2.14 shows the structural formulas of some isolated phytoncides from vegetables: *allicin* or allyl disulfide sulfoxide - (I); *alliin* or *S*-allyl L-cysteine sulfoxide - (II); *L-dihydro-alliin* or *S*-propyl-L-cysteine dihydrogenated sulfoxide – (III).

Allicin – is an organosulfur compound obtained from garlic, isolated in 1944 by Cavallito and Bailey. In fresh garlic, biologically active allicin is formed in the presence of the enzyme *alliinase* from alliin, it gives the garlic fresh specific smell.

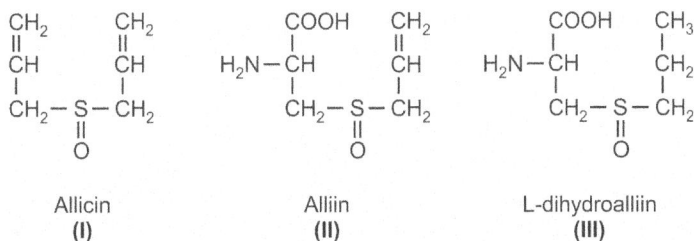

FIGURE 2.14 Phytoncides isolated from edible vegetables.

In an acidic environment (pH = 3), alliinase is irreversibly deactivated – a common situation in the stomach after eating fresh garlic. In the body allicin participates in defense processes in relation to microorganisms (bacteria, viruses).

Alliin and *dihydroaliin* – isolated from garlic and onions (*Allium cepa*). Allium is a powerful antioxidant that has the property of removing hydroxyl radicals.

Phytoncides isolated from garlic and onions are volatile substances. They have bactericidal action on numerous microorganisms in the digestive tract (e.g., microorganisms that produce staphylococci, diphtheria, cholera, etc.). In 1983, another phytoncide called ajoene (ajo) was isolated from garlic.

This compound is formed by the reaction of two molecules of allicin. Ajoene is a pharmacologically active substance with antioxidant effects – inhibits the release of superoxides. It also has antithrombotic action (prevents aggregation platelets) reducing the risk of heart attack and stroke (so it intervenes in the pathobiochemistry of cardiovascular and cerebrovascular diseases). Also, it has virucidal potential (destroys viruses), e.g., vaccinia virus, rhinoviruses, Herpes simplex, etc.

Allium was synthesized in the laboratory by alkylation of L-cysteine with allyl bromide. Hydrogenation gave dihydroalliin. Phytoncide substances have also been isolated from some fruits: blueberries, blackberries, currants, and raspberries (these are estimated to be derivatives of dyes anthocyanins). The effects of phytoncides reveal a chronobiochemical conditioning, which confirms the importance of chronobiology in the plant kingdom. In this regard, it is mentioned the fact that biochemical and biomedical effects are dependent on "chronophase" (growth, development, maturation). This concept is often called in plant biochemistry and plant physiology - "phenophase". In this case, the fact is exemplified that parsley, carrot, pepper that have phytoncidal effects only at the end of chronophases.

However, there are also cases in which phytoncides occur regardless of "chronophase". Characteristic examples are garlic, onion and horseradish. The issue of phytoncidal effects of "natural compounds of plant origin" interest various classes of biologically active substances. For this reason, the names of some phytoncides – with mention of effects and even with implications in pathobiochemistry, appear in other chapters on "biochemical effectors". Such compounds have been found in various vegetables, e.g., allyl-senevol – in seeds and in plants of mustard, horseradish, radish, etc.; tomatine – in tomato leaves (has antibacterial and antifungal effects, but toxic to humans); β-phenethyl-isocyanate – present in horseradish and gully roots, as well as in mustard seeds, etc.

2.4.4.3 Phytoncides in Biochemistry

Phytoncides, as shown, are antibiotics synthesized by plants. These compounds have bacteriostatic, bactericidal, insecticidal and even toxic properties to some invertebrates (worms) or vertebrates (rodents and birds).

Important compounds in the phytoncide group are found in vegetables, e.g., *allicin* – from garlic; *dihydroalliin and dimethylthiosulfonate* – from onions; *benzoic acids, p-hydroxybenzoic, vanillic, caffeic, ferulic, chlorogenic* acids – carrot; *ally-senevol* – from white mustard, horseradish, radish; sinigrin – from black mustard, etc.

These phytoncides have antibiotic attributes preventing the growth/development of many Gram-positive and Gram-negative bacteria. The best known are the phytoncides present in Alliaceae: onion and garlic – the compound of major interest is allicin. Garlic and onion extracts destroy the diphtheria bacillus, staphylococci, cholera vibrio, etc. Phytoncides such as benzoxazine heterosides in wheat, rye, maize – by hydrolysis – form substances with antibiotic action against mold.

The plants of the Brassicaceae family contain some phytoncides with structure of sulfonic acid type that have bactericidal and fungicidal properties. Phytoncides of biochemical interest with phytosanitary use for forestry have been isolated from various wood species. The wood of some trees contains antibiotics with bacteriostatic, fungistatic and fungicidal action.

Lichens contain phytoncides with an antibacterial effect that act on Gram-positive bacteria. The best known is the *usnic acid* from lichens of the species *Usnea barbata* and *Cetraria islandica*, which prevent the growth and multiplication of tuberculosis bacteria.

Phytoncides present in mustard plants and tomatoes are also important. First, such phytoncide is *sinigrin*, isolated from black mustard. Sinigrin is hydrolyzed in the presence of myrosinase enzyme, and allylic compounds with bactericidal action are released.

It is also mentioned that from the white mustard, a heteroside called sinalbin was isolated. An important phytoncide was obtained from tomatoes, known as tomatidine. It is biologically active, destroying mold (being antifungal) and yeasts. More than 50 phytoncides are currently known.

The chemical structure is very different. Phytoncides play an important role in protecting against phytopathogenic microorganisms. Seen in a biochemical context, phytoncides – isolated from fruits and vegetables – they structurally belong to various classes of compounds, e.g., heterosides, alkaloids, essential oils, phenols, etc. The phytoncidal effects – seen in terms of human nutrition – also show conditioning dependent on culinary processing or industrial processing of food.

For example, in the case of dill and eggplant, the phytoncidal effects increase by heating them (at about 50°C), a fact found in food chemistry and microbiology studies. Finally, it should be noted that phytoncides are of interest to industrial processing in the case of canned vegetables and fruits. Being natural products of vegetal origin, phytoncides do not cause side effects (often difficult to control), ensuring more efficient conservation.

2.4.5 HETEROSIDES

2.4.5.1 Synoptical Data

Chemical compounds that contain a carbohydrate component – *ide* (monosaccarides) or oligoholosides (oligosaccharides) bound by hemiacetal hydroxyl (glycosidic) of a non-carbohydrate component – called aglycone, are known by the generic name of *heterosides* or *glycosides*. According to the approach of glycosides in order to diversify the study of the *biologically active substances*, a special way of evaluating these compounds can be accredited.

Thus, glycosides can be considered as physiologically active substances with interest in nutrition or pharmacologically active substances, interesting pharmacology with therapeutic and toxicological implications (Liener, 1969; Strasburger, 1971; Goodwin and Mercer, 1983; Macrae et al., 1992).

In nature, heterosides are especially widespread in the plant kingdom and isolated from various organs of plants: leaves, flowers, fruits, seeds, stems and roots (Gârban, 2018; Kytidou et al., 2020; Yulvianti and Zidorn, 2021). A more detailed study of heterosides is carried out together with other biologically active substances.

2.4.5.2   Representative Compounds

The classification of heterosides is made – in general – by constitution, having in sight or the carbohydrate component, or the non-carbohydrate component (aglycone). Considering the nature of the aglycone, the following groups of heteroside are distinguished:

a. *O-heterosides* (oxygen glycosides) – results from condensation with alcoholic, phenolic, sterol or heterocyclic hydroxyl;
b. *S-heterosides* (sulfur glycosides) – formed by condensation with thiols;
c. *N-heterosides* (nitrogen glycosides) – derived by condensation with amine groups of some organic compounds.

Heterosides can exist in α and β forms, corresponding to two aldose isomers that bind to aglycone. Almost all natural heterosides are β-heterosides. Numerous heterosides have been obtained synthetically. The carbohydrate component of the heterosides is as shown, mono- or oligosaccharide. In many cases this is D-glucose (D-Glc), resulting glucoside. However, other monosaccharides may occur, e.g., D-galactose (D-Gal), Rhamnose (L-Rha), D-fucose (D-Fuc), D-ribose (D-Rib), D-deoxyribose (DdRib), as well as oligosaccharide, e.g., gentiobiosis, glucoarabinosis, etc.

The study of glycosides in biochemistry is usually done in the relevant chapter to carbohydrates because structurally they fall into their class.

In the field of human nutrition, as well as in pharmacology, heterosides are also studied along with carbohydrates. Heterosides (glycosides) have no reducing properties, lacking the free semi-acetal (glycosidic) hydroxyl. Hydrolysis can be accomplished under the action of specific acids and enzymes called and glycosidases.

Depending on the anomerism of the carbohydrate, there may be α- and β-glycosidases. From the observations made on glycoside-containing plants have been found to contain specific hydrolytic enzymes. The various subclasses of glycosides will be briefly presented below.

2.4.5.2.1   Group of O-Heterosides

They are compounds formed by the condensation of the hydroxyl group of the component non-carbohydrate (aglycone) with the hemiacetal (glycosidic) hydroxyl of a mono- or oligosaccharide. Depending on the nature of the aglycone, there are several subgroups of O-heterosides: (i) *aliphatic*; (ii) *aromatic* (phenolic, cyanogenic, anthracene); (iii) *sterolic* and (iv) *heterocyclic*.

Oxygen heterosides (O-heterosides) have been isolated from various plants and are important because they can enter into the human body by food consumption, by pharmaceuticals and, accidentally, as toxic substances.

2.4.5.2.1.1 Aliphatic Heterosides These heterosides have a glycosidic bond between the hemiacetal –OH of an "-ose" or "-oside" and a –OH function of a secondary or tertiary alcohol acting as aglycone (Figure 2.15). The bond can also be established in the –OH group of cyclical polyols. From the subgroup of these glycosides are exemplified: *methyl-glucoside*(I) and *mesoinositol-galactoside*(II).

> **Methyl-glucoside** – is the simplest aliphatic glycoside. Aglycone is represented by the methyl group which may be in the α or β positions of the D-Glc_p. It has been isolated from various plant tissues.
>
> **Mesoinisitol-galactoside** – consists of D-galactopyranose (D-Gal_p) and has as its aglycone mesoinositol (one of the isomers of hexane-hexol). It was isolated from sugar beet. This compound is present in plant and animal tissues. D-Glc_p binding is done in one of the –OH groups (binding position is not defined).

2.4.5.2.1.2 Aromatic Heterosides Aromatic heterosides are compounds in which the aglycone consists of one or more aryl nuclei. Depending on the nature of the aglycone, heterosides are *phenolic, cyanogenic* and *anthracene* (see Figure 2.16). In all formulas presented in this chapter, the dotted line indicates the separation of carbohydrate residues relative to the aglycone.

Phenolic heterosides are substances in which aglycone is phenol or other hydroxylate aryl derivative. Examples are: *salicin*(III) and *vanyllin glucoside*(IV).

> **Arbutin** – this heteroside contains β-D-glucopyranose and has aglycone *hydroquinone*. It is found in the hairy tree (*Pyrus communis*) and in the bear grapes (*Arctostaphylos uva-ursi*).
>
> **Salicin** or **saligenin** – this heteroside consists of β-D-glucopyranose (β-D-Glc_p) with salicylic alcohol as aglycone. It was isolated from leaves and willow bark (*Salix helix*).
>
> **Vanyllin glucoside** – is formed by binding D-Glc_p to vanilla alcohol. Isolated from green vanilla fruit.
>
> **Coniferin** – contains β-D-gluco-pyranose residues and has as aglycone coniferyl alcohol (p-hydroxy-m-methoxy-cinnamic alcohol). It was isolated

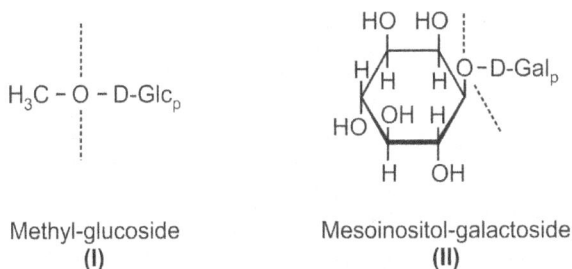

Methyl-glucoside
(I)

Mesoinositol-galactoside
(II)

FIGURE 2.15 Oxygen heterosides from the group of aliphatic compounds.

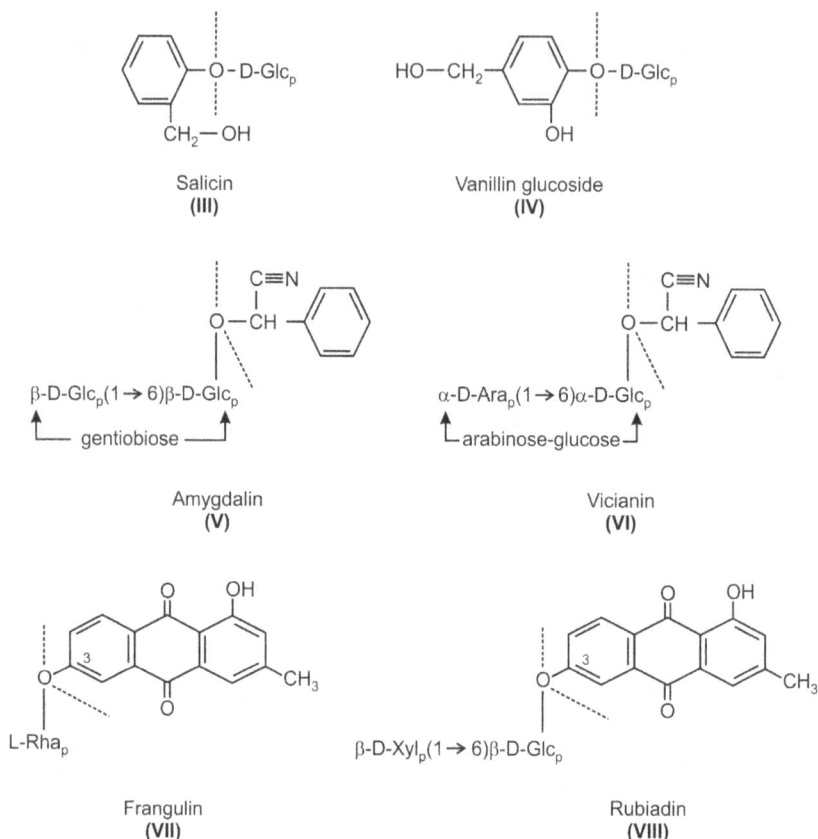

FIGURE 2.16 Oxygen heterosides from the group of aromatic compounds: phenolic (III, IV); cyanogenic (V–VI); anthracene (VII–VIII).

from the juice of conifers, then asparagus. Coniferyl alcohol is a highly reactive substance, which is easily polymerized.

Syringin – is a glycoside in which the aglycone is syringingenin. It was isolated from the bark of the lilac plant (*Syringa vulgaris*).

2.4.5.2.1.3 Cyanogenic Heterosides   In these heterosides, the aglycone is represented by a hydrocyanic compound. The structural formulas of *amygdalin* (V) and *vicianin* (VI).

Amygdalin – is a heteroside that has a disaccharide in its structure – gentiobiosis consisting of –β-D-Glc$_p$ (16) β-D-Glc$_p$ and an aglycone – mandelic acid nitrile. In nature, amygdalin can be found in bitter almond, peach, apricot, plum pits, apple and pears seeds, etc.

Vicianin – heteroside in which the carbohydrate component is O-α-Larabinopyranosyl - (1→6) - β-D-glucopyranose, and the aglycone component is also mandelic acid nitrile. It was extracted from the plant *Vicia angustifolia*.

Prunasin – isolated from the bark of wild cherries. Carbohydrate is D-Glc_p, and aglycone is, as in the present case, the nitrile of mandelic acid. It is generally in the leaves and bark of some fruit trees.

2.4.5.2.1.4 Anthracene Heterosides   In these heterosides, the aglycone is an oxidized or reduced derivative of anthracene: anthraquinone, anthrone or anthranol. They were isolated from the roots of **Rubia tinctorium** and **Rhamnus frangula** leaves. These glycosides have purgative effects. The best known are *frangulin*(VII) and *rubiadin*(VIII).

Frangulin – consists of the carbohydrate L-rhamnopyranose (L-Rha_p), which aglycone an anthraquinone derivative.

Rubiadin – is an anthracene heteroside in which the carbohydrate component is β-D-Xyl_p ($1\rightarrow6$) D-Glc_p and the anthracene component - aglycone is also an anthraquinone derivative.

Sennoside – contains bis (D-glucopyranose) and dantraquinone as aglycone.

2.4.5.2.1.5 Sterol Heterosides   They contain an aglycone from sterol class. These were obtained from plants and some have cardiotonic action (regulating the rhythm of cardiac activity), and others have diuretic effects. In high doses, however, they are cardiotoxic, stopping the heart in systole.

Among the most important sterol heterosides are: *digitaloids, strophantosides, saponins* and *solanine*. Aglycones in these glycosides are hydroxylated derivatives of *sterane*(IX) and are the followings: *genins* (X), *strophantidin* (XI), *sapogenin* (XII) and *solagenin* (XIII) - Figure 2.17.

Digitaloids – are also known as "cardiotonic heteroside drugs" were isolated since the Middle Ages from the leaves and seeds of Digitalis plants.

In these digitaloids, the aglycons are called "*genins*" (IX), and they have a steroid structure and an unsaturated lactone nucleus at the C_{17} a position of the steroid macrocycle. The first isolated cardiotonic glycoside in crystalline form was digitalin, realized by Nativelle (1869). Among the "specific genins" (R_{12} and R_{16} differ) present in heterosides, digitoxigenin, gitoxygenin and digoxygenin are mentioned. The general structure of digitaloids was formulated by Stoll et al. (1951) as follows:

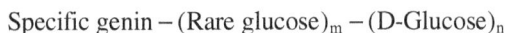

Specific genin – (Rare glucose)$_m$ – (D-Glucose)$_n$

The binding of the carbohydrate chain is done at –OH from the C_3 position of the macrocycle sterol, (so C_3 of the genin).

Strophantosides – are carbohydrates in which the carbohydrate component is D-glucose, L-rhamnose, but other 6-deoxyhexoses or 2,6-di-deoxyhexoses also appear. The aglycon of strophanthosides is *strophanthidin* similar to genin, but with a hydroxyl at the C_5 position and a carbonyl group at the C_{10}

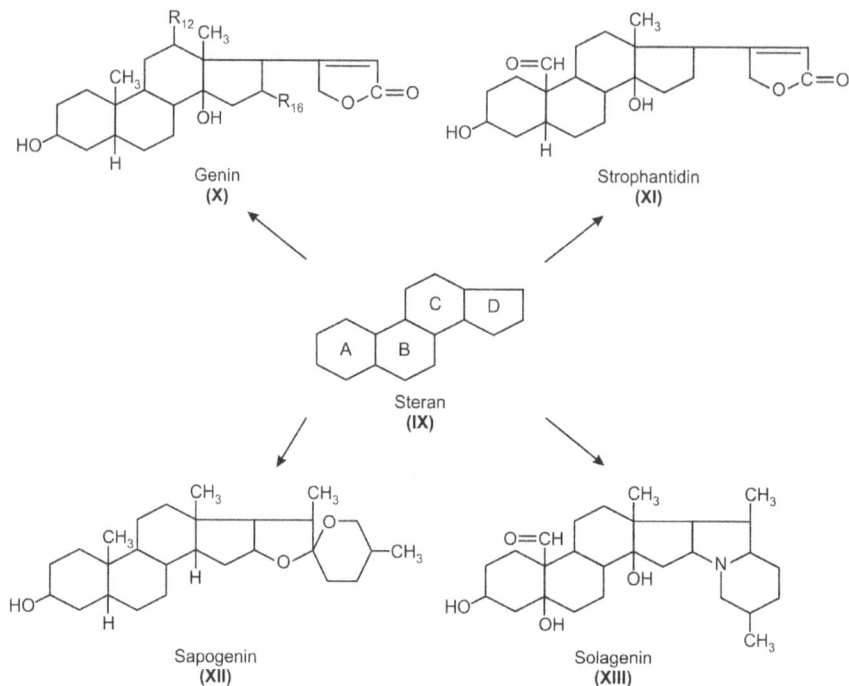

FIGURE 2.17   Aglycones detected in sterol heterosides.

position. The binding of the sterol derivative is done through the hydroxyl in the C_3 position.

Saponins – contain an oligosaccharide consisting of five monosaccharides (meaning a pentasaccharide) and an aglycone called *sapogenin*. Isolation of saponins was done from the leaves, roots and seeds of some plants. They also belong to the group of saponins heterosides: tigonine, gitonine, holoturine, etc.

Solanine – is a heteroside made up of a trisaccharide called solatriosis (consisting of L-rhamnose, D-galactose and D-glucose) and an aglycone called *solagenin*. Solanine was isolated from leaves, tubers and potato fangs (*Solanum tuberosum*), from eggplant and, in general, from plants of the Solanaceae family.

Tomatine – is a heteroside isolated from tomatoes (*Solanum lycopersicum*). The aglycone present herein is *tomatidine* to which a residue binds to C_3 tetrasaccharide (which by similarity to the previous one can be called "tomatetrose".

2.4.5.2.1.6 Heterocyclic Heterosides   These heterosides have heterocyclic aglycones in their structure. In this group, the following compounds are included: *anthocyanins* (anthocyanin heterosides); *flavones* (flavone heterosides).

2.4.5.2.1.7 Anthocyanin Heterosides Anthocyanins or anthocyanin het-erosides – have anthocyanidins as aglycones (Willstätter and Everest, 1913). Anthocyanins are red, purple and blue pigments, spread in the flowers and fruits of many plants. The parent compound is *anthocyanidin,* a heterocycle which tends to stabilize in the presence of the Cl⁻ ion. So anthocyanidin is stable only as *anthocyanidin chloride* (XIV) – shown in Figure 2.18.

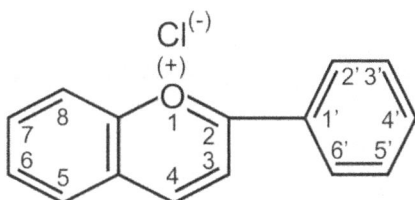

FIGURE 2.18   Precursor of anthocyanin heterosides.

This structure explains why anthocyanins (anthocyanin heterosides) can be iso-lated from plants only in the form of chlorides. From this group of compounds in Figure 2.19 are exemplified *oenin* (XV) and *prunicin* (XVI), but some mentions will be made for other anthocyanin heterosides, too.

Oenine – is a heteroside found in red grapes, in red wine. Aglycone is *oenidine* (reddish brown color) to which a residue of D-Glc$_p$ binds in position C$_3$.

Prunicin – is the diheteroside in which cyanidin binds in position to the agly-cone C$_3$ a residue of L-Ara$_p$, and in position C$_5$ a residue of D-Glc$_p$.

Fragarine – heteroside in which the aglycone is pelargonidine and the rest carbohydrate bound to C$_3$ of anthocyanidin is D-galactopyranose (D-Gal$_p$). The glycoside was isolated from the *Fragaria vesca* plant.

Keracyanin – is a heteroside in which the aglycone is cyanidin, bound in posi-tion C$_3$ with a residue of L-rhamnose (L-Rha$_p$). Isolated from black cherries and ripe olives.

Malvin – is the heteroside present in the flowers of the plant *Malva sylvestris* (color violaceous). Aglycone is the malvidin to which two residues bind in positions C$_3$ and C$_5$ of D-Glc$_p$.

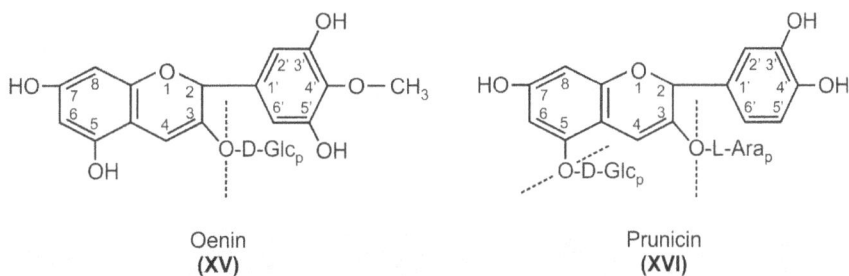

Oenin
(XV)

Prunicin
(XVI)

FIGURE 2.19   Oxygen heterosides from the group of anthocyanin heterocyclic compounds.

Mecocyanine – is the heteroside in which aglycone is cyanidin binds in position C_3 of gentiobiosis, i.e. β-D-Glc$_p$ (1→6) β-D-Glc$_p$. This glycoside isolated from *Papaver rhoeas* and *Prunus cerasus*.

Peonine – is the heteroside present in the leaves of the young plant of *Paeonia officinalis* (peony). Aglycone is the peonidine to which it binds at positions C_3 and C_5 two carbohydrate residues (the nature of the monoglycerides has not been established).

2.4.5.2.1.8 Flavone Heterosides   *Flavones* or *flavonic heterosides* have hydroxylated derivatives as aglycones of flavone. They are yellow pigments spread in the organs of plants (flowers, fruits, leaves and woody stems). Structurally, flavones are composed of *phenylchromone* type being known two isomers: 2-phenyl-chromone (*flavone*) - XVII and 3-phenyl-chromone (*isoflavone*) - XVIII (Figure 2.20). The first of these is more common in natural compounds.

There are hydroxylated derivatives in flavones or isoflavones: 3-hydroxyflavone (*flavonol*) – XIX and 2-hydroxy-isoflavone (*isoflavonol*) - XX. From these initial compounds derive various flavone aglycones (see Figure 2.20). Various carbohydrates are found in flavonic glycosides (flavones). Experimental or isolated: D-Glc$_p$, L-Rha$_p$, L-Ara$_p$, D-Xyl$_p$, glucuronic acid (GlcUA), and disaccharides, trisaccharides. Among the flavone glycosides are: cosmossine, propeonidin, robinin, rutin.

Rutin – an important flavone heteroside is found in buckwheat (*Fagopyrum esculentum*), gorse (*Ruta graveolens*), tobacco leaves (*Nicotiana tabacum*), etc. The glyceride of this hetroside is the disaccharide rutinose (consisting of L-rhamnose and α-D-glucose), and the aglycone is quercetin – a pentahydroxylated flavone derivative (3,5,7,3',4'-pentaoxyflavone) – XXI (see Figure 2.21). Rutin has a specific action on vascular permeability and blood vessel tone, being used as medicine.

In plant physiology, for example, it is assumed that by combining aglycone with a carbohydrate the toxic action of the aglycone is annihilated. Thus, experimentally, by injecting phenol into plants (e.g. corn) or keeping plants in a gaseous environment of o-chlorophenol or ethylene hydrochloride in plants, glycosides were

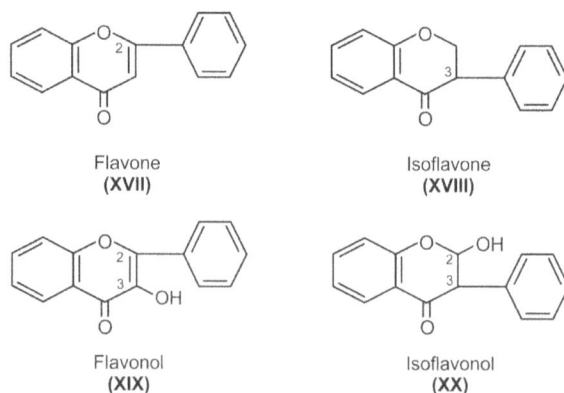

Flavone
(XVII)

Isoflavone
(XVIII)

Flavonol
(XIX)

Isoflavonol
(XX)

FIGURE 2.20   Flavone and isoflavone with their monohydroxylated derivatives.

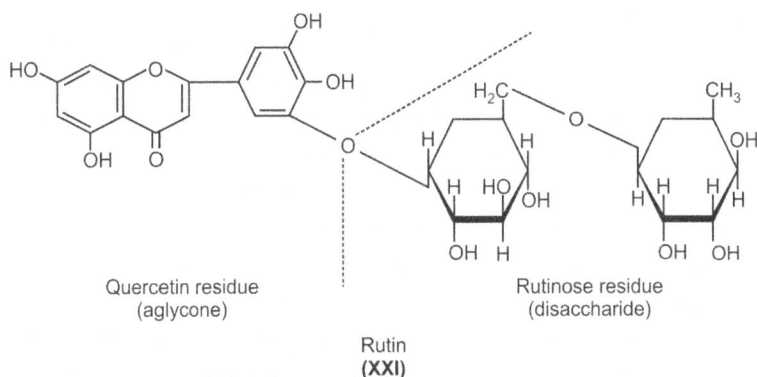

FIGURE 2.21 Rutin – structural formula.

synthesized. Carbohydrate synthesis was considered a specific mechanism for herbal detoxification.

2.4.5.2.2 Group of S-Heterosides

Compounds of the sulfur heteroside group, by hydrolysis, releases *oses* and *senevoles*. From a structural point of view, the senevoles are esters of isothiocyanic acid which have the general formula: R-N=C=S, in their structure the rest of the isothiocyane is noticeable.

In Figure 2.22 the structural formulas of some senevoles are presented: methyl-senevole (XXII); propenyl-senevole (XXIII); pentenyl-senevole (XXIV).

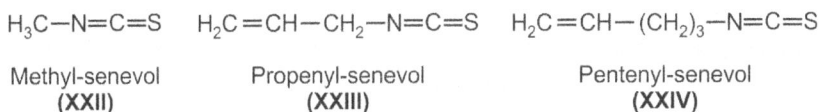

$$H_3C-N=C=S \qquad H_2C=CH-CH_2-N=C=S \qquad H_2C=CH-(CH_2)_3-N=C=S$$

Methyl-senevol   Propenyl-senevol   Pentenyl-senevol
(XXII)   **(XXIII)**   **(XXIV)**

FIGURE 2.22 Senevoles present in S-heterosides – structural formulas.

By enzymatic biochemical hydrolysis of S-heterosides, the following results: D-glucose and sulfate anion. On the formulation of the compound existed various opinions. The research undertaken by Ettlinger and Lundeen (1956) concluded that the true structure of a sulfur heteroside (thioheteroside) changes under the action of myrosinase enzyme (Figure 2.23), so that a transposition results in involuntary D-glucopyranose and sulfate anion.

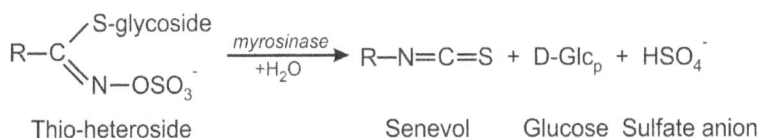

FIGURE 2.23 Biodegradation of a sulfur heteroside – details in text.

Compounds in the S-heteroside group also known as "sulfur glycosides" are found in the seeds of *Cruciferous, Liliaceae, Leguminosae, Capparidaceae* plants. They have been isolated mainly from seeds, roots or bulbs and very rarely from leaves or flowers. The resulting hydrolysis compounds are saturated, unsaturated and aromatic compounds. Sometimes they also contain a sulfur atom in the form of: thioether, sulfoxide, etc. In continuation the structure of some senevoles is exemplified.

> **Sinigrin** – is the first compound in the S-heteroside group that was isolated from black mustard seeds (*Sinapis nigra*). This S-heteroside present in seeds contain a residue of propenyl-senevole. It is volatile substance with the smell of mustard, potassium acid sulfate and β-D-glucopyranose.
>
> **Glucobrasicanapine** – sulfur heteroside containing *4-pentenyl-senevole*. It is an unsaturated substance isolated from *Brassica napus*.

2.4.5.2.3　Group of N-Heterosides

Compounds in this group result in condensation of carbohydrates with various amine derivatives and are named generically as N-heterosides (glycosides with nitrogen). In nature, the most important N-heterosides are formed by condensation of ribose or deoxyribose with purine and pyrimidine nucleobases.

So, purine nucleosides (e.g., adenosine nucleoside, guanosine nucleoside) and pyrimidine nucleosides, respectively (cytidine ucleoside, thymidine ucleoside, uridine nucleoside) are formed. The formed compounds are of importance for molecular biology and for biotechnologies, e.g., nucleosides, nucleoside monophosphates, nucleic acids, etc.

2.4.5.3　Heterosides in Biochemistry

Heterosides (glycosides) are biologically active compounds whose biosynthesis is produced by the interaction between the hydroxyl (–OH) attached to an anomeric carbon atom belonging to a monoglyceride with pyranose or furanose structure (at C_1 for aldoses or C_2 for ketoses) and a group possessing a nucleophilic atom. Nucleophilic atom may be oxygen (> O :), present in alcohols, phenols or in an acid. These will generate O-heterosidic compounds.

If the nucleophilic atom is the sulfur (> S :) present in the composition of some sulfhydryl groups or nitrogen (> N :) present in amines, S-heterosides and N-heterosides, respectively, are formed. Biochemical investigations have shown that the heterosides present in plants and isolated from them, are of interest both for nutrition and pharmacology, due to their biologically active properties. In this sense, it is quoted the fact that natural heterosides are widespread in the plant kingdom and in a few cases in the animal kingdom (e.g. in the skin of some species of frogs) which have found therapeutic uses.

Heterosides of therapeutic interest detected in plants of the genus Digitalis are compounds containing a monosaccharide group and a non-carbohydrate group (aglycone).

In the case of cardiotonic heterosides, aglycone is a compound called "genin". Derivatives of therapeutic interest are formed from genin such as digitoxigenin, gitoxigenin and digoxigenin. These compounds have cardiotonic action and are used as pharmaceuticals.

Other heterosides are important for their content in color pigments. In this regard, anthocyanin heterosides are cited as having anthocyanidins as aglycones, e.g., fragarine, malvine and oenine. There are also a number of glycosides that give rise to aglycones allow the generation of compounds of nutritional interest. These compounds are important both nutritionally and therapeutically.

Some heterosides, however, are notable for their toxic effects. Such are cyanogenic heterosides in which the aglycone is mandelic acid nitrile. Two are examples cases: amygdalin and vicianin. Upon hydrolysis, amygdalin releases: a molecule of amygdalin β-D-Glc$_p$, a molecule of benzoic aldehyde and a molecule of hydrocyanic acid. Amygdalin has been isolated from the first half of the last century by Robiquet and Boutron-Charlard (1830) from bitter almonds. Later, it was also obtained from peach, apricot, plum seeds, etc.

A general observation is that natural heterosides, almost without except, they are β - heterosides. Plants in which heterosides have been detected contain usually specific enzymes that can hydrolyze these compounds. For example, in the kernels of bitter almonds, an enzyme called emulsin has been detected, which has a specific action on the β-glycosidic bond. Heterosides often accompany other biologically active substances (e.g., essential oils) present in certain plant species such as the lilac (*Syringa vulgaris*), jasmine (*Jasminium officinale*).

These plants are also appreciated for their use in cosmetics (see essential oils). As a general conclusion, it can be concluded that heterosides contain: a residue carbohydrate and a non-carbohydrate residue, i.e., aglycone which is the biologically active compound characterized by physiological, pharmacological or even with toxicological effects (e.g. amygdalin – in excess after hydrolysis).

2.4.6 NATURAL PIGMENTS

2.4.6.1 Synoptical Data

Biologically active substances include natural pigments – detected especially in the plant kingdom – which differs greatly in terms of composition and chemical structures. Characteristic of natural pigments is that they intervene in the definition the composition of products and extracts of nutritional, pharmacological and cosmetic interest, giving them various colors (Britton, 1983; Goodwin and Mercer, 1983; Ciulei et al., 1993; Dewick, 1997).

The variety of colors, for example, characterizes the various vegetable food products (e.g., carrots, peppers, spinach, tomatoes, peas, etc.) and fruits (e.g., apples, apricots, cherries, etc.). It is mentioned that, in addition to the chromogenic properties of plant pigments gives fruits, vegetables and flowers a specific scent.

Obviously, the effects of pigments should also be considered in conjunction with other biologically active compounds, e.g., essential oils, alkaloids, etc. (Bodea et al., 1964–1966; Goodwin, 1976; Chadwich and Marsh, 1990).

From a chemical point of view, the issue of food color refers to both natural dyes (in this case involving vegetable pigments) and synthetic dyes (some of which are used as food additives).

Natural pigments of plant origin, called sometimes phytochromes, are organic compounds that contain in their molecule characteristic atom groups, known generically as *chromophore groups* (gr. *khroma*-color; *phoresis*-carrier). Among the most important chromophore groups are: *nitroso* (–N=O); *nitro* (–NO$_2$); *diazo* (–N=N–); thionic (= C=S), as well as double bond groups of the type (–CH=CH–) present in various compounds.

The problem of the color of the organic compounds in relation to the chromophore groups was studied by Witt (1876). It has been hypothesized that every substance that has a chromophore group is *chromogen*. To become a colorant of the environment in which is located the substance, the chromogen, must also contain *auxochrome* groups (gr. *khroma*-color; *auxo* to enlarge, to intensify). Auxochrome groups include phenolic *hydroxyl* –OH (C$_6$H$_5$–OH); monosubstituted (–NHR) or disubstituted (–NR$_2$) *amino* (–NH$_2$); carboxyl (–COOH); sulfonic acid (–SO$_3$H) groups, etc.

Development of knowledge in the field of theoretical organic chemistry (mechano-quantum theory, resonance theory, etc.) and physics (e.g., absorption, emission, fluorescence, spectroscopy, etc.) brought an important contribution to the knowledge of the particularities physicochemical properties of dyes.

Nutrition, pharmacology, cosmetics as well as experiments in biology and medicine contributed to the elucidation aspects of biochemistry and biochemical pathology related to dyes. Thus, it was made a delimitation between dyes of nutritional interest (present in food or introduced in food additives), pharmaceutical and cosmetic, compared to the dyes of industrial interest for textiles, leather, paper, etc.

Toxicological aspects have also been studied in the context of the structure-activity relationship. Natural pigments – which are the subject of this subchapter – are also called biochromes, i.e., "colored biological substances" or "colored natural substances".

These compounds have been isolated from various plant tissues, located in specialized cell organelles called "chromoplasts" present in leaves, flowers, fruits, pollen, etc. However, biochromes are not an exclusive prerogative of plant tissues. Such substances are also found in animal tissues, e.g., muscles, blood, skin, and plumage. Biochromes from the plant and animal kingdoms are involved in many biochemical reactions in the body, e.g., redox reactions, addition reactions, etc.

2.4.6.2   Chemical Structure-Chromogenic Activity Relationship

In biochemistry, nutrition, pharmacology and obviously organic chemistry books, the chemical structure biological activity relationship is frequently discussed, often with the revelation of physiological, pathophysiological and toxicological aspects. This approach is of interest to various bioconstituents and contributed, with real success, in explaining many process-specific feedback mechanisms of nutrient metabolization, characterized by various pathways (*biochemical pathways*).

It has also made possible to understand the mechanisms of the *biotransformation* of various xenobiotics, e.g., food and pharmaceutical additives, ingredients of some cosmetics, chemotherapeutics, pesticides, etc. The problem of the structure-activity relationship, known by the acronym SAR, is also of great interest to colored substances, in especially organic dyes. In the sense of approaching this subject, the aspect is mentioned particular of the problem of chemical structure-biological activity, which has physiological and/or pathophysiological connotation.

Before the chemical details, it is necessary to mention some physical aspects in the field of spectroscopy, which contributed to the elucidation of theoretical data. In this sense can be mentioned the importance of the Bouguer-Lambert-Beer law which postulates the relationship between the intensities of incident radiation (Io) and transmitted (It) when crossing an environment (Eq. 2.1):

$$It = I_o \cdot 10^{\varepsilon.c.l} \qquad (2.1)$$

where
  ε - extinction coefficient
  c - the concentration of the substance in the solution
  l - the distance (length) traveled by the radiation passing through the medium
     (i.e. the solution)

This law is also applied to the study of light radiation absorption in "spectrometry absorption". The method is based on changing the intensity of electromagnetic radiation (in case of light radiation) when crossing an environment, i.e. solution (sometimes gas or solid body).

The absorption spectra show the dark regions corresponding to the wavelengths absorbed in the form of *absorption lines* for single molecule substances, in gaseous state or by *absorption bands* to substances with large molecules and in the liquid state or dissolved in a certain transparent compound (solvent). The evaluation of absorption in spectrophotometry is done according to: wavelength (λ) expressed in nm; extinction (ε) - dimensionless size; transmittance (T) - expressed as a percentage. Measurements are made in ultraviolet (UV), visible (VIS) and infrared (IR). For color evaluation uses the visible and ultraviolet range.

To have an image of the distribution of the spectral domains, data on the corresponding wavelengths for the absorption spectra of the different colors are presented (Table 2.4). The wavelengths in the visible range to characterize the different colors absorbed are shown in more detail (350–750 nm).

TABLE 2.4
Spectral Ranges and Corresponding Wavelengths of Absorption Spectra

Domain	Wavelength (λ), (nm)	Adsorbed Color (Adsorption Spectrum)	Complementary Color
UV	Under 350	Ultraviolet	–
VIS	350–435	Violet	Yellow-green
	435–480	Blue	Yellow
	480–490	Blue-green	Orange
	490–500	Green-blue	Red
	500–560	Green	Purple red
	560–580	Yellow-green	Violet
	580–595	Yellow	Blue
	595–605	Orange	Blue-green
	605–750	Red	Green-blue
IR	Over 750	Infrared	–

Theoretical data on absorption spectroscopy allow an explanation of the notions used in the study of dyes.

Thus, in spectroscopy defines:

- *bathochrome* effect – characterized by dark color, due to movement the value of the maximum wavelength (λ_{max}) toward the IR range
- *hypsochrome* effect – lies in the lightness of the color and is due to movement λ_{max} value toward the UV range (so toward a shorter wavelength)
- *hyperchrome* effect – is at the origin of a color intensification and is caused by an increase in the extinction coefficient which reaches its maximum value (ε_{max}) achieving a specific bit in the absorption curve.

Based on these considerations, it is mentioned that over time, to explain the relationship between chemical structure and chromogenic activity several theories were accredited. The limited space allocated to this volume does not allow presentation of these theories.

2.4.6.3   Representative Compounds

Natural pigments can be classified according to various criteria, among which priority are: provenance, composition, structure, solubility, etc. The classification by composition and structure has a wider acceptance, according to distinguishing: (i) *non-nitrogenous* pigments (carotenoids, flavonoids and quinone) and (ii) *nitrogen* pigments (tetrapyrroles, indoles, pterins, melanins). A detail of this classification is made in Figure 2.24, following the main subclasses of natural pigments.

It is noticeable a great diversity, both regarding chemical structure as well as biological activity, of each subclass. Distinctive features reveal, in particular, the study of the structure-activity relationship.

2.4.6.3.1   Non-nitrogenous Pigments

These are a distinct group of compounds that include pigments: *carotenoids, flavonoids* and *quinoids*.

2.4.6.3.1.1   Carotenoid pigments   In terms of composition, this class of compounds includes:

- *hydrocarbon carotenoids* (carotenes) – one of the most important are carotenoids C_{40} (with forty carbon atoms) consisting of eight isoprene units;

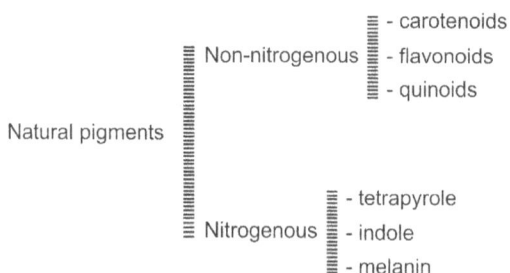

FIGURE 2.24   General classification of natural pigments.

- *oxygenated carotenoid* derivatives – in the constitution of which there are various functional groups: hydroxyl, aldehyde, ketone, epoxy, carboxylic (Goodwin, 1980; Britton and Goodwin, 1982; Neamţu, 1983; Krinsky, 1994).

Hydrocarbon carotenoids – These compounds have the general formula $C_{40}H_{56}$. The best known are: *lycopene* (I), *α-carotene* (II), *β-carotene* (III) and γ-carotene (IV). Initially, carotenes were isolated from carrots and lycopene from tomatoes and milk. Figure 2.25 presents the structural formulas of the main hydrocarbon carotenoids.

Oxygenated carotenoid derivatives – These compounds differ depending on the nature of the oxygenated functional group (Figure 2.26). Thus, in terms of structure, there are:

- *hydroxylated derivatives*, e.g., xanthophyll – isolated from plant and animal tissues; *lutein* (V) - isolated from plant tissues - yellow and red flowers and some products of animal origin (e.g., egg yolk, corpus luteum in the ovary); *zeaxanthin* (VI) - isolated initially from maize (*Zea mays*), then from various fruits; alloxanthin (from plant and animal tissues);
- *hydroxylated and epoxy derivatives*, e.g., *violaxanthin* (VII) – isolated for the first time given from the flowers of Viola spp.); *neoxanthin* (preferably present in tissues) vegetable);

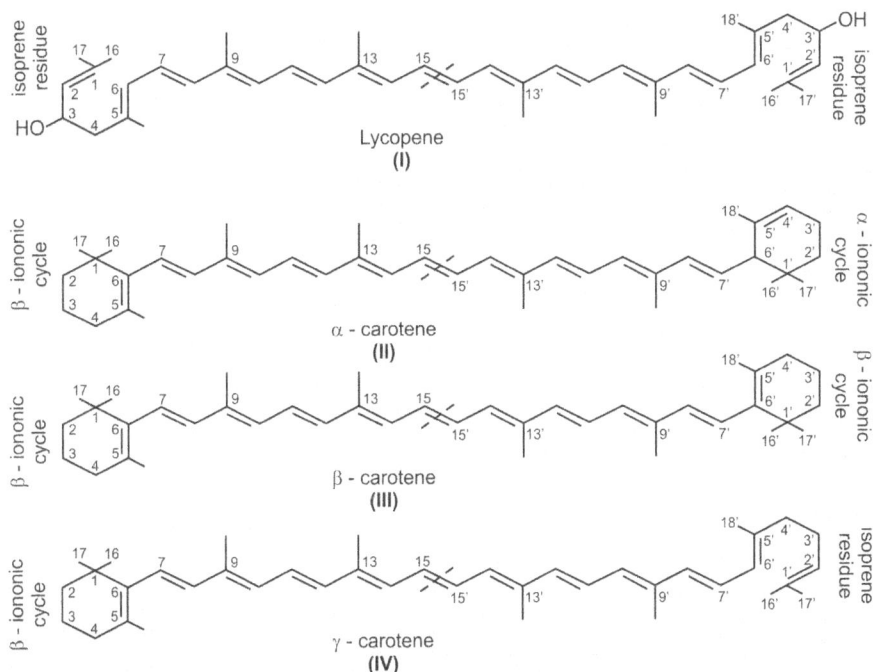

FIGURE 2.25 Hydrocarbon carotenoids – structural formulas.

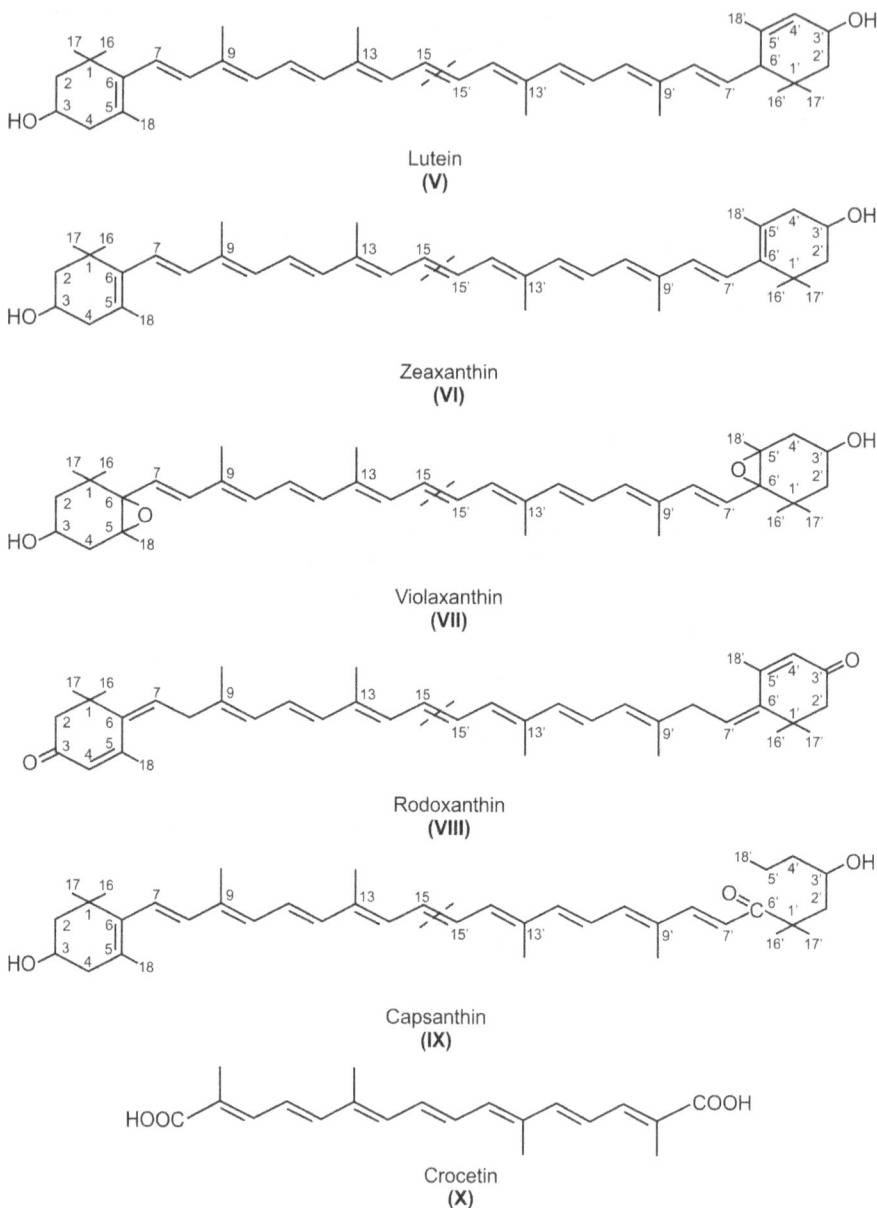

FIGURE 2.26   Oxygenated carotenoid derivatives – main representatives.

- *ketone derivatives*, e.g., *rhodoxanthin* (VIII) – isolated from the aquatic plant *Potamogeton natans,* the compound is reddish-brown; *astacin* - isolated from the lobster shell - *Astacus gammarus,* the compound is red-purple color; *astaxanthin* - isolated from green algae and animal tissues (e.g., crustaceans, protozoa); *capsanthin* (IX) - isolated from pepper red - *Capsicum* spp., has a carmine red color;

* *aldehyde derivatives*, e.g., *retinal* – isolated from plant tissues, are vitamin A precursors;
* *carboxylate derivatives*, e.g., *crocetin* (X) – isolated from the stigmas of *Crocus sativus*; *bixin* – isolated from the plant *Bixa orellana*, which grows in tropical (yellow, isolated compound called "orelline") used as a dye.

The structure of oxygenated carotenoid compounds reveals the great diversity of there are derivatives: hydroxylates (V) and (VI); hydroxylated and epoxy (VII); ketones (VIII) and (IX); carboxylates (X) – see Figure 2.26.

Carotenoid pigments are widespread in nature and are especially in the vegetable kingdom; however, it is also in the animal kingdom.

The chromogenic properties are due to the conjugated double bonds in the molecules, which gives the tissues in which it is found various colors: yellow, orange, red and in some cases blue. Carotenoid pigments in plant or animal tissues are free or associated with proteins, e.g. carotenoproteins, with lipids, e.g. esters, with carbohydrates, e.g. carotenoid glycosides.

Structurally, carotenoids belong to a wide range of compounds terpenoids with similarities to sterols, phytol, vitamins E, vitamins K, quinones, etc.

2.4.6.3.1.2 Flavonoid Pigments Flavonoid pigments or *flavonoids* are phenolic compounds having condensed heterocycle molecules (benzopyran or benzofuran) in which the heteroatom is oxygen. Benzene is another nucleus binds to the benzopyren or benzofuran. The compound thus formed may bind one or more hydroxyl phenolic groups (Goodwin, 1976; Das et al., 1994; Harborne and Williams, 1995; Neamţu, 1983).

Among the best-known flavonoid pigments are the derivatives that originate: flavan, flaven (from which anthocyanins are derived), flavanone, flavone, chalcone and aurone. The structural formulas of these nuclei of origin are shown in Figure 2.27 mentioning: *flavan* (XI), *flaven* (XII), *flavanone* (XIII), *flavone* (XIV), *chalcone* (XV) and *aurone* (XVI). Nuclei of origin XI–XVI, by hydroxylation, methoxylation reactions, etc., generates the great diversity of flavonoid pigments. The typical primary structures originate from benzopyran (XI–XV), respectively benzofuran (XVI).

Anthocyanins (a subgroup of flavonoids) are red, purple, blue or black pigments. They can be found in flowers, leaves and fruits. These compounds contain a *carbohydrate residue* – more commonly Glc, Gal, Rha and an *aglycone residue* – which is an anthocyanidin. Six molecular species of anthocyanidins are currently found in nature: *cyanidin, delphinidin, malvidin, pelargonidin, peonidin* and *petunidin*. By hydrolysis of anthocyanins the anthocyanidins are released from plant tissues. The sources used for the extraction of anthocyanins are: red cabbage, beetroots, blueberries, blackcurrants, cherries, sour cherries, raspberries, grapes, etc. More details on anthocyanins can be found in the specialized literature – (see Goto and Kondo (1991); Gould et al. (2009); Ockermann et al. (2021 a.o).

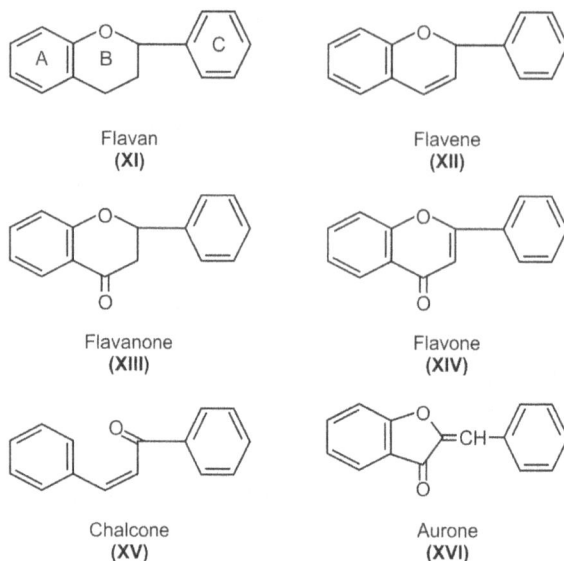

Flavan
(XI)

Flavene
(XII)

Flavanone
(XIII)

Flavone
(XIV)

Chalcone
(XV)

Aurone
(XVI)

FIGURE 2.27 Structural formulas of the main nuclei of origin of flavonoid pigments.

Flavonoids are pigments that are found mostly in the vegetable kingdom: flowers, fruits, leaves, stems, roots, bark, etc. Topobiochemically, these pigments are located in the chromoplasts and vacuoles in the cells from which can be isolated free or in the form of phenolic glycosides. Flavonoids have been isolated also from various microorganisms, algae and even insects.

In nature, the detected flavonoids, belonging to different groups, have dissimilar forms with varied structural features, number and position of phenolic hydroxyl groups, presence of some methoxyl groups etc., resulting compounds in different colors. Thus, *flavans* – are compounds that give the environment a yellow color; *anthocyanidins* – color red or blue (found mainly in flowers and fruits); *flavones* and *isoflavones* – they are yellow pigments; *flavanones and chalcones* - are colorless; *aurones* - are golden yellow colored. Chromophore system is burdened by the discontinuity of double conjugate bonds.

Data regarding the above-mentioned group were also discussed in the section Heterosides where a distinction was made between anthocyanin heterosides and flavonoid heterosides.

2.4.6.3.1.3 Quinone Pigments This class of pigments includes compounds with a quinone structure: derivatives of benzene, e.g., *p*-benzoquinone and other di- and tricyclic aromatic hydrocarbons, e.g.. naphthoquinone, anthraquinone, phenanthroquinone and even tetracyclic, e.g., tetracenchinone. Structures of the compounds of origin from which they are formed the quinone pigments are shown in Figure 2.28.

These are: *benzoquinone* (XVII); *naphthoquinone* (XVIII); *anthraquinone* (XIX); *phenanthroquinone* (XX) and *tetracenoquinone* (XXI). By grafting hydroxyl groups (–OH), methoxy (–O–CH$_3$) etc. – the during metabolic processes – various

p - Benzoquinone (XVII) Naphthoquinone (XVIII) Anthraquinone (XIX)

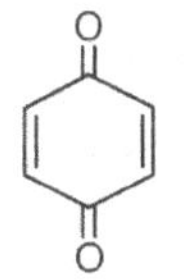

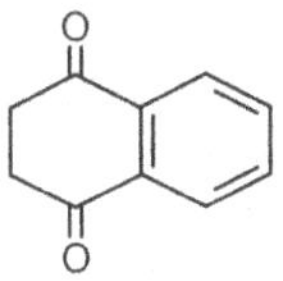

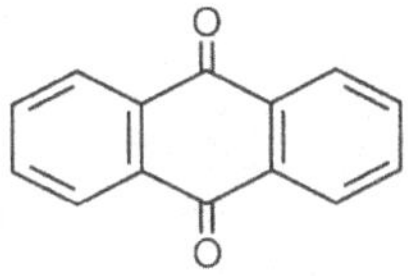

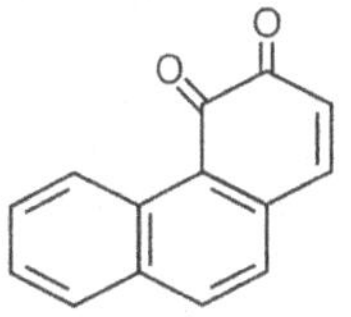

Phenanthraquinone (XX) Tetracenoquinone (XXI)

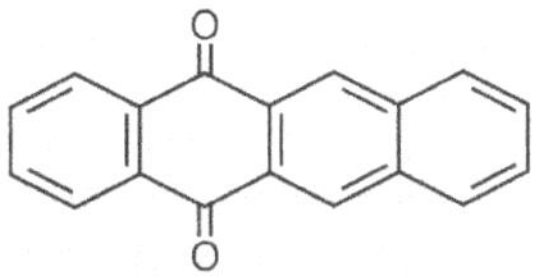

FIGURE 2.28 Structural formulas of mono- and polycyclic compounds of quinone-type.

quinone pigments are formed (Goodman and Gilman, 1966; Goodwin and Mercer, 1983; Bruneton, 2009).

In addition to their role as a coloring compound, these pigments are remarkable that participates in the redox processes in the body, ensuring the non-enzymatic transfer of hydrogen and electrons transfer. For these reasons, compounds of this group are involved in the breathing process (e.g., ubiquinones).

Some aspects of the structure and the natural distribution of various quinone pigments will briefly present.

Benzoquinone pigments – There are in microorganisms (bacteria) in the soil, in fungi, in superior plants. More important are two compounds (shown in Figure 2.29): one disubstituted −2,6-dimethoxybenzoquinone (XXII), another tetrasubstituted - ubiquinone (XXIII).

Ubiquinone –also known as coenzyme Q (abbreviated CoQ) is a compound of the benzoquinone group with almost universal distribution in living organisms. For this reason, ubiquinone is often also mentioned in writing "Ubiquinone" (lat. *ubique* – everywhere).

Naphthoquinone pigments – They were isolated from plant tissues. The best known are: juglone (5-hydroxy-1,4-naphthoquinone) isolated from the green skin of

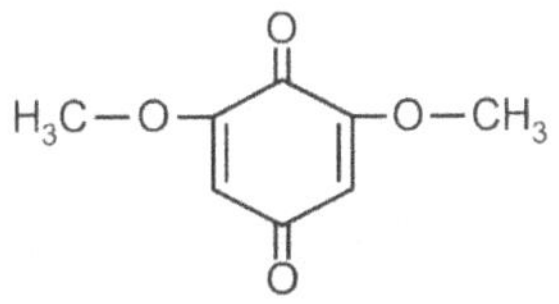

2,6 - Dimethoxybenzoquinone (XXII) Ubiquinone (XXIII)

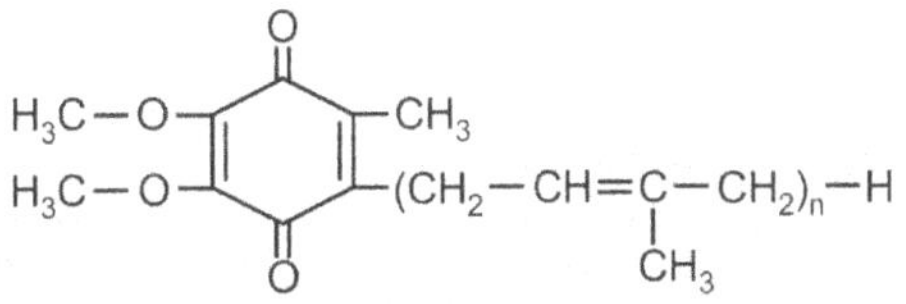

FIGURE 2.29 The structure of the main benzoquinones.

the walnut (*Juglans regia*); menaquinone (vitamins K) - compounds of the class of fat-soluble vitamins important in redox reactions from the body.

Anthraquinone pigments – They were found in mushrooms and superior plants of the Rubiaceae, Ramnaceae, Polygonaceae families. Among these pigments are better known: a derivative anthraquinone dihydroxylate - alizarin and a trihydroxylate derivative – purpurin; both used in coloring textiles since antiquity.

Phenanthroquinone pigments – They have a lower natural distribution. Compounds in this class generally have toxic effects. Study of compounds derived from tricyclic hydrocarbons with angular disposition (e.g., phenanthrene) and polycyclic (e.g., tetracene, pentacene, etc.) may interact with nucleic acids, being involved in the mutagenesis, oncogenesis and are studied in pathobiochemistry.

The compounds in the group of quinone pigments give the environment different colors, from yellow, orange to red. Study of the chromogen relationship between chemical structure and activity revealed that the presence of ketone groups in ortho position confer red color, whereas the presence of ketone groups in the para confer yellow color.

In the case of metabolic reduction reactions, quinones – compounds whose structure determines the color of natural pigments, turns into hydroquinones – which are colorless compounds. In nature – in the plant kingdom – this reaction exemplifies biological evolution in the sense that in young plants there are compounds with a structure of hydroquinone, and as plants grow and mature, following processes oxidants form quinones, which give color to the flowers, leaves, fruits, etc.

2.4.6.3.2 Nitrogenous pigments

This group includes substances belonging to various classes of compounds representing the following nitrogenous plant pigments (nitrogenous phytochromes): *tetrapyrrole, indole* and *melanin*.

2.4.6.3.2.1 Tetrapyrrole Pigments   The group of vegetable tetrapyrrole pigments includes colored compounds mostly from leaves. At their origin is a basic "macrocycle" called porphin, which consists of four pyrrole rings (noted A, B, C and D) linked in the α positions by methyne groups (=CH–). In the β positions of the pyrrole rings are hydrogen atoms whose substitution by various radicals leads to the formation of derivatives called *porphyrins* (Strasburger, 1971; Neamţu, 1983; Rawn, 1991; Gârban, 2018).

Addition of a carboxylic ring to the porphin ring results in the formation of *phorbin* (XIV) - considered as the unsubstituted structure (precursor) of chlorophyll. As known, *chlorophyll* (XV) is the most widespread tetrapyrrole pigment in the plant kingdom and contains Mg^{2+} ion (Vernon and Seely, 1966). Figure 2.30 presents the structures of phorbin and chloropphyll.

It is mentioned that the group of tetrapyrrolic pigments includes also compounds from animal tissues (see Chapter 3).

Depending on the nature of the substitutes in the alpha positions of pyrrole nuclei several molecular species of chlorophyll are distinguished (e.g., chlorophylls a, b, c, d). Chlorophyll is the prosthetic group from the chromoprotein called chloroglobin.

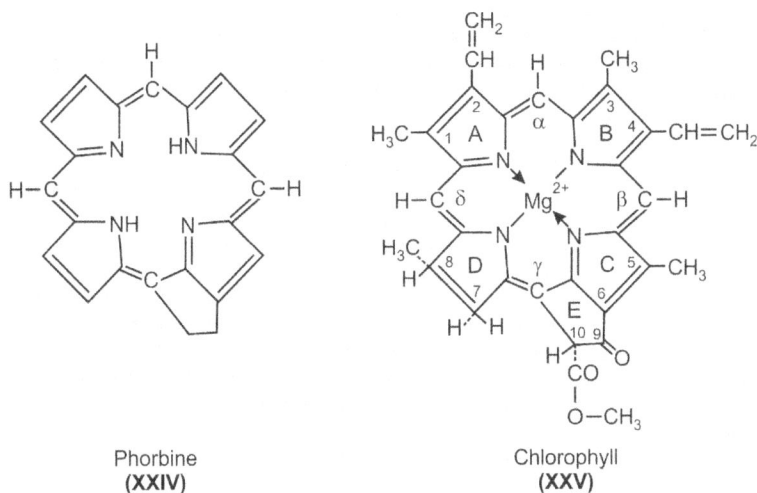

Phorbine
(XXIV)

Chlorophyll
(XXV)

FIGURE 2.30 Chemical structures of phorbin and chlorophyll.

For information, it is mentioned that porphyrins have been isolated alo from petrol and originated from the plant and animal kingdom during geological evolution. The content of various porphyrins in petrol is approx. 0.004–0.02 mg/100 g. The presence of these compounds is considered an indication of the origin of the oil. From the point of view of geology, porphyrins from oil are indicators of the geothermal gradient.

2.4.6.3.2.2 Indole Pigments Structurally, indole pigments contain the indole heterocycle. They are naturally bound to proteins – indole chromoproteins or carbohydrates – indole glycosides (Bodea et al., 1964–1966; Goodwin, 1976; Britton, 1983; Waterman, 1993).

Structurally, *indole* is a free, but naturally occurring benzopyrrole, especially in the form of protein, carbohydrate derivatives, etc.

The structural formulas of *indole* (XXVI) and *indigo* (XXVII) the main indole pigment are shown in Figure 2.31.

In the free state it was detected in small quantities in some essential oils, e.g., essential oils from orange leaves. In the form of indole derivatives in living matter, there are: tryptophan – an amino acid present in all proteins; and tryptamine – biogenic amine derived from it.

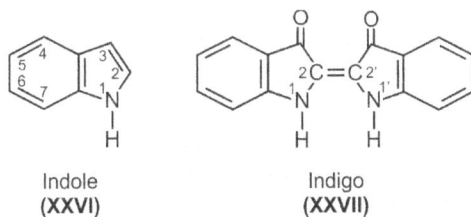

Indole
(XXVI)

Indigo
(XXVII)

FIGURE 2.31 Indole and indigo – chemical structures.

Also, the indole core is in numerous biologically active substances, e.g., alkaloids – some of therapeutic importance, vegetable pigments – present in tissues, from where they can be extracted. It is mentioned that a number of indole derivatives were extracted from coal tar.

Indigo or *indigotine* was known in antiquity in India and Egypt. Originally it was obtained from plants belonging to the genus Indigofera – later from the plant *Isatis tinctoria* (woad) grown in Europe. Indigo is found in plants in the form of a glycoside called *indican* whose indole residue is *indoxyl*. Indigo was studied by Adolf Baeyer (1860) who extended his observations on the model and its derivatives.

Indigo is obtained by macerating plants in aqueous medium. So, enzymatic hydrolysis occurs (under the action of enzymes present in plants) after to which indoxyl is released, which in the presence of oxygen in the air is oxidized to indigo.

2.4.6.3.2.3 Melanin Pigments   Melanin nitrogen pigments – known generically as "melanins" – are formed from aromatic amino acids, i.e., tyrosine and hydroxyphenylamine. Structurally, they are in the form of quinoid polymers. Melanins were isolated from plant tissues, mainly from superior plants and animal tissues taken from invertebrates (e.g., insects) and from vertebrates.

These pigments cause in plants browning until blackening after heavy rainfall. In animals, melanin is disposed in cells called "melanocytes" (Prota, 1995). Melanin pigments give various colors to the environment: yellow, red, brown to black. Melanins are mainly involved in expressing hair and skin color. Melanin gives various colors, such as yellow, red, brown and black, to tissues. The color black (gr.melanos-black) is dominant.

In the plant tissues, the color is highlighted quite quickly by the action of water stress (after prolonged rainfall).

2.4.6.4   Natural Pigments in Biochemistry

Phytochromes (plant pigments) are photoreceptors with the role of capture and use of light energy to stimulate or inhibit biological processes and are spread in all green plants, except mushrooms.

They are located intracellularly, associated with the biomolecular system from plasmalemma and the inner membrane of chloroplasts. Phytochromes in plant tissues are found in higher amounts in meristematic tissues and in the storage bodies of reserve substances.

From a chemical point of view, they are chromoproteins with molecular mass of 60,000 Da. They contain three chromophores of tetrapyrrolic nature whose structure is close to phycocyanin and bile pigments.

Physiologically, these compounds act as photon receptors in all reversible photoreactions that occur in red and dark red in green plants, intervening in: greening of plants; seed germination; synthesis of anthocyanins, chlorophylls, carotenes, flavones; induction of gibberellins biosynthesis; hormonal regulation of plant flowering; organ morphogenesis; photoperiodicity; tuber formation, stem growth, leaves growth, etc.

In food processing, and even in their culinary preparation, one of the major objectives is to ensure superior organoleptic qualities. Among these qualities, color plays an essential role.

Color provides a criterion to estimate presenting nutritional (physiological and biochemical), psychological and even economic interest. This criterion is concerned with the issue of so-called "biochemical barriers" that correlate with food consumption.

A simple approach to these issues can be seen that the chromatic aspect is perceived as a superior quality aspect of food. Color changes relative to a certain standard (existing in consumer information) are considered to be effects of unhealthy food.

The use of natural dyes in food preparation is known from very long time. In the beginning, food colorings were used in the form of raw materials extracted from vegetable, animal and mineral products. Natural dyes were used to give the food a tint chromatic or to restore the natural chromatic aspect that has been lost through industrial processing (Britton, 1983; Bruneton, 2009; Di Salvo et al., 2023). In practice, they are used to color various sugar products (candy and sherbet), fruit juices, alcoholic beverages, butter, margarine, pastry, etc.

It is to mention that in food industry not only extraction dyes (natural) but also synthetic dyes (artificial) are used. Among the *natural dyes* can be mentioned: *carotenoids* (yellow-orange or red) E 160; *chlorophylls* (green) E 140; *anthocyanins* (various colors: red, purple, and blue) E 163; *riboflavin* (yellow) E 101, etc. The group of *synthetic dyes* includes: *azo dyes*, e.g., tartrazine (yellow) E 102, *azorubin* (red) E 122; *triaryl-methane dyes*, e.g. green bright acid BS (green S) E 142; *xanthine dyes*, e.g. erythrosine (red) E 127 and so on. They are generally used to restore the color of a food after processing or to impart color of a colorless product.

Prior to use, coloring agents in food industry are necessary to obtain the approval of the authorities based on special studies starting from stability assessment, animal model research and even complex toxicology research. In fact, for all substances used as food additives, there are Codex Alimentarius prescriptions.

As natural dyes – in the food industry – pigments used are: (i) *carotenoids*; (ii) *flavonoids* (e.g. anthocyanins); (iii) *anthraquinones* and (iv) *porphyrins*.

Carotenoid pigments are obtained by extraction from various plants, e.g., saffron, red pepper, carrots, tomatoes, annatto plants, etc. The main natural extraction preparations from this group are: annatto-carotenoid mixture obtained from *Bixa orellana* (Annatto plant) has yellow color and is used for coloring butter, bakery products, etc.; oleoresin obtained from the fruits of ripe peppers *(Capsicum annuum)* – color from pink to carmine red, used to color meat products, salads, etc. The oleoresin extract contains 35–57 carotenoid derivatives, including higher concentrations of capsanthin and capsorubin; tomato extracts – yellow to pink – are used for coloring meat products. It contains lycopene and lutein; saffron extract obtained from the stigmas of the flowers of the plant *Crocus sativus* (saffron) – dark yellow – used for coloring cheeses, soft drinks, meat products, soup concentrates, etc. The main pigment in its composition is crocetin; xanthophyll extracts are obtained from alfalfa and are used in the coloring of some dairy products. The main compounds are lutein and β-carotene.

The mostly used flavonoid pigments in food industry are compounds from anthocyanin group, which have red, purple and blue colors.

Porphyrin pigments. From their group, the chlorophyll extracted from plant tissues is used in coloring oils and beverages, as well as the recoloration of plant foods discolored by processing.

It is to mention that even achromatic flavonoids (colorless), e.g., flavones, reveal biological effects characterized by intense absorption of UV radiation and are detectable by insects, considering that in this way they ensure their participation in flower pollination.

Studies on flavonoids in the field of *nutrition* and *pharmacology* revealed that these compounds play a beneficial role in acting as antioxidants. Flavonoids detected in red wine (quercetin, kaempferol and anthocyanidins) and in tea (catechins and catechin gallate esters) proved to be effective antioxidants against free radicals. It helps to reduce the oxidative stress caused by biogenesis of free radicals harmful to health, e.g., superoxide radical; hydroxyl radical. The antioxidant role of flavonoids allows them to provide protection against cardiovascular disease and certain forms of cancer (Dewick, 1997).

With reference to plant pigments and biochemical aspects of the cosmetic action of these, it is noted that derivatives of aldehydes and ketones interact with components of the keratinous layer of the skin. This interaction changes the skin color, characterized by the appearance of a brownish hue. It is supposed to originate from Maillard's reaction.

A remarkable feature in interactions of cosmetic interest and limitation of excessive staining, for example, in the case of dihydroxyacetone, concomitantly vitamins A, vitamins E and to a very small extent B vitamins (Raab, 1985) are used.

2.4.7 TANNINS

2.4.7.1 Synoptical Data

Tannins are a class of organic compounds with a heterogeneous polyphenolic structure. They were extracted from plants in whose tissues accumulate secondary metabolites. These substances give reduction reactions (e.g., with Fehling, Tollens solutions, etc.), precipitation (e.g., with alkaloids, metal ions or organic substances). They also give color reactions with ferric chloride. These reactions attest the phenolic character of tannins.

In the laboratory, the tannins were extracted from various plants, being present in peel, leaves and fruit. They have an astringent, characteristic taste – found in some fruits that contain these substances. The notion of *tannin* was taken from technology. The word has its origin from the Latin term (derived from the Celtic language) *tan* – which means oak.

Studies undertaken since 1908 by Freudenberg and continued in the period 1910–1930, led to the acceptance of the existence of two tannin subclasses, based on structural criteria (Figure 2.32) and distinct chemical properties: (i) *gallotannins* (hydrolyzable tannins) – based on *gallic acid* (4-pyrogalol-carboxylic acid) - (I); (ii) *catechin tannins* (condensed tannins) – are considered to have, formally, a common precursor, i.e., *catechin* (3,5,7,3', 4'-pentahydroxy-flavan) – (II).

The isolation of the tannins was made from oak bark (*Quercus sessilifolia* and *Quercus pedunculata*), spruce bark (*Picea excelsa*), willow bark (*Salix viminalis*), birch bark (*Betula alba*), rosehip marrow (*Rosa canina*), etc. Tannins have a molecular weight 500–3000 Da.

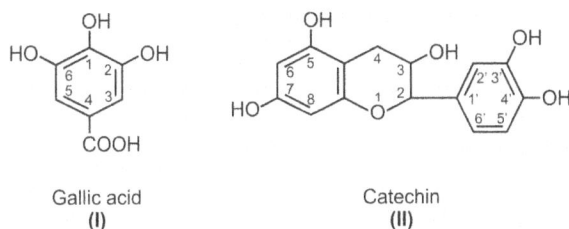

Gallic acid
(I)

Catechin
(II)

FIGURE 2.32   The structure of the main compounds presents in various types of tannins.

There are numerous hydroxyl phenolic groups in the molecule. These groups make possible protein binding (Goodwin and Mercer, 1983).

Tannin compounds are preferred in the leather industry, having the property of turning raw skin into tanned skin. Tannins are polyphenolic compounds. In the case of gallic acid, for example, there is also a carboxylic group. They have a weak acid character.

The main chemical properties are: reduction reactions (e.g., with Fehling's solutions, Tollens, iodine-iodide, potassium permanganate, etc.); precipitation with alkaloids, metal ions (e.g., Cu^{2+}, Zn^{2+}, Al^{3+}, etc.) or some organic substances (e.g., phenyl hydrazine, albumin); color reactions (with ferric chloride – it forms blue-violet combinations).

2.4.7.2   Representative Compounds

In plant, tannins accumulate in leaves, fruits and generally in tissues with intense physiological activity. Detection of tannins in tissues was made in specialized cells – "tannin idioblastic cells" but also in the vacuolar fluid of some parenchymal cells, which they fulfill a protective role against attacks by viruses and microorganisms.

Tannins are mainly found in plants belonging to *Rosaceae, Amentaceae, Leguminosae, Myrtaceae, Geraniaceae* families. Due to the phenolic groups, the tannins oxidize slightly in contact with air, forming reddish-brown or dark red colored compounds. Vegetable tannins are preferred in the leather and textile industries. The food and pharmaceutical industries are also interested. In the following, general data on subclasses of mentioned tannins are presented.

2.4.7.2.1   Gallotannins

Gallotannins also known as *hydrolyzable tannins* are natural esters of glucose with gallic acid (see Figure 2.32) and/or of some derivatives of it (Neamţu, 1983; Ciulei et al., 1993; Haslam and Cai, 1994; Okuda et al., 1995; Braghiroli et al., 2019).

Gallic acid (4-pyrogallol-carboxylic acid) is a trihydroxylated phenolic acid, which is found in the free form in plant tissues – small amounts in risky donuts, oak bark, tea leaves and combined in the form of gallotannins – compounds which decompose by acidic or enzymatic hydrolysis.

Acid hydrolysis can be performed in the presence of dilute H_2SO_4, and enzymatic hydrolysis can be performed in the presence of *tannase* enzyme present in mycelium (e.g., *Penicillium glaucum, Aspergillus niger*).

In its pure state, gallic acid is in the form of aciform and colorless crystals. It is soluble in hot water, alcohol and ether. It tastes sour and astringent. It melts at 222°C when it decomposes to carbon dioxide and pyrogallol. Gallic acid reactivity is characterized by intense reducing properties, e.g., precipitate Au and Ag from their salt solutions. In an alkaline environment, it absorbs oxygen from the air – at same as pyrogallol and turns brown.

Gallic and *ellagic* acids were more frequently detected in the composition of hydrolyzable tannins. For this reason, the older classification of these compounds even discusses the existence of two subgroups of compounds: *gallotannins* and *ellagitannins*.

Gallic acid and ellagic acid – the main constituents of "gallotannins" – are formed in the biochemical hydrolysis process of (under the action of the enzyme *tannase*) or chemical. Studies on ellagic acid biosynthesis have shown that it comes from two molecules of gallic acid in several successive stages (Figure 2.33).

Thus, initially it has an oxidative dehydrogenation with hexahydroxydiphenic acid formation. Then, luteolic acid is formed by dehydration followed by partial lactamization, and by total lactamization, the ellagic acid is formed. In the laboratory, the synthesis of ellagic acid is performed by oxidative dehydrogenation in the presence of ferric chloride ($FeCl_3$) or potassium permanganate ($KMnO_4$).

However, since a single compound is at origin – gallic acid – it is becoming more common to define all their esters with glucose (e.g., glucogallicin) by the term "gallotannins".

A remarkable fact, in structural terms, in the case of gallotannins, is that at the same molecule of cyclized glucose (having one glycosidic hydroxyl and four alcoholic hydroxyls available for esterification) can bind different compounds, e.g., gallic acid and *m*-digallic acid (metadigallic).

Such a structure features Chinese tannin – a representative compound for gallotannins. The presence of gallic acid derivatives explains why in some tannins the ratio remains glucose: gallic acid residues exceed 1:5 (value 5 corresponding to the total number of groups hydroxyl).

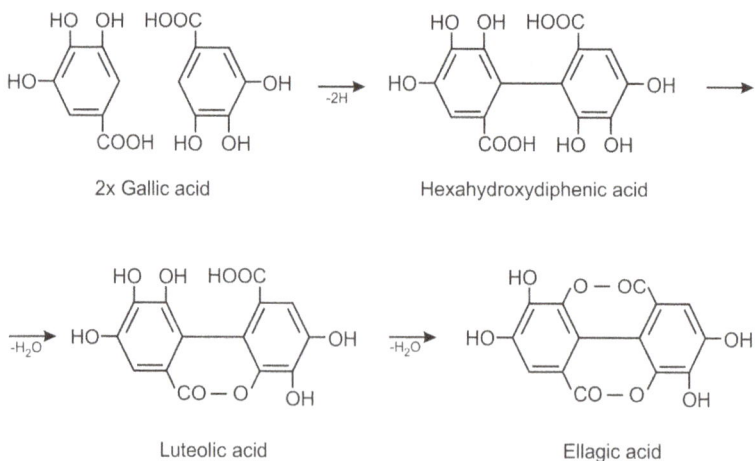

FIGURE 2.33 Biosynthesis of ellagic acid and intermediate compounds derived from gallic acid.

Gallotannins – derived mainly from shikimic acid biosynthesis – they are mainly used for industrial purposes, suitable for tanning hides and skins. Among the following best-known gallotannins are:

Turkish tannin – with the molar ratio of glucose residue: gallic acid residues 1: 5; to hydrolysis D-Glc is formed, a gallic acid residue, four acid residues m-digallic and even alginic acid residues are formed at the end of hydrolysis.

Chinese tannin – in which the ratio of glucose residue: gallic acid residues is 1: 9, which proves the presence of gallic acid derivatives; hydrolytic was D-Glc, one gallic acid residue and 4 m-gallic acid residues each with other related gallic acid derivatives; the structure described reveals that the initial report is as at Turkish tannin 1:5 (but four of the gallic acid residues are bound to other derivatives of gallic acid).

Chebulinic acid – is another tannin compound that has a residue of glucose, three gallic acid residues and one m-digallic acid residue.

Chebulic acid – is a tannin in which the ellagic acid derivative predominates; hydrolysis releases glucose, gallic acid, ellagic acid and other derivatives.

Hamamelis tannin – obtained from the *Hamamelis virginica* shrub, which has the particularity that the monosaccharide is D-ribose (D-Rib) methoxylated to C_2; esterification is done with only two gallic acid residues.

Gallotannins were originally studied for technological reasons – given applications to *skin tanning* (actually the transformation of collagen from the dermal layer into a material resistant, rot-proof), for the purpose of processing for industry. Vegetable tannins used for tanning were obtained from various species of plants: oak bark (*Quercus alba*), chestnut (*Castanea vesca*), myrobalan plum (*Prunus cerasifera*), etc.

For tanning, various processes have been used: (i) *tanning with vegetable tannins* – using substances from the gallotannin group; (ii) *tanning with mineral salts* – based on the use of aluminum salts (white tanned leather used for gloves), iron salts, chromium salts, etc.; (iii) *tanning with fats* – by using oil fish which contains unsaturated fatty acids which polymerize in the air and preserve collagen fibers (thus obtained soft, water-permeable tanned skin), etc.

The tanning process – It is made explicit by the formation of covalent bonds – inside three-stranded collagen fiber. In this case, condensation of phenolic groups or ketone groups with free amine groups, e.g., diamine amino acids, amino acids that are N-terminals in polypeptide chains. Collagen is a protein with a helical, three-stranded structure. In its composition, there are in proportion of 50% three amino acids: glycocolle, proline and hydroxyproline.

The implications of these interactions have been used in the leather (tanning) industry, described briefly above. Later, synthetic tannins were also obtained for industrial purposes starting from sulfonated phenols and cresols. The essential characteristic of tannins lies in their ability to give interactions with protein compounds, which results in distortion structure-activity relationship in proteins. This interaction explains the action of tannins on the human body, the effects of nutritional and pharmacological interest, respectively.

Reaction mechanism – Tannin interacts with polypeptide chains from proteins. This reaction is at the origin of both the tanning process and the astringent taste of some vegetal food products (Goodwin and Mercer, 1983; Wagner, 1985; Ciulei et al., 1993; Fraga-Corral et al., 2020).

The astringent effect is also based on the interaction between tannins and proteins, the situation in which they can form:

a. ionic bonds – between polar hydroxyl groups in tannins (Tannin–O–) and polar amine groups derived from proteins (H_3N+ - Protein).
b. hydrogen bonds – between hydroxyl groups in tannin (Tannin–OH) and hydroxyl groups (–OH) in proteins belonging to hydroxyl amino acids (AA) or even acyl residues (R–CO–) from peptide bonds (R–CO–NH–R'). From the above data, it can be concluded that the tanning process is irreversible, unlike the astringent effect is reversible.

2.4.7.2.2   Catechin Tannins

Catechin tannins, also called "condensed tannins", are macromolecular compounds obtained through the polycondensation of several catechin molecules (see Figure 2.33). In some cases, compounds have also been obtained by polycondensation of catechin derivatives (Neamţu, 1983; Chadwick and Marsh, 1990; Devi, 2021).

From the group of catechin tannins, one of the first extracted products called *Catechu* was isolated from the leaves taken from the *Acacia catechu* plant (fam. *Leguminosae*), which is commonly grown in East Africa and India. The crude extract, in a dry state, actually contains two compounds: catechins and phlobaphenes.

With regard to phlobaphene, it is stated that such compounds have been isolated from various species of oak (*Quercus* sp.) and peanuts, i.e. American hazelnuts (*Arachis hypogaea*). An *arachi-phlobaphene* (III) was extracted from the seeds of Arachis. The compound consists of two catechin molecules linked in positions C_4–C_4 and is presented in Figure 2.34.

In general, catechin tannin forms small polymers (oligomers). Another typical representative, given in Figure 2.35, is the trimer known *as epicatechin* (IV). C_8–C_8 bonds appear in this compound.

The compounds of the catechin tannin subclass do not hydrolyze, they are hardly soluble in water - especially those with a high degree of condensation.

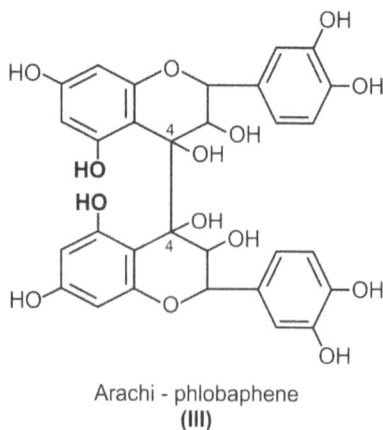

Arachi - phlobaphene
(III)

FIGURE 2.34   Arachi-phlobaphene – dimer.

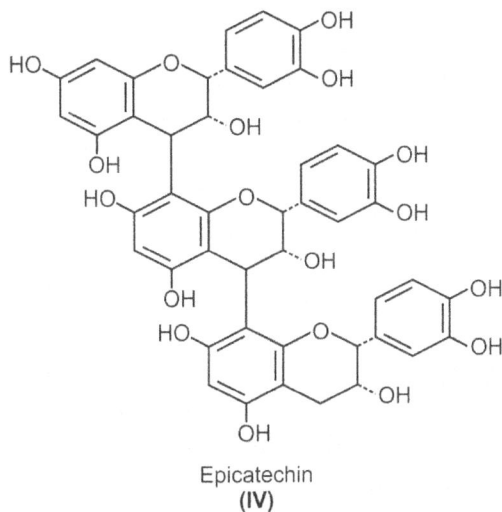

Epicatechin
(IV)

FIGURE 2.35   Epicatechin - the trimer molecular structure.

They have reddish-brown color. When heated, these tannins form pyrocatechin. Among the best-known condensed tannins are:

Quebracho tannin – obtained from *Quebracho colorado* tree, South American wood in which it is in the proportion of 14%–26% compared to dry material.

Acacia tannin – obtained from *Robinia pseudacacia* wood in which it is in the proportion of 3.5%–4.0% of the dry matter.

Oak tannin – taken from oak wood of different species (*Quercus* sp.) in which it is in the proportion of 3%–10%.

Phlobaphene type tannin – extracted from cocoa beans, has crude macromolecular structure $(H_{16}O_7)_n$ has red or reddish brown color.

Tannins are formed in various plant organs such as tree bark, fruits, leaves, stems or even roots. In many cases, catechin tannins biogenesis occurs along with that of gallotannins.

2.4.7.3   Tannins in Biochemistry

Tannins are accompanied by carbohydrates, as well as other nutritional principles in food products. They were especially isolated from fruits and vegetables but also from hops, tea leaves, etc.

The presence of tannins in *fruits* (e.g., apples, pears, quinces, etc.) and vegetables (e.g., potatoes) is explained by the browning process, commonly called in the food industry "browning". This process is due to the oxidation of tannins with oxygen in the air and under the action of enzymes called tannases, which exist in plant tissues.

The browning process is also encountered in the industrial processing of fruits and vegetables as a result of cutting them into thin slices. Contact with metal parts accelerates this process due to catalytic effects.

Tannins are also important in the field of *oenology*. It is known that during vinification, *grapes* (husks, seeds and bunches) release tannins, especially *enotannin*. These tannins are involved in redox processes contributing to the coloring of the wine and the formation of its bouquet.

Iron excess, in some cases, is incriminated by the production of a blackening of the wines. In foods and beverages (especially wines), some catechins (e.g. epicatechin) give a slightly astringent character (Dewick, 1997; Watrelot and Norton, 2020).

The role of catechin tannins in protecting capillaries of blood vessels and ensuring their permeability has also been noted. Tanning substances present in hops (the amount of which may be up to 5%) influence the chemical processes specific to beer technology.

In this sense, the tannins in hops precipitate AA and, in general, the proteins in hops beer wort, helping to clear the beer. The important role of catechin tannins in tea leaves is stated, it is known that during the preparation of teas, an intense brown coloration takes place as a consequence of the oxidation of tannins in green tea leaves.

It should also be noted that there are some tannins in the tobacco leaves that, after the oxidation processes during drying, give the chestnut color until it turns brown.

Gallotannins are found in plant products of food and phytopharmaceutical interest. Compounds including gallic acid, ellagic acid, etc. are released from gallotannins during various enzymatic hydrolysis (Okuda et al., 1995; Amarowicz and Janiak, 2019). Ellagic acid, which is present in raspberries, strawberries, pecans and blueberries, has been studied for decades.

From a pharmacological point of view, it has been found that by the interaction of tannins, in moderate doses, with proteins, *tannin-protein complexes* are formed. Such complexes provide anti-infective protection (contributing to the destruction of microorganisms by coagulation effects); have antidiarrheal action; limits or stop fermentation processes; intervene in the coagulation processes of the bleeding wounds at intestinal level; decrease the intestinal secretions. At high doses, the tannins have laxative and vomiting effects.

Discussing the biochemical and pharmacodynamic role of tannins, it is stated that research over the past two decades has shown their antitumor action. So, in pathobiochemistry and especially in oncology, there are opinions about the existence of alternative and complementary therapies in which compounds containing ellagic acid and gallic acid are used.

In this respect, there are experimental data obtained from laboratory animals (mice). Studies on animals have shown that ellagitannins are precursors that hydrolytically release ellagic acid with effects on neoplasms remission, detoxifying effects on the colon and limiting abnormal cell proliferation.

Experiments in laboratory animals have been shown that ellagic acid inhibits malignancy in esophageal, lingual, lung, colon, liver and skin cancers. The mechanism of oncostatic action has shown that ellagic acid: (i) inhibits the transformation of carcinogenic compounds (e.g., polycyclic aromatic hydrocarbons, food additives - preservatives, aflatoxins); (ii) forms adducts with some carcinogens that are thus inactivated; (iii) forms adducts with DNA, thus occupying malignancy-specific receptors.

Remarkable is the observation regarding the fact that in small quantities the ellagitannins slow down the proliferation of cancer cells, and in large quantities induce their thanotocytosis.

The relationship between apoptosis and necrosis is currently being studied extensively specific to molecular biology and pathobiochemistry. There are also studies on the antimutagenic action of ellagic acid, tested in vitro for mutagens in some foods.

2.4.8 ESSENTIAL OILS

2.4.8.1 Synoptical Data

Essential oils, within the biologically active substances that accompany the natural products are included. Essential oil compounds present a great diversity of composition and of application in nutrition, pharmacology and cosmetics (Gildemeister and Hoffmann, 1959; Wiss, 1960; Neamțu, 1983; McGarvey and Croteau, 1995; Mookherjee and Wilson, 1996; Bhavaniramya et al., 2019).

Essential oils – also called *volatile oils* or *essential oils* – are substances of plant origin whose biogenesis takes place in flowers, fruits, leaves, bark and seeds of many plants. Essential oils extracted from vegetable products have a density of 0.80–1.07. The refractive index (n) of these compounds is 1.45–1.60. Boiling temperature is 150–300°C. So, it can be concluded that there are wide areas of extension of physicochemical constants due to the diversity of composition.

From a biochemical point of view, knowledge of biogenesis in plant tissues; chemical composition, topobiochemical distribution, diversity molecular species and biologically active features is important.

Essential oils are free or in combination with other biologically active compounds, as an expression of common biogenesis in plant tissues. Thus, in plant tissues, one can find: (i) essential oils; (ii) resins; (iii) balms; (iv) copals and (v) lignans. Storage of these metabolic products in plants is made in various cells, e.g. cells from leaves, flowers, roots (in cell vacuoles), as well as in cells from glandular pores and from scales or intercellular channels.

  i. *Essential oils* – are chemical compounds with various structure (hydrocarbons, oxygenated, heterocyclic derivatives) present in flowers, fruits, leaves, seeds, etc.
 ii. *Resins* – are solid or semi-solid substances represented by excretion products of woody plants (e.g., conifers, but also fruit trees). Often, they are produced as a consequence of a mechanical injury to the tissues that release excretion products. The resins are yellow, yellow-brown, insoluble in water and have a fragrant smell. In contact with air, they solidify. In their composition, resin acids, resinols, etc. are found. Such substances are rosin and incense.
iii. *Balsams* – are liquid or semi-solid substances that have a high content of volatile oils (about 40%). In contact with air, the oils volatilize, take place oxidation-reduction reactions, polymerization, so that a solidification occurs.
 iv. *Copals* – are residual resins (from various plants), which self-deposit in the ground. For this reason, they are also called *fossil resins.*
  v. *Lignans* – have the physical characteristics of resins, but from a chemical point of view, they contain compounds with lignin-like structures. One such substance is guaiacum resin.

The name "volatile oils" (used in conjunction with "essential" or "ethereal") is motivated by some authors for the consideration that the characteristic properties of these compounds are: high vapor pressure and volatilization in ambient conditions (at room temperature).

For "essential oils", the use of the term "ethereal oils" is considered to be improper. Various authors point this out for the reason that the name refers to the "ethereal" character in the sense of "smelling" of these compounds. The term ethereal oils makes indirectly, reference to the notion of ethers, which in chemistry and biochemistry is a good circumscribed class of chemical compounds.

2.4.8.2 Representative Compounds

Compounds in the composition of essential oils are acyclic hydrocarbons (saturated or unsaturated), aromatic hydrocarbons, alcohols, aldehydes, ketones, organic acids, heterocyclic compounds (containing O, N, S), ethers, esters, etc. From the plethora of substances found in essential oils, there will be mentioned the group of: (i) *monoterpenes* – which originate from isoprene and are presented in the form of hydrocarbons and their derivatives with oxygen, sulfur or nitrogen and have an acyclic or homocyclic (mono- or polycyclic) structure; (ii) *sesquiterpenes*; (iii) *tropolone compounds* (tropolones) – with a hepta atomic nucleus unsaturated homocycles which derives from tropone; (iv) *miscellaneous compounds* in essential oil; (v) *other terpenes*.

In case of terpenes, the number of isoprene units (with formula C_5H_8) differs so that several types (Table 2.5), i.e., hemiterpenes (C5); monoterpenes (C10); sesquiterpenes (C15); diterpenes (C20); triterpenes (C30); tetraterpenes (C40) and superior terpenes (Cn) can be distinguished.

Different methods were used in the chemical technology of extracting essential oils, e.g., entrainment with water vapor (a process specific to distillation), extraction with liquid solvents (ethyl ether, petroleum ether) or solids (lipids) and mechanical extraction (by pressing).

In general, the idea of a classification dependent on provenance is also accredited in the study of essential oils. Thus, for example, in the field of cosmetics (Thiers, 1962; Goodwin and Mercer, 1983; Heymann, 1994), a classification according to which odorous substances *of plant origin* are grouped into: (i) *simple natural odorants* substances, e.g., citronellol, geraniol, nerol, linalool, menthol, terpineol, benzyl alcohol, phenyl-ethyl

---

TABLE 2.5
Structural Characteristics of Terpenes

No.	Type of Terpenoids	Usual Notation	Number of Isoprene Units	General Formula	Natural Distribution (Examples)
1	Hemiterpenes	C5	1	C_5H_8	Essential oils
2	Monoterpenes	C10	2	$(C_5H_8)_2$	
3	Sesquiterpenes	C15	3	$(C_5H_8)_3$	
4	Diterpenes	C20	4	$(C_5H_8)_4$	Resins and balms
5	Triterpenes	C30	6	$(C_5H_8)_6$	
6	Tetraterpenes	C40	8	$(C_5H_8)_8$	

alcohol, enanthol, citronellal, citral, benzoic aldehyde, vanillin, camphor, ionone, anethole, thymol, eugenol, eucalyptol, diphenyl ether and methyl anthranilate; (ii) *complex natural odorants,* e.g., essential oils, bergamot oil, olive oil, clove oil, eucalyptus oil, orange blossom oil, orange peel oil, olive oil geranium, lemon oil, lavender oil, peppermint oil, rose oil, resins like: Tolu balsam, Peru balsamm, Styrax and Benzene resin; (iii) *substances extracted from wood,* e.g., essence of Santal, fir, vetiver and guaiac.

2.4.8.2.1 Monoterpenes

Monoterpenes are compounds that belong to a subgroup of terpenoids. As known, tepenoids are derived from isoprene (2-methyl-butadiene) that differ in physical and chemical properties and are included in two distinct subclasses: *terpenes* - substances in the molecular structure include only isoprene units that appear as acyclic or cyclic compounds (so they contain only C and H) and *terpenoids* - substances whose structure includes O atoms (oxygenated derivatives) and even S or N atoms. In some terpenes there is also the phenyl (aromatic) nucleus.

The notion of terpenoids characterizes isoprene-derived compounds (2-methylbutadiene): $H_2C=C(CH_3)–CH–CH_2$

In organic chemistry and biochemistry, there are extensive studies on natural products with polyisopentene skeleton (in this case polyisoprene). These compounds differ in the number of "isoprenoid groups" (also called "isoprene units"), by properties and by natural distribution (Chappel, 1995).

In the chemistry of natural organic compounds, there are four major classes of compounds with polyisopentene skeleton, i.e., *terpenes, terpenoids, carotenoids and steroids.*

The name "terpenoids" has been accredited by analogy with "carotenoids", "steroids" known for a long time. The group of terpenes and terpenoids include hydrocarbon compounds named terpenes and their oxygenated derivatives – alcohols, carbonyl compounds (aldehydes and ketones) and carboxylic compounds. Terpenes C_{10}, C_{15} are mostly in volatile essential oils, and terpenes C_{20}, C_{30} are found in vegetable resins, balsams, etc.

Terpenes and terpenoids of nutritional, pharmaceutical and/or cosmetic interest are present in superior plants, being located in seeds and fruits. Table 2.6 shows data on essential oils (main representatives) highlighting plant sources, storage tissues and types of compounds and areas of use.

Monoterpenes are isoprene derivatives, which contain two isoprene residues. Their formula is $(C_5H_8)_2$, and structurally they can be acyclic, monocyclic or dicyclic. The structure of isoprene and monoterpenoids more commonly found in essential oils are shown in Figure 2.36. So, it can be seen that the subgroup of monoterpenoids includes several structurally distinct types of compounds: (i) acyclic monoterpenoids; (ii) cyclic monoterpenoids, which can be of two types: monocyclic monoterpenoids and dicyclic monoterpenoids (Bodea et al., 1964–1966; Bonner and Varner, 1976; Grayson, 1996; Masyita et al., 2022).

2.4.8.2.1.1 Acyclic Monoterpenes and Monoterpenoids In the subgroup of acyclic monoterpenes and monoterpenoids in essential oils, compounds with two isopentane (isoprene) units with hydrocarbon structure are included, e.g., *myrcene*

TABLE 2.6
Essential Oils Containing Terpenoids – The Main Representatives

Essential Oil Type	Plant (Taxonomic Name)	Sampled Plant Tissue	Oil Content (%)	Predominant Constituents (Composition in %)	Main Uses
Anise	*Pimpinella anisum* *Foeniculum vulgare* (Umbelliferae)	Ripe fruits	2–5	Anethole (50–70); fenchone (10–20); estragole (3–20)	Flavoring; antispasmodic aromatherapy
Bergamot	*Citrus aurantium* ssp. *Bergamia* (Rutaceae)	Orange peels	0.5	(+) – Limonene (42); (−) – linalyl acetate (27); γ-terpinen (8); linalool (7); bergapten; bisabolene	Flavoring perfumery aromatherapy
Caraway	*Carum carvi* (Umbelliferae)	Ripe fruits	3–7	(+) – Carvone (50–70); (+) – limonene (47); dihydrocarvone	Flavoring (spice) carminative flatulence cramps aromatherapy cosmetics
Thyme	*Thymus vulgaris* (Labiatae)	Flowers in inflorescence stage	0.5–2.5	Thymol (40); *p*-cymene (30); linalool (7); carvacrol (1)	Flavoring antiseptic aromatherapy
Coriander	*Coriandrum sativum* (Umbelliferae)	Ripe fruits	0.3–1.8	(+) – Linalool (60–75); γ-terpenes (5); α-pinene (5); camphor (5)	Flavoring carminative
Cloves	*Syzygium aromaticum* (Myrtaceae)	Dry buds	15–20	Eugenol (75–90); eugenyl acetate (10–15); β-caryophyllene (3)	Flavoring antiseptic aromatherapy
Eucalyptus	*Eucalyptus globulus* (Myrtaceae)	Fresh leaves	1–3	Cineole (70–85); α-pinene (14)	Flavoring antiseptic
Eucalyptus (lemon flavor)	*Eucalyptus citriodora* (Myrtaceae)	Fresh leaves	0.8	Citronellal (65–85)	Flavoring (spice) carminative flatulence, colics aromatherapy cosmetic
Orange blossom	*Citrus aurantium* ssp. *Amara* (Rutaceae)	Fresh flowers	0.1	Linalool (36); α-pinene (16); (+) – limonene (12); (−) – linalyl acetate (6)	Flavoring antispasmodic aromatherapy cosmetics

(Continued)

TABLE 2.6 (*Continued*)
Essential Oils Containing Terpenoids – The Main Representatives

Essential Oil Type	Plant (Taxonomic Name)	Sampled Plant Tissue	Oil Content (%)	Predominant Constituents (Composition in %)	Main Uses
Ginger	*Zingiber officinale* (Zingiberaceae)	Dry rhizomes	1.5-3	Zingiberene (34); β-sesquifelandrene (12); β-felandrene (8); β-bisabolene (6)	Flavoring
Juniper	*Juniperus communis* (Cupressaceae)	Dried ripe grains	0.5–2	γ-Pinene (20); limonene (9); mircene (9); (−) – borneol (8)	Flavoring antiseptic diuretic aromatherapy
Lemon	*Citrus limon* (Rutaceae)	Dry shells	0.1–3	(+) – Limonene (60–80); β-pinene (8–12); γ-terpinene (8–10); citral (2–3)	Flavoring perfumery
Dill	*Anethum graveolens* (Umbelliferae)	Ripe fruits	3–4	(+) – Carvone (40–65)	Flavoring carminative
Peppermint	*Mentha spicata* (Labiatae)	Fresh leaves	1–3	Menthol (30–50); menthone (15–32) menthyl acetate (2–10); menthol (1–9)	Flavoring antispasmodic aromatherapy cosmetics
Pine	*Pinus palustris* (Pinaceae)	Pine needles, pine branches		α-Terpineol (65)	Antiseptic disinfectant aromatherapy
Orange	*Citrus sinensis* (Rutaceae)	Dry shells	0.3	(+) – Limonene (90–95); mircene (2)	Flavoring (flavor and smell come from oxygenated derivatives)
Sage	*Salvia officinalis* (Labiatae)	Flowers in the inflorescence stage	0.7–2.5	Thujone (40–60); (−) – camphor (5–22); mircene; cineole (5–14); β-caryophyllen (10); (+) – limonene (6);	Flavoring flatulence dysmenorrhea biliary dyskinesia aromatherapy
Rose	*Rosa damascena Rosa cenifolia* (Rosaceae)	Fresh flowers	0.02–0.03	(+) – Citronellol (36); geraniol (17) 2-phenyl-ethanol (3); eugenol	Perfumery
Valerian	*Valeriana officinalis* (Valerianaceae)	Dry roots		Butyric acid, valerenic acid, (−) – camphor, (−) – borneol	Cardiac arrhythmias, hyperactive gastritis, helminthiases

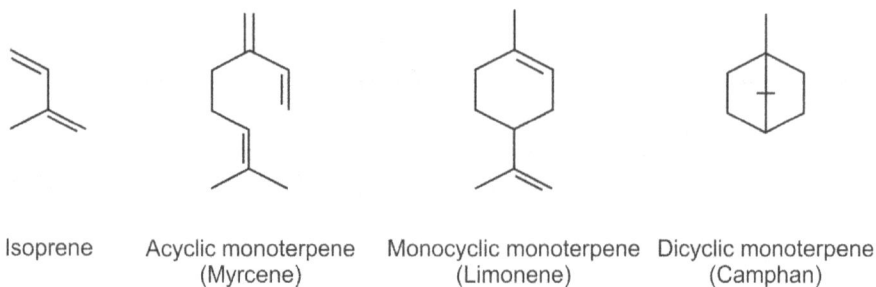

Isoprene   Acyclic monoterpene   Monocyclic monoterpene   Dicyclic monoterpene
                    (Myrcene)   (Limonene)   (Camphan)

FIGURE 2.36 Isoprene and representative monoterpenes.

(I), or with a structure of oxygenated derivatives thereof, e.g., *geraniol* (II), *linalool* (III), *citral* (IV), etc. (Figure 2.37).

> *Myrcene* – is a hydrocarbon compound isolated from hops (*Humulus lupulus*), a cereal used in beer production.

Oxygenated compounds of this class, more commonly found in essential oils are:

> *Geraniol* – a hydroxylated compound to a secondary C, is found in rose, eucalyptus, geraniums, lemons oils (volatile), etc.
> *Linalool* – a hydroxylated compound to a primary C, isolated from teardrop flowers, lavender, coriander and orange oils;
> *Citral* – is an aldehyde present in orange oil.

2.4.8.2.1.2 Cyclic Monoterpenes and Monoterpenoids   These compounds constitute another distinct subgroup of hydrocarbons compounds (terpenes) and their oxygenated derivatives(terpenoids) which include: monocyclic monoterpenes and monoterpenoids; dicyclic monoterpenes and monoterpenoids; iridoid type of mono- and dicyclic monoterpenoids.

2.4.8.2.1.3 Monocyclic Monoterpenes and Monoterpenoids   There are found in fruits, flowers, fruit peel, even in woody tissues (in the case of conifers). This subgroup includes compounds that have a structural hexacycle structure (i.e. hexane

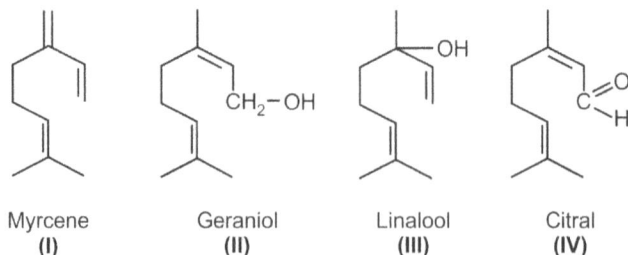

Myrcene   Geraniol   Linalool   Citral
   (I)   (II)   (III)   (IV)

FIGURE 2.37 Acyclic monoterpenes and monoterpenoids – representative structural types: hydrocarbon (I); oxygenated derivatives (II, III, IV).

or hexene), to which side chains are grafted: limonene (V), menthol (VI), menthone (VII), and carvone (VIII). In general, the monocyclic compounds (Figure 2.38) may be hydrocarbons or oxygenated derivatives (hydroxylates and carbonyls)-terpenoids.

Limonene or carvene is a hydrocarbon present in the cumin, dill, turpentine (obtained by dry distillation of coniferous resins), in citrus peel extracts (oranges, tangerines and lemons) oils. It is also present in essential oils obtained from anise, cumin, dill and celery (see Table 2.6).

The most well-known oxygenated compounds are:

Menthol – hydroxylated and dialkylated derivative of cyclohexane present in mint and dill essentials oils;

Mentone – a dialkyl ketone derivative of cyclohexane present in mint, along with menthol;

Carvone – present in cumin, dill, thyme, etc.;

Terpenol - hydroxylated compound present in lilac oil;

Eucalyptol – hydroxylated compound isolated from eucalyptus extract, lavender.

α-phellandrene – was identified in the essential oil obtained from the species *Anethum graveolens, Foeniculum vulgare, Eucalyptus phellandra*. It has antitumor activity. It is used as a biopesticide and repellent. In the food field, it is used as a preservative.

2.4.8.2.1.4 Dicyclic Monoterpenes and Monoterpenoids They consist of tetra-, penta- or hexatomic cycles. They are hydrocarbons, oxygenated derivatives (hydroxylated and carbonyl). The best-known structures are: camphane (IX), pinene (X), borneol (XI) and camphor (XII). Figure 2.39 presents their structural formulas.

Camphane – consisting of two pentacyclic nuclei. It has been isolated from lavender and rosemary oils.

Pinene – consisting of a hexa nucleus - and a tetra atomic one. It can be found in eucalyptus, lavender, lemon, coriander, parsley and of turpentine essences.

From the point of view of biological activity, it stands out for its antioxidant, anti-inflammatory, nematocidal, neuroprotective and gastroprotective activity.

Camphene – was detected in the essential oil of *Ocimum gratissimum* (basil clove) and *Thymus algeriensis*. It has hypolipidemic, antioxidant activity, inhibits superoxide radicals. At the same time, it can be used in the treatment of fungal skin infections.

Two the best-known oxygenated compounds are from camphane:

Borneol – an alcohol found in lavender, rosemary and rosemary oil fir oil;

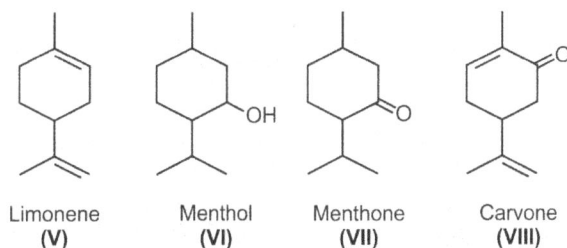

| Limonene | Menthol | Menthone | Carvone |
| (V) | (VI) | (VII) | (VIII) |

FIGURE 2.38   Monocyclic monoterpene and monoterpenoides - structural formulas.

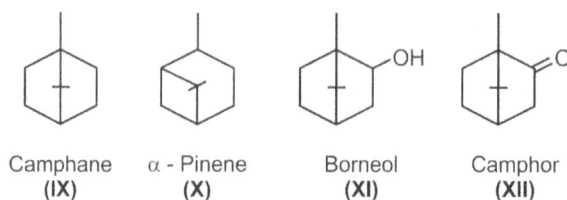

Camphane   α - Pinene   Borneol   Camphor
(IX)   (X)   (XI)   (XII)

FIGURE 2.39   Dicyclic monoterpenes and monoterpenoids – structural formulas.

Camphor – a ketone that has been isolated from the leaves of the shrub *Cinnamomum camphor* and wormwood leaves (*Artemisia absinthium*). In pure form, both compounds are solid.

2.4.8.2.1.5   Iridoid Type Mono- and Dicyclicterpenoids   The group of monoterpenoids also includes iridoids (Junior, 1990; Dewick, 1997). At the origin of the iridoids is the skeleton of the "iridiane nucleus" (XIII) which contains a pentacycle (pentane). It "fuses" with an oxygenated heterohexacycle resulting iridoid (XIV), nepetalectone (XV), secoiridoid (XVI), presented in Figure 2.40.

Iridoid – contains a pentacyclic nucleus and an oxygen hexacycle. from biologically active derivatives are formed: hemiacetal - iridoid; loganin - which can be glycosylated, etc. (Dewick, 1997).

Nepetalactone – a compound in which the heterocycle has a ketone bond. It was isolated from the *Nepeta cataria* plant (family Labiatae). The compound is a powerful attractive and stimulating for cat.

Secoiridoid – this compound has the characteristic that the pentane cycle is open. Secoiridoid nucleus – with the characteristics of a monocyclic monoterpenoid (with oxygen heteroatom) – is found in *secologanin*, a compound with biologically active properties that can bind to carbohydrates – forming glycosides or alkaloids – forming indole alkaloid complexes.

In these situations, it acquires biologically active properties of newly formed compounds (Harborne, 1989).

Regarding iridoids, research conducted on the mechanisms of their biosynthesis, led to the observation that biosynthesis starts from geraniol. Through successive steps – via hydroxylation and oxidation – various iridoid derivatives are obtained, e.g. iridodial, iridotrial, loganin, secologanin, etc. (Dewick, 1997).

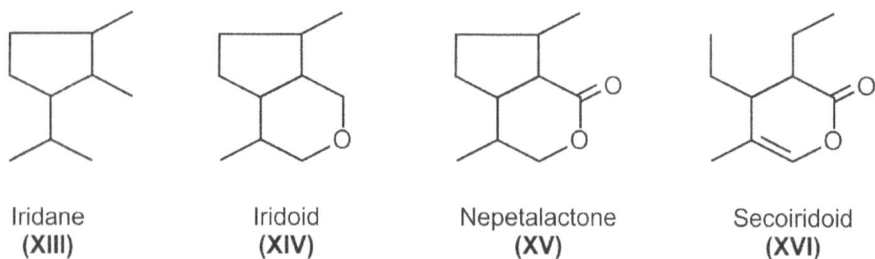

Iridane   Iridoid   Nepetalactone   Secoiridoid
(XIII)   (XIV)   (XV)   (XVI)

FIGURE 2.40   Iridiane-derived monoterpenoids – representative structural types.

2.4.8.2.2   Sesquiterpenes

Sesquiterpenes and sesquiterpenoids have the basic structure $(C_5H_8)_3$ and are spread in various essential oils (Fraga, 1996; Neamţu, 1983). Structurally compounds of this class may be acyclic and cyclic. Figure 2.41 shows examples of **acyclic substances** - *farnesol* (XVII), **monocyclic** - *bisabolene* (XVIII) and **dicyclic** with similar cycles, e.g. *cadinene* (XIX) or different, e.g. *azulene* (XX) acyclic, monocyclic and dicyclic sesquiterpenoids. The tricyclic sesquiterpene *santonin* (XXI) was isolated from *Artemisia* spp. Next the characteristic representatives of these sesquiterpenoids are described.

Farnesol – acyclic compound with a hydroxyl group, isolated from various essential oils. Oils from flowers were especially detected, e.g., linden flowers, pearls, tears, etc.

Bisabolene or *limene* – is a very monocyclic sesquiterpenoid widespread in nature, present in rose, lemon (generally citrus) oils, carrot oils, oils obtained from spruce needles.

Cadinene – has a dicyclic structure, with two hexamonocycles, each having a double connection. It was originally isolated from vegetable oil from the *Piper cubeba* plant. Later it was obtained from other plants present in the spontaneous flora.

Azulene – is a dicylic compound consisting of heptacyclotriene and pentacyclodiene. It was extracted from the upper fractions of the oil obtained from one chamomile species (*Matricaria recutita*). The volatile oil called *chamazulene* is dark blue-colored. Azulene and its derivatives have antispasmodic, anti-inflammatory and antiulcer properties. It is suitable for use in low concentrations (1%) in cosmetics (lotions and creams).

Santonin is a tricyclic sesquiterpenoids, which have also been isolated from plants. These are compounds which are generally formed by the appearance of lactone structures. Santonin (XXI) was isolated from wormwood (*Artemisia* spp., e.g., *A. cinae, A. maritima*). In the structure of santonin (Figure 2.42), a lactone bond and a quinoline nucleus can be identified. It has therapeutic uses as a dewormer.

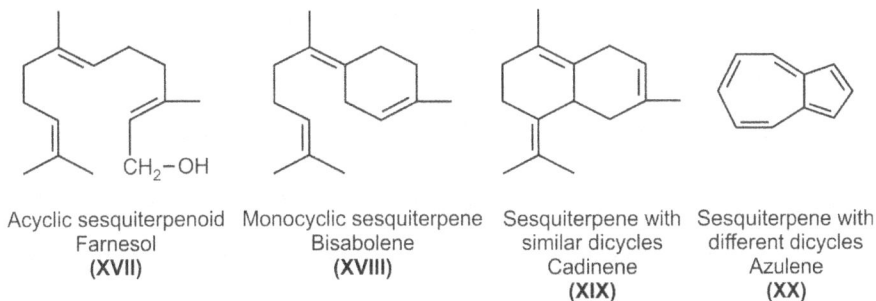

Acyclic sesquiterpenoid   Monocyclic sesquiterpene   Sesquiterpene with   Sesquiterpene with
Farnesol   Bisabolene   similar dicycles   different dicycles
(XVII)   **(XVIII)**   Cadinene   Azulene
                                                                    (XIX)   **(XX)**

FIGURE 2.41   Sesquiterpenes and sesquiterpenoids – representative structural types: acyclic, monocyclic and dicyclic.

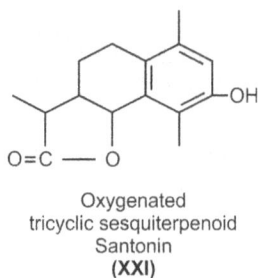

Oxygenated
tricyclic sesquiterpenoid
Santonin
(XXI)

FIGURE 2.42   Tricyclic sesquiterpenoid.

2.4.8.2.3   Tropolone Compounds –Present in Essential Oils

Tropolones are biologically active organic compounds in the structure of which it can find an unsaturated hepta atomic nucleus, known generically as a *tropolonic nucleus*. The basic compound is tropone – a ketone substance – from which, "theoretically", various derivatives are formed. Tropolones have a pronounced aromatic character. From the tropolone group in Figure 2.43, the following compounds are mentioned: *stipitatic acid* (**XXII**), *procerin* (**XXIII**) and *thujaplicin* (**XXIV**).

These tropolones are present in essential oils obtained from cultivatated mush-rooms or shrubs. There may be hydroxyl, carbonyl (especially ketone) or even car-boxylic groups in the structure of the tropolones,

Stipitatic acid – extracted from Penicillium stipitatum cultures;
Procerin – isolated from Juniperus procera;
Thujaplicin – has α,β,γ isomers and has been isolated from essential oil Thuja spp.

In general, tropolones are substances of lesser nutritionally interest. In some cases, however, their derivatives may have toxic action. There are some tropolones with insecticidal action manifested in the plant kingdom and through this is especially protective for wood essences.

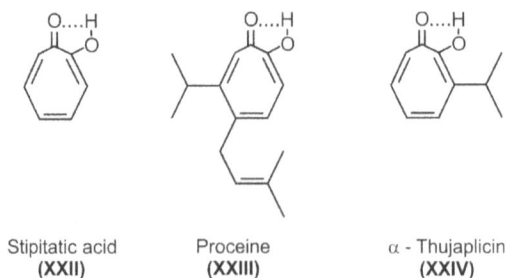

Stipitatic acid   Proceine   α - Thujaplicin
(XXII)   (XXIII)   (XXIV)

FIGURE 2.43   Tropolones.

2.4.8.2.4 Miscellaneous Compounds in Essential Oils

Among the essential oils, as shown, along with terpenoid compounds and tropolone compounds, there are "various compounds" whose essential characteristic resides in an atypical structure. Although there are a variety of acyclic compounds – the presence of similar functional groups stands out, i.e., (i) *carbonyl functional groups* (aldehydes and ketone); (ii) *aliphatic structure* (methyl and ethyl) of some organic acids; and (iii) *aromatic structure*. A brief statements is made about the essential oils belonging to the group of "various compounds".

2.4.8.2.4.1 Essential Oils with Carbonyl Functional Groups
This class includes various aldehydes and ketones. Obviously, the carbonyl compounds presented in the terpenoid and tropolone groups will not be mentioned. Among the aldehydes more frequently present in essential oils, benzaldehyde (XXV) – isolated from various fruits and vegetables; cinnamaldehyde (XXVI) – preferred in cinnamon and cloves are mentioned. Both compounds have an aromatic nucleus (Figure 2.44).

2.4.8.2.4.2 Essential Oils with Aliphatic Structure
In fruits and vegetables, there are essential oils made up of aliphatic esters from formic, acetic, propionic, butyric, valerenic, capronic acids and their methyl, ethyl, amyl esters, etc. (see Figure 2.44). Among the ester, the following compounds are mentioned: isoamyl formate (XXVII) and isoamyl acetate (XXVIII) – isolated from apples; octyl propionate (XXIX) and octyl butyrate (XXX) – isolated from parsnips; isocapronylmethyl ester (XXXI) – isolated from pineapple fruit.

2.4.8.2.4.3 Essential Oils with Aromatic Structure
This group of compounds includes many free substances – phenols or esters. Their identification was made from vegetables, fruits, flowers and spontaneous flowering plants. In Figure 2.45, the structural formulas of the main essential oils from the phenol class are presented.

Among these, more important are: *thymol* (XXXII) – gives an odor the thyme (it also has antiseptic action); *carvacrol* (XXXIII) – present in thyme and pepper; *guaiacol* (XXXIV). Various alkyl, hydroxylated or aldehyde side chain etherified phenolic compounds have also been isolated.

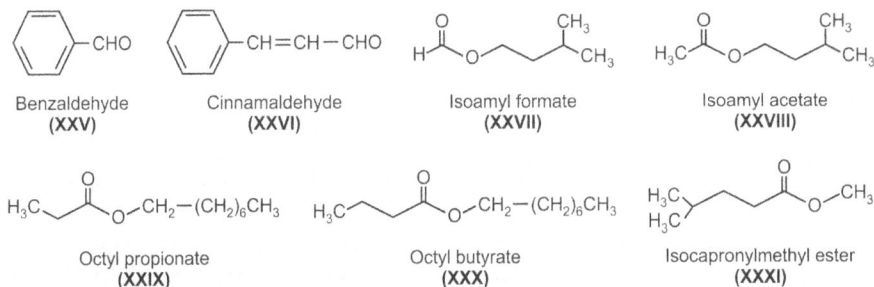

FIGURE 2.44 Aryl-aldehyde compounds and aliphatic esters isolated from essential oils.

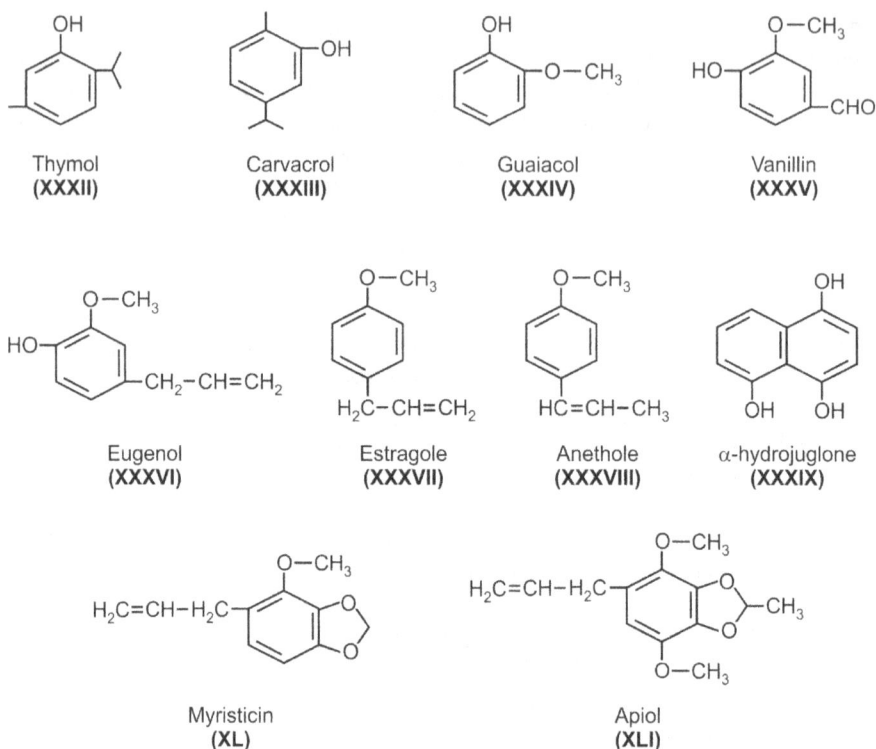

FIGURE 2.45   Phenolic compounds isolated from essential oils.

Monocyclic compounds are mentioned in this group: *vanillin* (**XXXV**) is obtained from vanilla fruit (this compound is synthesized in large quantities for food industry); *eugenol* (**XXXVI**) is found in the essence of geranium, carnations, bananas, bay leaves, cloves, etc.; *estragol* (**XXXVII**) is present in tarragon essence; *anethole* (**XXXVIII**) is isolated from parsley roots.

The essential oils with phenolic hydroxyl (free or esterified), dicyclic compounds can be found, e.g., α-*hydrojuglone* (**XXXIX**) - or even heterocyclic, e.g., *myristicin* (**XL**); *apiol* (**XLI**) - present in parsley roots. Numerous essential oils containing carbonyl, phenolic compounds and esters have applicative use.

2.4.8.2.5   Other Terpenes

Diterpenes/diterpenoids and triterpenes/triterpenoids are found in resins, especially in resins from trees and shrubs (Hanson, 1995; McGarvey and Croteau, 1995; Connolly and Hill, 1996). It should not be overlooked that certain derivatives of compounds from this class have been isolated from natural plant products and are studied in organic chemistry and in plant biochemistry having connotations regarding the biologically active character of them.

2.4.8.3 Essential Oils in Biochemistry

Essential oils, as shown, are mixtures of organic compounds isolated from plants. They are in a liquid state, have an average subunit density (about 0.9), have an aromatic odor, are colorless or yellow to brown. It is possible to solidify by crystallizing some of the main components. They were isolated in the form of mixtures of compounds, sometimes exceeding the number of 50 molecular species.

The diversity of composition is explained by the presence of acyclic hydrocarbons, aromatic hydrocarbons, alcohols (lower and upper), of aldehydes, ketones, organic acids, their ester derivatives, etc. Many of them may also contain ethers or heterocyclic compounds containing oxygen, nitrogen or sulfur. Essential oils are of particular interest in various fields of application: nutrition, e.g. nutrients in the category of physiologically active substances (Guenther, 1950; Heller, 1977–1978; Alais and Linden, 1993); pharmacology and pharmacognosy, e.g. pharmacologically active substances used as and pharmaceuticals for aromatherapy, carminative, antispasmodics effects, etc. (Gildemeister and Hoffmann, 1959; Charlwood and Banthorpe, 1991; Ciulei et al., 1993; Dewick, 1997); cosmetics e.g., soaps, perfumes, creams, brillantines, face lotions (Thiers, 1962).

Some comments on essential oils and their uses (therapeutic and/or nutritional) have been made since antiquity. In this regard, there are information about the therapeutic use of essential oils by Chinese for approx. 4000 years (BC) ago, by Egyptians 1555 years (BC). Data on essential oils was revealed in medical works left by Hippocrates of Kos (460–375 BC) and botanical writings from Theophrastus of Eresos (370 BC). There are 188 references to essential oils in the Bible. In Europe, the first crops of aromatized herbs appeared in the Middle Ages, which is pleading to expand their use.

The therapeutic utility of essential oils was noted by the French chemist René Maurice Gattefosse since 1920. He had a laboratory accident – a burn on his hand – and he hurriedly dipped his hand in the lavender oil, thinking it was water. To his surprise, after a few minutes, the pain subsided and soon the wounds healed.

With the development of knowledge in analytical chemistry and biochemistry, essential oils have found uses in phytotherapy (aromatherapy, fruit therapy and vegetable therapy). They started from considering that the recommended plants also contained essential oils. Obviously, they have also found uses in the field of nutrition. Their expansion was gradual, with the diversification of food processing methods.

Essential oils are the source of the characteristic odor of various species of plants. They are found in certain organs of the plant predominating in flowers, leaves, fruits, and seeds. However, small amounts are also found in roots or even woody stems.

In the vegetable kingdom, there are representative families of plants for the oil content, e.g., *Pinaceae, Labiatae, Umbelliferae, Mirtaceae, Compositae*, etc.

In general, the name "herbs" is attributed to those species that contain a significant amount of essential oils, approx. 0.1%–0.2%, have an odor sufficiently perceptible and suitable for economic exploitation.

The study on essential oils allowed the approach of biochemical taxonomy problem, following the existence of some genera and even plant species among which there is a certain homogeneity in the composition of essential oils. This was not possible; however, there was a predominance of anethole in *Umbelliferae* and glycosyl oleate in *Cruciferae*. In general, plants belonging to very distant families in taxonomically relation contain similar or even identical essential oils. Such oils present in many plants are: anethole, borneol, camphor, carvone, citral, etc.

The physiological role of essential oils in plants is not fully elucidated. However, they facilitate pollination by attracting insects to the flowers, which have a characteristic odor. Also, some essential oils are produced by protection against animals.

In terms of application, essential oils have found their use in the cosmetic industry (perfumes and soaps), in the pharmaceutical industry (ointments and lotions), in the food industry (sugar products, soft drinks and even alcoholic beverages).

Biogenesis of essential oils. Biogenesis was a constant concern of botanists (morphology, taxonomy), analytical chemists, biochemists and specialists in processing technologies (pharmaceuticals, cosmetics and food).

In this context, the aspects of *topobiochemistry* were followed, trying to locate the tissue or organ in which the essential oils are produced or stored. Another interesting aspect, from a biochemical point of view, was that of tracking aspects of plant *chronobiochemistry* by which periods of time are established (referring to phenophases), when the amount of essential oils is higher. These data are used in practice for the purpose of collecting plant tissues and extraction of essential oils.

Topobiochemical studies have shown the existence in tissues of so-called oleiferous cells, studied more in the case of terpene compounds detected in certain tissues of plant families, e.g., *Magnoliaceae, Laureaceae, Zingiberaceae, etc.*

The oleiferous cells have an arrangement that converges into lined tissues with single-cell layers. Oleiferous cells are generally larger and are yellow due to the inside oil. The extension of topobiochemical studies also highlighted the possibility of disposing of essential oils in the intercellular space. Such situations are found in *Mirtaceae, Coniferae.*

In coniferous, for example, there is even talk of the existence of resin channels specific to this family of plants. Resiniferous channels are in fact secretory channels lined with parenchymal cells with intense metabolic activity.

Another biogenesis way take place in the intercellular space and accumulation is characteristic of fruits, e.g., the pericarp of citrus fruits of the *Aurantiaceae* family, has "*spherical cavities*" lined with small cells in which essential oils are produced. They are then removed as exudates in a central cavity. Often, in plant morphology, this biosynthetic system has been compared to that characteristic of glands with internal secretion, present in superior animals and in humans. A special feature of topobiochemistry in the biogenesis of essential oils is represented by the existence of a biosynthesis process in the glandular cilia present in some plants. So, glandular cilia that represent taxonomic characteristics are found in *Labiatae* and *Compositae*. In their case, the essential oils are developed as cellular exudates that cross the canaliculi of cellulose wall and accumulates in the space underlying the cuticle that covers the glandular cilia. Glandular cilia change gradually, in terms of morphology, due to the increase in the amount of essential oils from the underlying tissues.

Obtaining essential oils. The methodology used for this purpose represents another important issue of biochemical analysis. In this regard, it is mentioned that there are various specific procedures for extracting essential oils: (i) *water distillation*; (ii) *steam distillation*; (iii) *extraction with volatile solvents*; (iv) *extraction with solid solvents* and (v) *sampling by physical*, i.e. mechanical procedures (*squeezing* and *pressing*).

Chemically, essential oils are organic substances, usually liquid. In terms of composition in essential oils are: saturated and unsaturated aliphatic hydrocarbons (ethylene, acetylene), aromatic hydrocarbons, alcohols, aldehydes, ketones, acids, esters, phenolic ethers, lactones, nitrogen compounds, sulfur compounds, pigments, etc.

The biological action of the essential oils used in aromatherapy is conditioned by their chemical structure. The main groups of chemical compounds are mentioned in this framework, the effects produced and various essential oils are exemplified in which they are present are also mentioned.

Monoterpenes –Effects: antiviral, antiseptic and bactericidal. They are irritating for the skin. Example: lemon oil.

Aldehydes – Effects: antiseptic and sedative. Examples: lemon balsam, citronella oil.

Ketones – Effects: in congestion, helps the elimination of mucus. It can be toxic. Examples: anise, fennel, sage oils.

Alcohols – Effects: strong antiseptic, antiviral. Examples: geranium, rose oils.

Phenols – Bactericidal effects, strongly stimulating. May be irritating to skin. Examples: clove oil, thyme oil.

Esther – Effects: fungicides, sedatives. They have a pleasant aroma. Examples: bergamot oil, lavender and sage.

Oxides – Effects: expectorants and bactericides. Example: rosemary oil.

Absorption of essential oils – Essential oils are substances with small molecules, they are liposoluble – fact that makes possible to pass them through skin. If the essential oils are diluted and applied directly to the skin they are absorbed completely and penetrates deep: into the tissues, into the interstitial fluid and into the circulation blood. Different oils have varying absorption rates, ranging from 20 to 120 minutes.

In aromatherapy, it is known that the essential oils reach to skin contributes to the regulation of capillary activity and the restoration of tissue vitality. The chemicals in the essential oils (used for body massage) are transported to organs and system, in which it contributes to the stimulation of natural functions.

When inhaling an essential oil, there is an *olfactory* response and it produces passage, by absorption, into the bloodstream. Smell is achieved through the olfactory membrane, which is the only place in the human body at which the CNS is in direct contact with the environment. When a cellular olfactory receptor is stimulated, the impulse in the form of chemical mediators passes along the olfactory nerve to the "limbic system".

Thus, before conscious knowledge, we are in contact with aroma, our subconscious already receives and reacts. The limbic system is a part of the brain in which memory, hunger, response sexuality or emotion are evoked.

Purity control – In the case of essential oils, control is required due their use in pharmacy, nutrition and cosmetics. Given that the oils interact with tissue bioconstituents – so they participate in biochemical reactions – it is necessary to establish certain: *physical characteristics*, e.g., olfactory examination, density, refractive index, specific rotation, melting point, freezing point, etc.; *chemical characteristics*, e.g., acidity index, esterification index, index of saponification, acetyl index, presence of total alcohols, percentage of total alcohol, the presence of aldehydes, the presence of ketones, phenols, etc. In general, specific determinations are made according to standards accredited in this field.

Use of essential oils – Request of these oils in pharmacy, nutrition and cosmetics have led to the development of a plant processing industry in view of oil extraction. An impressive expansion of the extraction was found by the amount of various oils obtained. For example, worldwide, statistics data from 1978 show that oil extraction has reached tons, e.g., peppermint oil 800 t, orange oil 1,100 t, lavender oil 1,250 t, cinnamon oil 5,000 t; cloves 12,000 t, sage oil 25,000 t, etc. (Ciulei et al., 1993).

Fruit-flavored essential oils are formed during ripening when there are significant changes in the concentration of AA. Decreasing their concentration is accompanied by enzymatic transformations that lead to chemical precursors specific to volatile oils. Experimentally, using enzymatic extracts (banana, citrus), it was found on the "sliced slices" of the respective fruits that transformations also take place especially for branched amino acid chains. Thus, carbonyl compounds (e.g., ketones), alcohols, esters appear, which gives to fruits, specific aromas.

Essential oils can have immunostimulant, antiviral, antibacterial, antifungal, antioxidant and even antitumor effects. Advances in knowledge of essential oils have led to the establishment of practical methods of their use in herbal medicine and aromatherapy.

Owing to its content in essential oils, some plants are used in the mixture as spices for obtaining processed foods, e.g., concentrates for soups (powders, cubes), for pudding, for juices, etc. Also, certain oils are used to preserve meat, e.g. sage oil.

There is even a specific analysis for the so-called essential oils *aromagram* indicating the odor properties of the species aromatic plant and at the phenophase in which the tissue subjected to oil extraction was taken.

Aromagrams are determined using the gas-chromatographic method in which the peaks highlight and define the molecular species in the studied essential oil. To get an idea of the diversity of molecular species that compete with the formation of oils extracted from various plants is generally done in complex studies based on physico-chemical analyzes (chromatography, spectrophotometry, polarimetry, etc.).

The study of biologically active substances of plant origin is done by systematic approach, following stages: (i) documentation on the class of chemical compounds; (ii) documentation on the current state of knowledge in various aspects of biochemistry, physiology, chemistry clinic, pharmacology, toxicology, etc.; (iii) toxicological examinations: acute, subacute effects, chronic (on laboratory animals); (iv) special tests (teratogenicity, mutagenicity and carcinogenicity); (v) clinical-biological tests, following the mechanism of action, the effects on homeostasis; (vi) biologically active specificity (physiological effects, pharmacological effects, etc.); (vii) observations and preventive examinations: effect on consumers/users); (viii) studies based on epidemiological data - investigations limited to a certain usual area, to a certain category of users, etc.; (i) establishing programs for medical trials, and so on.

It is relevant to stage the research and the accuracy of the investigation of all substances to be used as food and/or pharmaceutical ingredients but and for cosmetic ingredients.

REFERENCES

Alais C., Linden G. - *Biochimie alimentaire*, 2-ème édition, Masson, Paris-Milan-Barcelone-Bonn, 1993.

Ahmed T., Wang C.-K. - Black garlic and its bioactive compounds on human health diseases: a review, *Molecules*, 2021, 26, 5028 https://doi.org/10.3390/molecules26165028.

Amarowicz R., Janiak M. - Hydrolysable Tannins, pp. 337–343, in *Encyclopedia of Food Chemistry* (Melton L., Shahidi F., Varelis P., Eds.), Academic Press, 2019.

Bhambhani S., Kondhare K.R., Giri A.P. - Diversity in chemical structures and biological properties of plant alkaloids, *Molecules*, 2021, 26, 3374 https://doi.org/10.3390/molecules26113374.

Bhavaniramya S., Vishnupriya S., Al-Aboody M.S., Vijayakumar R., Baskaran D. - Role of essential oils in food safety: antimicrobial and antioxidant applications, *Grain Oil Sci. Technol.*, 2019, 2(2), 49–55.

Bodea C., Fărcăşan V., Nicoară E., Sluşanschi H. - Treatise on vegetable biochemistry (in Romanian), Editura Academiei R.S.R., Bucureşti, Part I, Vol. I, 1964, Vol. II, 1965, Vol. III, 1966.

Bonner J., Varner J.E. - *Plant Biochemistry,* 3rd edition, Academic Press, New York- London, 1976.

Borza Al. - *Ethobotanical Dictionary* (in Romanian), Editura Academiei R.S.R., Bucureşti, 1968.

Braghiroli F.L., Amaral-Labat G., Boss A.F.N., Lacoste C., Pizzi A. - Tannin gels and their carbon derivates: a review, *Biomolecules*, 2019, 9, 587 https://doi.org/10.3390/biom9100587.

Braquet P., Esanu A., Buisine E., Hosford D., Broquet C., Koltai M. - Recent progress in ginkgolide research, *Med. Res. Rev.*, 1991, 11, 295–355.

Britton G., Goodwin T.W. - *Carotenoid Chemistry and Biochemistry*, Pergamonn Press, Oxford - New York -Toronto-Sydney-Paris-Frankfurt, 1982.

Britton G. - *The Biochemistry of Natural Pigments*, Cambridge Univ. Press, Cambridge, 1983.

Brossi A. - *The Alkaloids, Chemistry and Pharmacology*, Academic Press, San Diego, 1989.

Bruneton J. - *Pharmacognosie - Phytochimie, plantes médicinales*, 4-ème édition, revue et augmentée, Tec & Doc - Éditions médicales internationales, Paris, 2009

Bullen C., McRobbie H., Thornley S., Glover M., Lin R., Laugesen M. - Effect of an electronic nicotine delivery device (e cigarette) on desire to smoke and withdrawal, user preferences and nicotine delivery: randomised cross-over trial. *Tob Control.*, 2010, 19(2), 98–103.

Cavallito C.J., Bailey J.H. - Allicin, the antibacterial principle of Allium sativum. I. Isolation, physical properties and antibacterial action, *J. Am. Chem. Soc.*, 1944, 66(11), 1950–1951.

Chappel J. - Biochemistry and molecular biology of the isoprenoid biosynthetic pathways in plants, *Ann. Rev. Plant Physiol. Plant Mol. Biol.*, 1995, 46, 521–547.

Chadwich D. J., Marsh J. (Eds.) - *Bioactive Compounds from Plants*, John Wiley and Sons, New York, 1990.

Charlwood B.V., Banthorpe D.V. (Eds.) - *Methods in Plant Biochemistry*, Vol. 7, Academic Press, London, 1991.

Ciulei I., Grigorescu E., Stănescu U. - *Medicinal Herbs. Phytochemistry and Phytotherapy: Pharmacognosy treatise, Vol. I–II* (in Romanian), Editura Medicală, Bucureşti, 1993.

Connolly J.D., Hill R.A. - Triterpenoids, *Nat. Prod. Rep.*, 1996, 13, 151–169.

Das A., Wang J.H., Lien E.J. - Carcinogenicity, mutagenicity and cancer preventing activities of flavonoids: astructure-system-activity relationship (SSAR) analysis. *Prog. Drug Res.*, 1994, 42, 133–166.

Devi S. - A Centrum of Valuable Plant Bioactives, pp. 525–544, in *A Centum of Valuable Plant Bioactives* (Mushtaq M., Anwar F., Eds.), Academic Press, Imprint of Elsevier, Cambridge, MA, 2021

Devlin T. M. (Ed.) - *Textbook of Biochemistry with Clinical Correlations*, 3rd edition, Wiley and Sons Inc., Publication, New York-Chichester-Brisbane-Toronto-Singapore, 1992.

Dewick P.M. - *Medicinal Natural Products: A Biosynthetic Approach*, John Wiley and Sons, Chichester-New York-Weinheim-Brisbane-Singapore-Toronto, 1997.

Di Salvo E., Lo Vecchio G., De Pasquale R., De Maria L., Tardugno R., Vadala Rossella, Cicero N. - Natural pigments production and their application in food, health and other industries, *Nutrients*, 2023, 15, 1923 https://doi.org/10.3390/nu15081923.

Ensminger A.H., Ensminger M.E., Konlande J.E., Robson J.R.K. - *The Concise Encyclopedia of Foods and Nutrition*, 2nd edition, C.R.C. Press, Boca Raton, 1995.

Ettlinger M. G., Lundeen A.J.-The structures of sinigrin and sinalbin; an enzymatic rearrangement. J. A. C. S. 1956, 78, 16, 4172–4173

Fleurentin J., Cabalion P., Mazars G., Dos Santos J., Younos C. - Ethnopharmacologie : Sources, Méthodes, *Objectifs, Premier colloque européen d'Ethnopharmacologie*, Metz 23–25 mars, Édition de l'ORSTOM et Société Française d'Ethnopharmacologie, Paris-Metz, 1990.

Fraga B.M. - Natural sesquiterpenoids, *Nat. Prod. Rep.*, 1996, 13, 307–326.

Fraga-Corral M., García-Oliveira P., Pereira Antia G., Lourenço-Lopes C., Jimenez-Lopez C., Prieto M.A., Simal-Gandara J. - Technological application of tannin-based extracts, *Molecules*, 2020, 25, 614 https://doi.org/10.3390/molecules25030614.

Funayama S., Cordell A.G. - Alkaloids. *A Treasury of Poisons and Medicines*, Academic Press, Elsevier, Amsterdam-Boston-Heidelberg-London-New York-Oxford-Paris-San Diego-San Francisco-Singapore-Sydney-Tokyo, 2015.

Gârban Z. - *Human Nutrition, Vol. I* (in Romanian), Editura Didactică şi Pedagogică R.A., Bucureşti, 2000.

Gârban Z. - *Molecular Biology: Concepts, Methods, Applications* (in Romanian), 6th edition, Editura Solness, Timişoara, 2009.

Gârban Z. - *Biochemistry: Comprehensive Treatise, Vol. II, Biochemical Effectors*, 5th edition, Editura Academiei Române, Bucureşti, 2018.

Gildemeister E., Hoffmann F. - *Die Ätherischen Öle, Vol. V*, Akademie Verlag, Berlin, 1959.

Goodman L.S., Gilman A. - *The Pharmacological Basis of Therapeutics, Vol. III*, McMillan Comp., New York, 1966.

Goodwin T.W. - *Chemistry and Biochemistry of Plant Pigments*, Academic Press, London - New York, 1976.

Goodwin T.W. - *The Biochemistry of the Carotenoids, Vol. I*, Chapman and Hall, New York - London, 1980.

Goodwin T.W., Mercer E.I. - *Introduction to Plant Biochemistry*, Pergamon Press Ltd., Oxford - New York - Toronto - Paris - Frankfurt, 1983.

Goto T., Kondo T. - Structure and molecular stacking of anthocyanins flower colour variation, *Angew. Chem. Int. Ed. Engl.*, 1991, 30, 17–33.

Gould K., Davie K., Winefield Ch. (Eds.) - *Anthocyanins. Biosynthesis, Functions and Applications*, Springer, New York, 2009

Grayson D.H. - Monoterpenoids, *Nat. Prod. Rep.*, 1996, 13, 195–335.

Grinţescu G. - *The Culture and Harvest of Pharmaceutical Plants*, Editura Universul, Bucureşti, 1945.

Guenther E. - *The Essential Oils, Vol. IV*, Van Nostrand, New York, 1950.

Guthrie Helen A. - *Introductory Nutrition*, 3rd edition, C.V. Mosby Comp., Saint Louis, 1975.

Hanson J.R. - Diterpenoids, *Nat. Prod. Rep.*, 1995, 12, 207–218.

Harborne J. B. (Ed.) - *Methods in Plant Biochemistry, Vol. 1*, Academic Press, London, 1989.

Harborne J.B., Williams C.A. - Anthocyanins and other flavonoids, *Nat. Prod. Rep.*, 1995, 12, 639–657.

Haslam E., Cai Y. - Plant polyphenols (vegetable tannins): gallic acid metabolism, *Nat. Prod. Rep.*, 1994, 11, 41–66.

Hasler C., Blumberg J. - Phytochemicals: biochemistry and physiology, *J. Nutr.*, 1999, 129(3), 756S–757S.

Heller R. - *Abregé de physiologie végétale, Vol. I. Nutrition, Vol. II. Development*, Masson, Paris-New York-Barcelone-Milan, 1977–1978.

Heymann E. - *Haut, Haar und Kosmetik*, S. Hirzel Verlag, Stuttgart, 1994.

Hudson B.J.F. - *Food Antioxidants*, Elsevier Applied Sicence, London-New York, 1990.

Jacob, M.C.M., Teixeira, C.D., Bautista, D.A., Ramos, V.A.N. - Ethnonutrition. *Ethnobiol. Conser.*, 2021, 10. https://doi.org/10.15451/ec2021-10-10.35-1-8

Junior P. - Recent developments in the isolation and structure elucidation of naturally occurring iridoid compounds, *Planta Med.*, 1990, 56, 1–13.

Khan F., Qidwai T., Shukla R.K., Gupta V. - Alkaloids derived from tyrosine: modified benzyltetrahydro-isoquinoline alkaloids, pp. 405–460, in *Natural Products* (Ramawat K., Mérillon J. M., Eds.), Springer, Berlin-Heidelberg, 2013.

Krinsky N.I. - The biological properties of carotenoids, *Pure Appl. Chem.*, 1994, 66, 1003–1010.

Kritchevsky D., Binfield C., Anderson J.W. - *Dietary Fiber: Chemistry, Physiology and Health Effects*, Plenum Press, London, 1990.

Kutchan T.M. - Alkaloid biosynthesis - The basis for metabolic engineering of medicinal plants, *Plant Cell*, 1995, 7, 1059–1070.

Kytidou K., Artola M., Overkleeft H.S., Aerts J.M.F.G. - Plant glycosides and glycosidases: a treasure-trove for therapeutics, *Front. Plant Sci.*, 2020, 11, 357 https://doi.org/10.3389/fpls.2020.00357.

Lawson L.D. - Bioactive organo-sulfur compounds in garlic and garlic products: role in reducing blood lipids. *Human Med. Agents Plants. ACS Symp. Series*, 1993, 534, 306–330.

Liener E. (Ed.) - *Toxic Constituents of Plant Foodstuffs*, Academic Press, New York, 1969.

Macrae R., Robinson R.K., Sadler M.J. - *Encyclopedia of Food Science and Nutrition*, Academic Press, London, 1992.

Masyita A., Sari R.M., Astuti A.D., Yasir B., Rumata N.R., Emran T.B., Nainu F., Simal-Gandara J. - Terpenes and terpenoids as main bioactive compunds of essential oils, their roles in human health and potential application as natural food preservatives, *Food Chem.: X*, 2022, 13, 100217 https://doi.org/10.1016/j.fochx.2022.100217.

McGarvey D.J., Croteau R. - Terpenoid metabolism, *Plant Cell*, 1995, 7(7), 1015–1026.

Mookherjee B.D., Wilson R.A. - Oils, essential. pp. 603–674, in *Kirk-Othmer Encyclopedia of Chemical Technology, Vol. 17*, 4th edition, John Wiley, New York, 1996.

Müller B.W. - *Kosmetik aus der Apotheke*, Govi Verlag, Frankfurt, 1989.

Nativelle C.A.- Sur la digitaline cristallisée. J. Pharm. Chim.,1869, 4, 255–262.

Neamțu G. - *Ecological Biochemistry* (in Romanian), Editura Dacia, Cluj-Napoca, 1983.

Ockermann P., Headley L., Lizio R., Hansmann J. - A review of the properties of anthocyanins and their influence on factors affecting cardiometabolic and cognitive health. *Nutrients*, 2021, 13(8), 2831. https://doi.org/10.3390/nu13082831.

Okuda T., Yoshida T., Hatano T. - Hydrolizable tannins and related polyphenols, *Prog. Chem. Org. Nat. Prod.*, 1995, 66, 1–117.

Pelletier S.W. (Ed.) - *Alkaloids, Chemical and Biological Perspectives, Vol.6*, John Wiley, New York, 1998.

Pelletier P.J., Caventou J.B.- "Recherches Chimiques sur les Quinquinas" (Chemical Research on Quinquinas). *Annal. Chim. Phys.*, 1820,15, 337–365.

Prota G. - The chemistry of melanins and melanogenesis. *Prog. Chem. Org. Nat. Prod.*, 1995, 64, 93–148.

Pugliese T. - Concepts in aging and skin, *Cosmet. Toiletries*, 1987, 102, 19–44.

Raab W. - *Hautfibel. Medizinische Kosmetik*, 3.Auflage, Gustav Fischer, Stuttgart, 1985.

Radu A., Andronescu E., Firi I. - *Pharmaceutical Botany* (in Romanian), Editura Didactică și Pedagogică, București, 1981.

Rawn L.D. - *Biochemistry*, Neil Patterson Publishers, Burlington-North Carolina 1991.

Runge F. - Ueber einige produkte der steinkohlendestillation (On some products of coal distillation). Ann. Phys. Chem., 1834, 31. 65–78.

Robiquet J.P., Boutron-Charlard A.-Nouvelles expériences sur les amandes amères et sur l'huile volatile qu'elles fournissent. *Ann. Chim. Phys.,* 1830, 44, 352–382.

Sandermann H. - Plant metabolism and xenobiotics, *Trends Biochem. Sci.*, 1992, 17, 82–84.

Schröder E., Balansard G., Cabalion P., Fleurentin J., Mazars G. - Médicaments et aliments. Approche ethnopharmacogique, Deuxième colloque européen d'Ethnopharmacologie/ Onzième conférence internationale d'Éthnomédicine, Heidelberg 24–26 mars, 1993, Publ. Société Française d'Éthnopharmacologie, Paris, 1996.

Shang A., Cao S.-Y., Xu X.-Y., Gan R.-Y., Tang G.-Y., Corke H., Mavumengwana V., Li H.-B. - Bioactive compounds and biological functions of garlic (*Allium sativum* L.), *Foods*, 2019, 8, 246 https://doi.org/10.3390/foods8070246.

Shi Y., Po D., Zhou X., Zhang Y. - Recent progress in the study of taste characteristics and the nutrition and health properties of organic acids in foods, *Foods*, 2022, 11, 3408 https:// doi.org/10.3390/foods11213408.

Simmonds R.J. - *Chemistry of Biomolecules: An Introduction*. Royal Society of Chemistry, Cambridge, 1992.

Stoll A., Angliker E., Barfuss F., Kussmaul W., Renz J.- Trennung und bestimmung herzwirksamer glykoside und deren aglykone durch chromatographie an silicagel. 27. Mitteilung über herzglykoside. *Helv. Chim. Acta.*, 1951, 34, 1460–1477.

Strasburger E. - *Lehrbuch der Botanik*, 3. Auflage, VEB Gustav Fischer Verlag, Jena, 1971.

Süntar I. - Importance of ethnopharmacological studies in drug discovery: role of medicinal plants. *Phytochem. Rev.*, 2020, 19, 1199–1209. https://doi.org/10.1007/ s11101-019-09629-9

Thangaleela S., Sivamaruthi B.S., Kesika P., Bharathi M., Kunaviktikul W., Klunklin A., Chanthapoon C., Chaiyasut C. - Essential oils, phytoncides, aromachology and aromatherapy - a review, *Applied Sciences*, 2022, 12, 4495 https://doi.org/10.3390/ app12094495.

Thiers H. - *Les Cosmétiques: Pharmacologie et Biologie*, Masson et Cie, Paris, 1962.

Troll W. - *Allgemeine Botanik*, 3rd edition, Enke Verlag, Stuttgart, 1973.

Vernon L.P., Seely G.R. - *The Chlorophylls*, Academic Press, New York, 1966.

Wagner H. - *Pharmazeutische Biologie, Drogen und ihre Inhaltsstoffe*, Gustav Fischer Verlag, Stuttgart, 1985.

Waterman P.G. (Ed.) - *Methods in Plant Biochemistry, Vol. 8*, Academic Press, London, 1993.

Watrelot A.A., Norton E.L. - Chemistry and reactivity of tannins in *Vitis* spp.: a review, *Molecules*, 2020, 25, 2110 https://doi.org/10.3390/molecules25092110.

Wiss R.F. - *Lehrbuch der Phytotherapie*, Hippokrates Verlag, Stuttgart, 1960.

Willstätter R: Ueber die constitution der spaltungsproducte von atropin und cocain. *Ber. Dtsch. Chem. Ges.*, 1898, 31, 1534–1553.

Willstätter R., Everest A. E. - Anthocyanins. I. Pigment of cornflowers. *Liebigs Ann.*, 1913, 401, 189.

Witt O. N.- Zur kenntniss des baues und der bildung färbender kohlenstoffverbindungen. *Ber. Dtsch. Chem. Ges.*, 1876, 9, 522–527.

Yulvianti M., Zidorn C. - Chemical diversity of plant cyanogenic glycosides: an overview of reported natural products, *Molecules*, 2021, 26, 719 https://doi.org/10.3390/ molecules26030719.

Ziolkowsky H. - *Kosmetikjahrbuch*, Verlag für chemische Industrie, Augsburg, 1995.

3 Natural Compounds of Animal Origin

3.1 INTRODUCTION

Among the biologically active substances found in natural products that are used in the processing of food and medicine, there are many compounds of animal origin. These are defined by distinct features in terms of chemical structure and biological activity.

This chapter presents some special aspects regarding the intake of biologically active substances (de facto physiologically active, pharmacologically active and toxicologically active) of animal origin called *zoochemicals* – a term accredited in the literature in this field.

Zoochemicals are specifically "biologically active substances" present in various *food products of animal origin* or obtained from *animal tissues*. They can also be used in the form of extracts as ingredients in food supplements and/or pharmaceuticals.

The name *zoochemicals* and often *zoonutrients* has been more frequently used by nutritionists (Macrae et al., 1992; Daria et al., 1996; Ward and Bruce, 2003; Peterson et al., 2009; Kathuria et al., 2019). The name "zoonutrients" derives from their use in their natural state and/or extraction products from biological tissues and liquids (e.g., meat and milk). Such substances, in a general context, are of interest for the processing of "food supplements" and "functional foods" (Goldberg, 1994; Prates and Mateus, 2002; Maher, 2007).

There are also biologically active substances specific to zoochemicals that are suitable for use in the processing of pharmaceuticals and cosmetics. There are, for example, zoochemicals with pharmacologically active properties in which anti-inflammatory, antihypertensive, anticoagulant and antimicrobial effects have been observed by stimulating the development of beneficial microorganisms for food digestion.

Over time, research on zoochemicals has been disparate and mostly oriented to food products of animal origin following the physiological and pharmacological effects. For this reason, the classification of zoochemicals was possible taking into account both the food groups (i.e., milk and dairy products, meat and viscera, eggs and bee products) and certain compound groups extracted from tissues.

As phytochemicals were discovered centuries ago and there were performed numerous studies that targeted a certain plant, a plant family or a certain chemical compound, their classification was possible into structurally well-circumscribed classes.

DOI: 10.1201/9781032702520-3

3.2   CHEMICAL COMPOSITION AND NATURAL DISTRIBUTION

Natural compounds in animal food products characterized by attributes of biologically active substances, can be included in two distinct categories: (i) zoochemicals present in various *food groups* (see Section 3.3) and (ii) zoochemicals present in *animal tissues* (see Section 3.4).

3.2.1   ZOOCHEMICALS PRESENT IN VARIOUS FOOD GROUPS

Compounds from milk and dairy products – The following compounds may be included in this group: colostrinin, immunoglobulins, colostrum lactoferrin, specific "encrypted peptides" (in milk, whey, casein), whey lactoferrin, etc.

Compounds from meat – These compounds are mainly represented by: carnitine, coenzyme Q (CoQ), choline and lipoic acid.

Compounds from eggs – Zoochemicals are in a smaller amount in eggs but, given their food consumption (including processed foods), it has caught the attention of nutritionists and pharmacologists. Among the egg zoochemicals: (i) compounds present in the egg yolk: phosvitin, livetin; (ii) compounds present in egg white: ovolysozyme, ovotransferrin, ovomucin, flavoprotein, etc.

Compounds from bee products – From the bees are collected/taken various products represented by: honey, propolis, royal jelly, apitoxin, wax, pollen, etc. In these products there are many biologically active substances, e.g., flavonoids, phenols, terpenoids, peptides, alkaloids, etc.

A general look at the biologically active substances of animal origin gives us interesting information on their spectrum of action.

Thus, *milk and dairy products* contain protective substances against some diseases both in children and adults. They also contain peptides with antibacterial action, lipids and oligosaccharides that assure the normal functioning of the gastrointestinal tract and improve the action of probiotics. Milk casein has antihypertensive and anticarcinogen action. Alfa lactalbumin and betalactoglobulin are immunomudulator and reduce colesterolemia. Lactoferin is antioxidantand antimicrobian, while lysozyme is an efficient antimicrobian. Conjugated linoleic acid is an immunomudulator and anticarcinogen.

In *meat and viscera*, certain compounds considered as "endogenous antioxidants" were found. Such substances are coenzyme Q10, lipoic acid, glutathione, etc. Meat also contains other zoochemicals having physiologic activity, such as carnitine, carnosine, conjugated linolcei acid and essentil fatty acids – omega 3 (known as polyunsaturated acids).

Thus, meat L-carnitine intervenes in lipid metabolism and prevents cardiovascular disease. Lipoic acid is antioxidant, carnosine – antiaging.

Eggs are associated with biological activity by the presence of choline, phospholipids and some carotenoids a.o. in their composition. Eggs phospholipids have anti-inflammatory and anticarcinogenic action. Lutein and zeaxantin are antioxidants and have a role in the prevention of macular degeneration (ophthalmologic disease).

Bee products are considered as natural therapeutic agents by the presence in their composition of compounds with phenol, flavonoid, etc., structures. Derivatives of various bee products have protective action against certain chronic diseases and are used for health maintenance.

In this framework, we discuss the food groups represented by milk, meat, bird eggs and bee products. Details on the biologically active specificity are mentioned together with general information regarding the chemical structure and biological activity.

3.2.2 ZOOCHEMICALS PRESENT IN ANIMAL TISSUES

The natural compounds of animal origin include various substances that have detected a certain biologically active effect, of particular interest with specific *physiologically active* in nutrition, respectively *pharmacologically active* in pharmaceutical products and even *toxicologically active* in the case of chemical xenobiotics with toxic effects (West, 1991; Ensminger et al., 1995; Clark et al., 1999; Ward and Bruce, 2003; Maher, 2007; Devlin, 2010; Alasalvar et al., 2011; Gârban, 2018b). Some zoochemicals can be obtained as extracts (e.g., unsaturated fatty acids). These will be presented separately for the reason that substances extracted from them can be of nutritional interest (used in the processing of food supplements or fortify some foods), others of pharmaceutical or cosmetic interest and some of toxicological interest.

3.3   ZOOCHEMICALS PRESENT IN VARIOUS FOOD GROUPS

Data regarding zoochemicals found in various food groups are exposed in Table 3.1 with details on representative compounds and their natural distribution.

3.3.1 ZOOCHEMICALS FROM MILK AND DAIRY PRODUCTS

3.3.1.1   Synoptical Data

The biologically active specificity of zoochemicals in milk and dairy products is of interest to human nutrition and pharmacology. In general, with regard to milk, it is known that it contains various nutrients belonging to the group of macronutrients (proteins, lipids and carbohydrates), micronutrients (vitamins and minerals) and other nutrients. More detailed data can be found in the literature in the field of food chemistry, nutrition, pharmaceutical chemistry, etc. (Macrae et al., 1992; Gârban, 2000; Guyton and Hall, 2006; Kanwar et al., 2009).

3.3.1.2   Representative Compounds

In this group of compounds, the main biologically active substances detected in milk (preferably in colostrum), dairy products and derivatives resulting from processing are presented. It is necessary, ab initio, a clarification regarding the biologically active substances in milk and dairy products (fermented), known for their beneficial effects, but not distinctly defined in terms of chemical composition. The explanation of the composition was given by the introduction of an adjuvant concept for the characterization of so-called *encrypted bioactive peptides*.

TABLE 3.1

Zoochemicals with Attributes of Biologically Active Substances Detected in Food Groups

No.	Specification (Food Groups)	Representative Compounds (General Examples)	Natural Distribution (Animal Food Source)
1	Milk and dairy products	Colostrinin	Colostrum
		Lactoferrin	fresh milk
		Immunoglobulins (IgG, IgM, Ig A)	whey
		Milk polypeptides (α-lactalbumin, β-lactoglobulin)	natural cheeses
		Casein	
2	Meat and viscera	Carnitine	Land animals
		Coenzyme Q (CoQ)	(beef, pork, lamb,
		Choline	poultry, turkey, venison, etc.)
		Lipoic acid	
3	Eggs from birds	Phosvitin	Yolk
		Livetin	
		Lutein	
		Zeaxantin	
		Lysozyme	Egg white
		Ovotransferrin	
		Ovomucin	
		Ovoflavoprotein	
		Avidin	
4	Bee products	Flavonoids and derivatives	Honey, propolis, royal jelly,
		Phenols and derivatives	apitoxin, pollen
		Terpenoids	
		Fatty acids and esters	
		Alkaloids	

Another explanation stated that during the hydrolysis (in the body) or fermentation (during industrial processing) processes, "encrypted biologically active peptides" are released. These peptides are composed of 3–20 amino acids, which are released from original proteins.

A scheme (according to Möller et al., 2008) that suggests the release of these "encrypted biologically active peptides" during catabolic processes in the gastrointestinal tract, after consumption of milk or fermented dairy products is shown in Figure 3.1.

At the origin of the so-called "encrypted peptides" are protein precursors, e.g. lactalbumins, lactoglobulins, various proteins from processed dairy products, even post-processing residual products (e.g., whey).

3.3.1.2.1 Zoochemicals from Colostrum

Colostrum (*post-partum* milk) has been found to be rich in protein, which plays a crucial role in its beneficial effects in both newborns and adults. There are proteins (especially peptides) with an important role through the more strictly biologically active specificity called physiologically active (Golinelli et al., 2014).

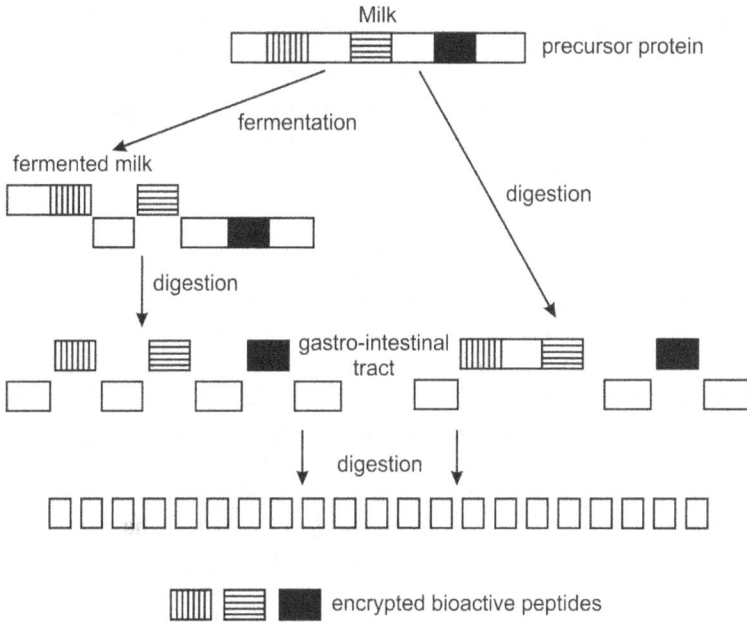

FIGURE 3.1 Release of "encrypted bioactive peptides" from protein precursors – scheme (according to Möller et al., 2008).

Various compounds present in milk are used in the treatment of asthma, diarrhea, hypertension, thrombosis, dental diseases, cancer and immunodeficiency. General data on biologically active substances (i.e., physiologically active and pharmacologically active) in colostrum are given below.

Colostrinin – A specific compound isolated from colostrum is colostrinin (CLN). The chemical composition of colostrinin is not rigorously defined. Biochemically, it is known to be a proline-rich peptide. Thus, it can be stated that it belongs to the type of "encrypted peptides" for which the amino acid sequence/sequences are not known. It is assumed to be a mixture of over 30 peptides with an extended molecular weight (possibly) over a range of 0.5–3.0 kDa.

Biological activity – In newborns, CLN activates the development of the immune system. Studies in cell cultures from mice have shown that there are changes in the mitochondrial level and a decrease in the intracellular level of oxygenated radical species – ROS (Reactive Oxygen Species). It has also been found that in the case of DNA, there is an increase in the efficiency of natural DNA repair.

In humans, it stimulates the development of the immune system in newborns, activates T cells and can inhibit multiple sclerosis disorders. It has also been tried to treat Alzheimer's disease (Leszek et al., 1999).

Lactoferrin – Another biologically active compound found in colostrum is lactoferrin (LF). It has enzyme attributes and is also called "lactotransferrin" (LTF).

Chemical structure – It corresponds to a globular glycoprotein. It has a molecular mass of 80 kDa. It is found in secretory fluids, e.g., milk, saliva, bile, nasal secretions, etc.

Biological activity – Lactoferrin is involved in the transfer of iron to cells and correlates Fe levels in the blood and in external secretions. Each lactoferrin molecule can bind reversibly two Fe, Zn and Cu ions. It has an increased affinity for Fe, hence the name.

Lactoferrin is generally thought to provide the immune response of cells. In cow's milk, the amount is 150 mg/L. Through histological/histochemical investigations, lactoferrin was also detected in certain regions of the CNS, in particular in dopaminergic neurons and microglial cells. Lactoferrin is sensitive to heat treatment (above 60° C). For this reason, it is sometimes recommended to eat fresh milk, foods rich in Fe (e.g., liver) or food supplements. *Biological activity.* In humans, LF has certain antibacterial, antifungal, antiviral effects (rotaviruses, enteroviruses, adenoviruses, etc.), antiparasitic and immunomodulatory properties (Orsi, 2004).

Colostrum immunoglobulins – In colostrum, there are mostly immunoglobulin (Ig) compounds of IgG, IgM and IgA classes. These ensure a more intense natural immunization of the newborn (provides passive immunity).

3.3.1.2.2 Zoochemicals from Milk/Dairy Products

Lysozyme – This is also known as "muramidase". It has the attributes of a hydrolytic enzyme with a polypeptide structure (129 amino acids) with bacteriolytic action and a molecular mass of 4.4 kDa. It is also found in the salivary and nasal secretions, in the egg white. It was isolated by Fleming (1922), who noted that nasal secretion inhibits the growth of bacterial cultures. The enzyme acts on the walls of bacterial cells (at the level of β-1,4-glycosidic bonds), for this reason, it is called "muramidase". This enzyme is effective in treating periodontosis and preventing dental problems. It is also known for its immune properties. It is used in the food industry and for pharmaceutical purposes.

Casein encrypted peptides – Various fractions were isolated from casein: α_{S1}-casein, α_{S2}-casein, β-casein (found in larger quantities) and κ-casein. Casein is involved in reducing bacteremia and in reducing colic in children. Casein hydrolysates reduce tumor growth. The composition of the "encrypted" peptides is not known, the biological effects exist, which is why their presence has been reported.

The biologically active properties of dairy zoochemicals provide a picture of the natural products involved in the prevention and treatment of certain diseases, replacing (in part) the use of synthetic drugs known for their toxic effects.

3.3.1.2.3   Zoochemicals from Whey

Zoochemical studies have also been performed on whey resulting from the preparation of cheeses. Whey – according to food chemistry – is defined as the "liquid part" left over after obtaining the sweet milk cheese (Madureira et al., 2007; Golinelli et al., 2014).

Investigations into milk protein have determined that the ratio of fractions: whey protein/milk protein is 20/80 in cow's, buffalo's and sheep's milk. In human breast milk, this ratio is 40/60.

Whey from cow's milk contains predominantly proteins. Serum studies have focused on proteins and biologically active substances, in fact, "specific peptides" in this product. In the whey, they are:

Whey lactoferrin – Lactoferrin was detected in whey in the amount of 1%–2% and is involved in inhibiting the growth of bacteria and fungi as well as iron retention.

Specific encrypted peptides – Serum-specific "encrypted peptides" include "cystine-rich peptides." These are important because they also derive from "protein precursors" in milk. In the body, cystine is the origin of glutathione synthesis.

Increasing the amount of glutathione reduces the risk of infection by improving the activity of leukocytes. At the origin of these peptides are "protein precursors" existing in milk, represented – mainly – by two compounds:

lactalbumin (123 amino acids) – present in the amount of 20%–25% of the proteins in whey, increased quantities in tryptophan – an essential amino acid; increases the amount of serotonin, intervenes in the regulation of sleep and reduces stress.

lactoglobulin (162 amino acids) – is in the amount of 50%–55% of whey proteins and is an important source of essential amino acids and branched chain amino acids; intervenes in the prevention of muscle depletion by avoiding excessive consumption of glycogen in the muscles.

Specific immunoglobulins – They were detected in quantities of 10%–15% in whey. Immunoglobulins (Ig), in general, belong to different classes, i.e. IgA, IgM, IgG.

Different enzymes – Whey contains approx. 70 enzymes. Among them can be mentioned: *hydrolases* (lipases, proteinases), *phosphatases* (alkaline and acidic), *ribonucleases, lysozymes, γ-glutamyl-transferase*, etc. These are found in greater proportion. They provide passive immunization.

If in the case of colostrum the importance of the physiologically active peculiarities was noticed, in the case of whey another specific one was noticed, namely the superior "nutritional value" due to the quality and quantity of the protein constituents (Clare and Swaisgood, 2000).

3.3.2 ZOCHEMICALS FROM MEAT AND VISCERA

3.3.2.1 Synoptical Data

The issue of meat and viscera nutrients has been extensively studied in relation to protein, lipid and carbohydrate macronutrients (Overvad et al., 1999; Guyton and Hall, 2006; Gârban, 2018a).

With reference to meat nutrients, more data are known about some minerals (Fe, Zn and Mg), about fat-soluble vitamin (E) and water-soluble vitamins, favorite B complex (B_2, B_6 and B_{12}).

The same cannot be said of investigating biologically active substances in meat, of which there is sporadic information.

3.3.2.2 Representative Compounds

There are biologically active substances in meat, such as carnitine, coenzyme Q10, choline, lipoic acid, conjugated linoleic acid, etc. In some biochemical treatises, these substances are discussed along with vitamins, considered as "vitamin-specific substances". The main zoochemicals in meat are listed below.

3.3.2.2.1 Carnitine

In the class of meat, zoochemicals are included carnitine (lat. *carnis* – meat). This compound is known under various names, which highlights the confusion that initially existed, with reference to its chemical composition and biologically active specificity.

This compound has been known by various names: "*torultine*" – isolated from yeasts belonging to the genus *Torula*; "*termitine*" – isolated from insects included in the termite family; "T factor" even "vitamin T" – derived from the substance isolated from flour worms included in the species *Tenebrio Molitor* (Gârban, 2018a).

Chemical structure – From a structural point of view, L-carnitine (I) can be considered: betaine of β-hydroxy-γ-amino-butyric acid (Figure 3.2).

Carnitine was originally isolated from meat extracts in 1905. The chemical structure was established in 1932. The physiological role of carnitine was for a long time unknown. Carnitine contains the carboxylic, hydroxyl and amine (trisubstituted) functional groups. The appearance of trisubstituted amine compound gives carnitine the character of "quaternary ammonium salt".

In specialized treatises, this compound is mentioned in either protein – in relation to amino acids metabolism, or lipids – in relation to fatty acid metabolism. On the other hand, it is known (from the biochemistry of vitamins) that nicotinamide is synthesized from another essential amino acid – tryptophan. This led to the accreditation of the idea that it may belong to the class of vitamins.

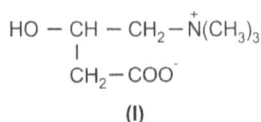

$$HO - CH - CH_2 - \overset{+}{N}(CH_3)_3$$
$$| $$
$$CH_2 - COO^-$$

(I)

FIGURE 3.2 Carnitine.

Biological activity – L-carnitine is currently being discussed in molecular biology, e.g. mitochondrial topobiochemistry and energogenesis, nutrition, e.g. dietary supplements (Galloway and Broad, 2005), and pharmacology, e.g. pharmacological effects of some chemotherapeutic products on lipid and carbohydrate metabolism (Ulvi et al., 2010).

Carnitine is absorbed in the intestine, so a maximum of 25% of the amount absorbed can be acylated in the intestinal mucosa. From here it is distributed throughout the body. Sodium present in the extracellular environment plays an essential role in the transport of carnitine.

This explains the fact that in the skeletal muscles and in the heart is approx. 95% of carnitine is concentrated, 4% in the liver, kidneys and other tissues and in 1% extracellular fluid.

During exercise, the concentration of carnitine in the "storage systems" decreases and a transition to acyl forms is made, i.e. acyl-carnitine. Elimination is mainly by renal excretion. More than 90% is reabsorbed in the kidneys – in the proximal tubules – and returns to the circulation (if there is a decrease of carnitine in plasma).

In certain situations of metabolic stress (e.g., diabetes, anoxia, etc.), the viability of the cell can be compromised by the accumulation of short and medium chain organic acids and by the concomitant decrease of carnitine. The biogenesis of L-carnitine starts from the essential amino acid lysine in the presence of the amino acid methionine, three vitamins (ascorbic acid, nicotinamide, pyridoxine) and a metallic bioelement – iron. The biosynthesis process is performed by enzymatic trimethylation of lysine – via S-adenosylmethionine – which is the methyl ($-CH_3$) donor group. It has been found that trimethyllysine can be cleaved (broken), resulting in L-carnitine.

The final stages of carnitine formation occur only in the liver, kidneys and brain because the necessary enzyme, butyrobetainhydroxylase, is present only in them. Trimethyllysine is present in some proteins, such as cytochrome C, histones, myosin, calmodulin, etc.

With regard to biological activity, it was found that the molecule of "acyl-CoA" which intervenes in lipid metabolism, more precisely in the metabolism of fatty acids, in the absence of carnitine can not cross the mitochondrial membrane. A carnitine deficiency causes a low concentration of fatty acids in the mitochondria, which results in depression in the process of energogenesis (Alhasaniah, 2023). The biochemical mechanism is not fully elucidated. It is known, however, that fatty acid is transferred from CoA to L-carnitine.

Thus acyl-carnitine is formed which can transfer the fatty acid molecule through the mitochondrial membrane (Sachan and Cha, 1994). Carnitine helps maintain an intramitochondrial ratio of acetyl-CoA/CoA. In the human body, as shown, carnitine is found in small amounts. Carnitine is also involved in protein metabolism. In this regard, the participation of carnitine in the conversion of analogous ketoacids in the branched chains of amino acids is mentioned, i.e. valine, leucine, isoleucine. This is extremely important during physical exercise. As major food sources with high carnitine content, zoochemicals are present in meat and dairy products. Cereals and fruits generally contain phytochemicals, but some also contain low amounts of carnitine.

Carnitine is present in the heart muscle, skeletal muscle and other tissues that have fatty acids as a source of energy. Their normal functioning depends on the proper transport of carnitine into the tissues (Paulson and Shug, 1981).

On the heart, the action of carnitine is manifested by reducing necrotic lesions, decreasing the incidence of arrhythmias, reducing the incidence of angina pectoris, limiting left ventricular hypertrophy, etc. (Rizzon et al., 1989). Carnitine is also recommended in the treatment of chronic heart ischemia (Orlando and Rusconi, 1986). Bioanalytical investigations have led to the detection of carnitine-specific proteins, which carry carnitine in the heart, skeletal muscle, epididymis, liver and kidneys.

The mechanism of carnitine transport activity allows the tissues to concentrate ten times more carnitine than the amount of plasma. Studying the aspects of homeorhesis – representing the biochemical mechanism specific to the maternal-fetal complex, it was found that in the umbilical cord the concentration of carnitine is higher than in the maternal blood. It was concluded that the placenta concentrates the carnitine needed by the fetus. A remarkable observation lies in the fact that there is a synergism between carnitine and coenzyme Q.

This synergism lies mainly in their antioxidant activity, as well as in their participation in the processes of intramitochondrial energogenesis (Barker et al., 2001).

In terms of carnitine deficiency, it was concluded that this concerns two aspects: *systemic deficiency* – affecting the whole body and *myopathic deficiency* – affecting the muscles.

The causes of carnitine deficiency are various, e.g., dietary deficiency of the amino acids precursors lysine and methionine; quantitative depression of a cofactor necessary in biosynthesis; a possible genetic defect that affects the biosynthesis of carnitine; intestinal absorption disorders of carnitine; defective transport of carnitine in tissues; liver and kidney problems, etc.

3.3.2.2.2   Coenzyme Q

Also known as "ubiquinone", is usually abbreviated as CoQ. This compound is found in mitochondria where it participates in redox reactions. It is a benzoquinone derivative with an isoprene side chain. The hydrocarbon chain gives liposolubility to ubiquinone. Thus, it is explained that ubiquinone was detected in the composition of lipoproteins in the inner mitochondrial membrane (Yamamura et al., 1980).

Chemical structure – Structurally, coenzyme Q (II) is a derivative of 1,4-benzoquinone, which has a side chain at C_5. This chain consists of a number "n" with the size of 6–10 isoprene residues (Figure 3.3). In humans and mammals in general, CoQ has 10 isoprene residues.

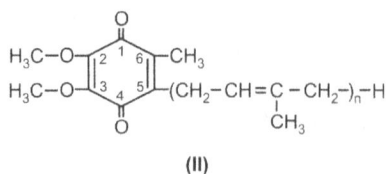

(II)

FIGURE 3.3   Coenzyme Q.

For this reason, it has the trivial name of coenzyme Q10 (denoted CoQ10). The substituent called decaisoprenyl – $[CH_2-CH=C (CH_3) -CH_2]_{10}-H$ at position C_5 of the benzoquinone ring is given as a radical – R.

The chemical name for CoQ10 is 2,3-dimethoxy-5-decaisoprenyl-6-methyl-1, 4-benzoquinone. Coenzyme Q (CoQ) is involved in certain redox reactions, as a transporter for H^+ and e^-.

The reaction takes place in two stages: first – with the formation of an anionic semi-quinone intermediate with free radical structure (CoQH); second – with the formation of the dihydrogenated reaction product called ubiquinol ($CoQH_2$) – Figure 3.4. The enzyme NADH-CoQ reductase, which transports H^+ and e^- from NADH (nico-tinamide adenine dinucleotide – reduced form) to CoQ, is involved in this reaction.

Biochemical interactions – involving transport for $2H^+$ and $2 e^-$ – occur at the mitochondrial level, preferably targeting the inner mitochondrial membrane (Pallotti et al., 2022). It is considered that there is a dicentric transport for electrons, in which the system [4Fe – 4S] intervenes – from the first center, which transfers e^- to CoQ, evolving at the same time to the system [2Fe–2S] – from the second center. Following the interactions that underlie transport, CoQ evolves into $CoQH_2$.

In turn, hydrogen is transported by other enzymes. There are other enzymes with NADH-CoQ reductase-like action. Examples include the enzymes: succinate-CoQ reductase; cytochrome-CoQ reductase; glycerophosphate-CoQ reductase, etc. These reductases work in mitochondria, but have also been found in the Golgi apparatus.

Biological activity – In metabolic processes specific to the "respiratory chain", coenzyme Q mediates the transport of hydrogen between NADH dehydrogenase and the cytochrome system. In the interaction, the oxidized form (CoQ) and the reduced form (CoQH2) alternate. In general, ubiquinone is considered to have antioxidant effects (Frei et al., 1990; Motohashi et al., 2017).

Ubiquinones are widespread in the plant and animal kingdom, being present in foods of plant and animal origin. Their widespread use is explained by the fact that they participate in the respiratory processes from tissues. Coenzyme Q is well known for its benefic effects and is used in case of the cardiovascular diseases (Lee et al., 2012; Martelli et al., 2020).

The increased distribution of ubiquinone in plant tissues explains – by itself – its higher plasma levels in the body of people with a vegetarian diet. In recent years, in various forms, CoQ10 has been introduced into therapy but is also used as an ingredi-ent in some dietary supplements.

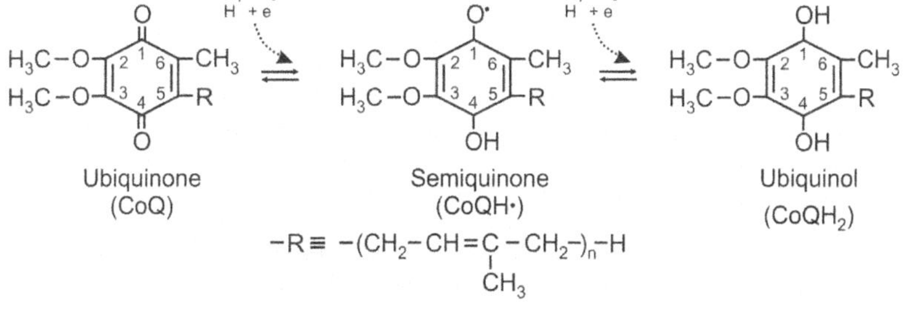

FIGURE 3.4   Tissue coenzyme Q reduction reaction.

3.3.2.2.3   Choline

Is the quaternary base of ammonium, fully methylated derived from cholamine. It was originally isolated from oxen and pigs bile (Strecker, 1862).

Later, it was also detected in the nervous tissue. However, in larger quantities, it is found in egg yolk (along with lecithin). It was obtained by synthesis, in 1868, by Würtz, starting from two compounds: ethylene oxide $[CH_2=CH_2]O$ and hydroxylated trimethylamine $[NH(CH_3)_3]^+ OH^-$.

It was found in complex lipids called choline phospholipids and in the composition of a chemical mediator called acetylcholine.

Chemical structure – From a chemical point of view, choline (III) is considered a derivative of cholamine (ethanolamine), also known as methylhydroxycholamine (Figure 3.5). The chemical name is 2-hydroxyethyl-trimethylamine-hydroxide or trimethyl- β-hydroxy-ethylammonium hydroxide.

There are a number of choline derivatives with a similar structure, e.g., *triethylcholine, diethylmonomethylcholine, betaine choline*, etc. In its pure state, choline presents itself as a solid, crystalline substance. In solution, it is colorless or viscous, being extremely hygroscopic. It is soluble in water, alcohol and formaldehyde. It behaves like a strong base. In alcoholic solutions with HCl, choline crystallizes, forming needle crystals. In alcohol-ether mixtures, choline crystallizes as platelets.

Biological activity – In animal and plant organisms, choline participates in many metabolic processes, being a donor of methylene groups (CH_3).

Thus, it participates in the biosynthesis of important metabolites. In this sense, the main metabolic activities will be discussed:

a. the quality of bioconstituent of some complex lipids (e.g., lecithins, sphingomyelins, etc.);
b. the bioconstituent quality of acetylcholine;
c. the role in detransmethylation reactions.

> *Choline as bioconstituent of lecithins* – In complex lipids choline was detected in choline phospholipids (lecithins). Depending on the arrangement of the phosphocholine group, α-lecithins and β-lecithins are distinguished.

> Choline also enters the structure of sphingophospholipids. A representative compound is sphingomyelin. Sphinomyelin contains choline as part of the phosphorylcholine residue in sphingomyelin.

> *Choline as bioconstituent of acetylcholine* – There are two precursors in the biosynthesis of acetyl choline: acetyl-coenzyme A (acetyl-CoA) and choline. The acetyl group is derived from the biodegradation of fatty acids by β-oxidation. In the presence of HS-CoA acetyl-CoA is formed. The endergonic reaction requires the presence of ATP and takes place under the action of the cholinetransacetylase enzyme (Figure 3.6).

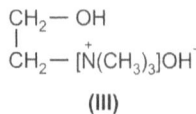

$$CH_2 - OH$$
$$|$$
$$CH_2 - [\overset{+}{N}(CH_3)_3]OH^-$$

(III)

FIGURE 3.5   Choline.

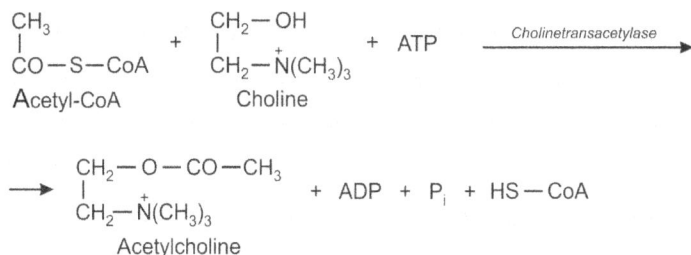

FIGURE 3.6   Acetylcholine biosynthesis under the action of cholinetransacetylase.

The reaction is more intense under the action of some metabolites, e.g., cystine, glutamic acid, lactic acid, acetylacetic acid, citric acid, K^+ ions and even ammonium ion.

Acetylcholine biosynthesis is also possible from acetic acid, choline, in the presence of coenzyme A (HS-CoA), ATP and Mg^{2+} ion. The reaction takes place in the presence of the cholinesterase enzyme (Figure 3.7).

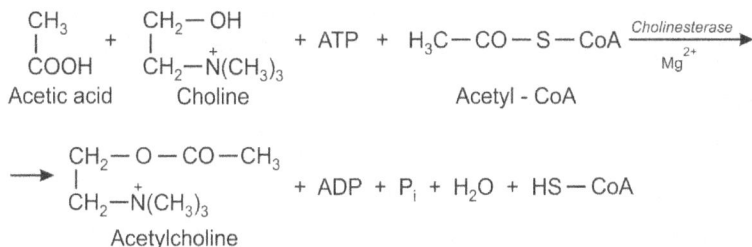

FIGURE 3.7   Acetylcholine biosynthesis under the action of cholinesterase.

In the body, acetylcholine acts as a chemical mediator ensuring the transmission of nerve impulses from cells (neurons) to muscles (Zhu and Zeisel, 2005; Guyton and Hall, 2006). Biodegradation of acetylcholine blocks its action and occurs in the presence of the acetylcholinesterase enzyme. Thus, acetylcholine is hydrolyzed to form choline and acetic acid (Figure 3.8). Decomposition of acetylcholine also results in its deactivation.

It is also mentioned that in the tissues, phosphorylcholine was detected – along with acetylcholine, choline and another derivative.

This compound is an intermediate in the biosynthesis of choline from acetylcholine. The interaction is correlated with the possible direct phosphorylation of choline.

FIGURE 3.8   Biodegradation reaction of acetylcholine.

This biosynthetic pathway has gained credibility with the discovery of the enzyme *choline phosphokinase* (in the liver, kidneys and intestines) – an enzyme that ensures the phosphorylation of choline.

Choline in transmethylation reactions – In the metabolic processes in the human and animal body, choline – through oxidation reactions – in the presence of two enzymes, *choline oxidase* and *betaine-aldehyde oxidase* are converted to betaine (Figure 3.9). Next, betaine – which is a donor of methyl groups ($-CH_3$) – will successively give up one methyl group, gradually forming dimethyl – glycine, sarcosine and, finally, glycochol. The uptake of methyl groups ($-CH_3$) will be done by homocysteine in the presence of the transmethylase enzyme. From reaction results methionine.

So, on the whole, it can be said that the transmethylation reaction has choline as donor and homocysteine as acceptor. In reality, transmethylation is a complex reaction, possible in the case of the binding of the methyl group (s) to N or S atoms.

Choline has been found in plant tissues, animals and microorganisms. It is composed of phosphate lipids (choline phospholipids, sphingophospholipids) and acetylcholine. Smaller quantity is free. Among the foods, higher amounts of choline are found in certain foods of animal origin, e.g., liver, meat and milk. In foods of plant origin, more significant quantities were detected in seeds, e.g., wheat grain (during the germination period it has a high choline content); in yeast, etc.

Choline contributes to the normal development of metabolic processes by participating directly in the transfer of methyl groups (Canty and Zeisel, 1994). In this way, it integrates with the biochemical pathways specific to lipid metabolism ensuring the normal functioning of the nervous system. It is known that acetylcholine causes vasodilation in the blood capillaries and thus lowers blood pressure. It has bradycardic effects (slows the heartbeat). It ensures the production of normal peristaltic movements of the intestine.

At the cellular level, choline phospholipids contribute to the formation of the lipid bilayer due to its polarity. The dipolar character is given by the acid function – the phosphorylated residue and the basic function – quaternary ammonium base (which includes methyl groups).

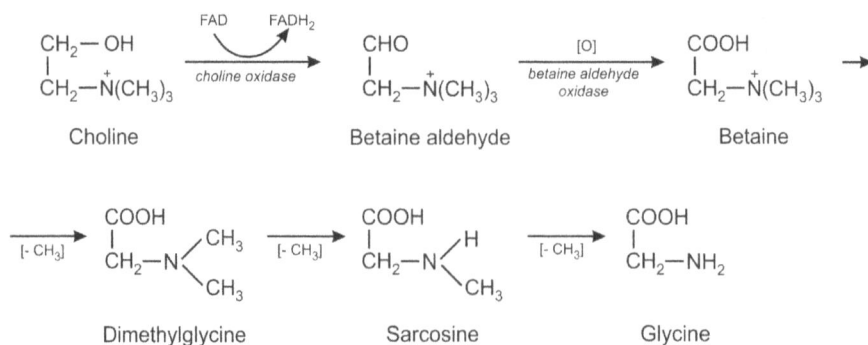

FIGURE 3.9   Choline demethylation – specific interactions.

Choline absorption occurs in the small intestine along with phospholipids. From here it is transported through the blood to the liver. The level of phospholipids in the blood is 4%–8% of the total blood lipids. The elimination of choline, in an average amount of 10 mg/24 hours, occurs through urine, feces and perspiration.

3.3.2.2.4   Lipoic Acid

The chemical compound called D-6,8-dithiooctanoic acid is known as *lipoic acid* or *thioctic acid*. This compound is considered as a "*growth factor*" for various microorganisms, especially *Lactobacillus casei, Streptococus faecalis* and for the protozoan *Tetrahymena gelei*.

Chemical structure. In literature, there is often discuss about a group of compounds called lipoic acids whose general formula is shown in Figure 3.10.

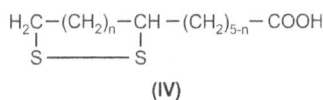

$$H_2C-(CH_2)_n-CH-(CH_2)_{5-n}-COOH$$
$$\underset{S\text{---------}S}{\vphantom{|}}$$

(IV)

FIGURE 3.10   Lipoic acids – general formula.

Among them, the best known is *α-lipoic acid*, with a chain of eight carbon atoms. If comparing the structure of α-lipoic acid to the general formula given above, it can be seen that the "n" value is unitary (Reed et al., 1953).

By organic synthesis, the compound with the racemic form L,D-lipoic acid was also obtained, which, however, has a reduced biological activity, about 50% compared to α-D-lipoic acid considered as the biologically active form (100%). α-lipoic acid was obtained in a crystalline state from liver extracts.

It can be noted that α-lipoic acid is an optically active compound, dextrorotatory (+97°) due to the asymmetric carbon in the C_6 position.

Biological activity – In nature, it can be found not only in plants, animals and the human body but also in microorganisms. It was isolated in an oxidized (disulfide) form, but also in a reduced (dithiolic) form. The acyclic, dihydrolipoic (dithiolic) form is often conjugated with various compounds, which are released during hydrolysis. The biological activity is dependent on the transformation capacity, so the reversibility of the redox reaction through which, from lipoic acid – the disulfide (oxidized) form, dihydrolipoic acid – the dithiolic (reduced) form is formed.

Their structural formulas are shown in Figure 3.12.

Lipoic acid intervenes in the transfer of elementary hydrogen within the cycle of tricarboxylic acids (TCA cycle, also called the Krebs cycle). The TCA cycle has an

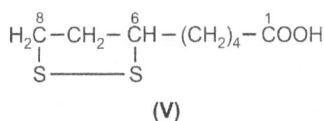

$$\overset{8}{H_2C}-CH_2-\overset{6}{CH}-(CH_2)_4-\overset{1}{COOH}$$
$$\underset{S\text{---------}S}{\vphantom{|}}$$

(V)

FIGURE 3.11   Alpha-lipoic acid.

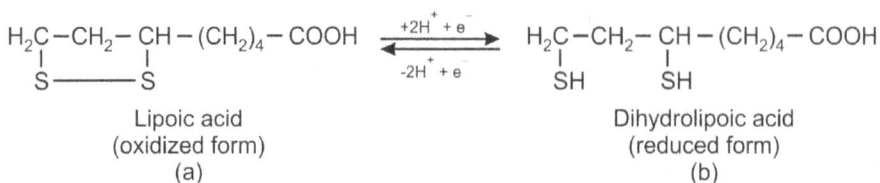

FIGURE 3.12   Lipoic acid and dihydrolipoic acid – structural formulas (a) disulfide structure (oxidized form); (b) dithiolic structure (reduced form).

essential role in ensuring transformations with "interactive specificity" that interest carbohydrate, lipid and protein metabolism.

Within the TCA cycle, lipoic acid in oxidized form interacts with succinyl-thiamino-pyrophosphate, forming succinyl-dihydrolipoic acid. It in the presence of coenzyme A (HS-CoA) forms succinyl-S-CoA and releases dihydrolipoic acid.

The intervention of lipoic acid in the TCA cycle reveals the interchangeability of the disulfide (oxidized) and dithiolic (reduced) forms mentioned above (Packer et al., 1995; Gârban, 2018a).

The reproduction of changes in the oxidation-reduction process is usually represented by symbolic forms – Figure 3.13.

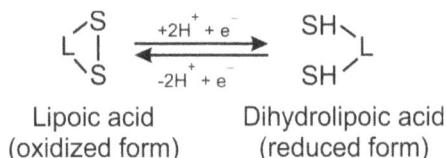

FIGURE 3.13   Symbolic representation.

Lipoic acid intervenes in redox processes and in the transport, we remember, by a non-enzymatic way of hydrogen. The biological activity is ensured by a so-called biologically active form represented by lipoic acid conjugated with thiamine pyrophosphate, a compound called lipoic-thiamine pyrophosphate abbreviated as L-TPP (Figure 3.14).

The lipoic-thiamine pyrophosphate represents, in fact, the formation of a coenzymatic form of lipoic acid, which is biologically active.

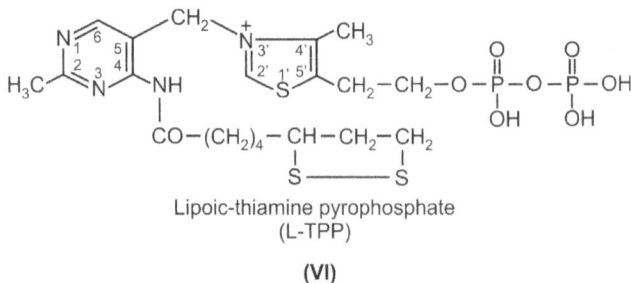

Lipoic-thiamine pyrophosphate
(L-TPP)

(VI)

FIGURE 3.14   Lipoic-thiamine pyrophosphate – structural formula.

Lipoic acid participates in metabolic interactions as a "coenzymatic factor" for certain enzyme systems in which vitamins B_1, B_2 and B_3 intervene. A redox system works in the body, represented by "lipoic-thiamine pyrophosphate" which highlights the reversible oxidation-reduction system.

The reaction is catalyzed by an enzyme called lipoyl dehydrogenase or lipoic dehydrogenase. However, its functioning requires the presence of NAD^+ and occurs at a pH of approximately 7.0 when the system becomes biologically active (Bustamante et al., 1998).

Lipoic acid intervenes in the decarboxylation processes of α-ketoacids, in the acylation processes and, very likely, in the processes specific to the visual function. In plants, it intervenes in the process of photosynthesis. The biochemical activity of lipoic acid in these processes correlates with that of thiamine and coenzyme A.

Alpha-lipoic acid, which intervenes in the function of various enzymes, participates in the oxidative mechanism in the mitochondria. Its redox specificity, as shown, consists of the conversion between lipoic acid and dihydrolipoic acid. From a biochemical point of view, alpha-lipoic acid acts as an antioxidant, intervening in the chelation of metals, and reducing the activity of other oxidants, e.g. vitamin C, vitamins E and glutathione. Alpha-lipoic acid is recognized as a therapeutic agent in the case of certain chronic diseases, such as cognitive dysfunction in Alzheimer's disease, obesity caused by non-alcoholic liver diseases, cardiovascular diseases, hypertension, some forms of cancer, glaucoma, etc. (Gomes and Negrato, 2014).

3.3.3 Zoochemicals from Eggs

3.3.3.1 Synoptical Data

The egg is structurally made up of egg yolk, egg white and shell. There are various macro- and micronutrients in yolk and egg white. Proteins and lipids predominate among macronutrients. The shell of the egg has a protective role. It is composed of 95–98% calcium carbonate, stabilized by an organic matrix, which facilitates the mineralization process. Also, small amounts of strontium, fluorine, magnesium and selenium were detected in the eggshell matrix. The biologically active substances in eggs are less well known than the nutrients in them. The detection of these substances was done by isolation from egg yolk and egg white, using specific laboratory techniques, known in food chemistry and pharmaceutical chemistry (Anton et al., 2005; Huopalahti et al., 2007; Zdrojewicz et al., 2016; Réhault-Godbert et al., 2019).

3.3.3.2 Representative Compounds

3.3.3.2.1 Yolk-Containing Compounds

Egg yolk contains two more important biologically active substances: phosvitin and livetin. In yolk, the carotenoid compounds such as lutein and zeaxanthin are also present. They have also been isolated from the ovary in humans and animals. However, it is predominant in the vegetable kingdom (see Chapter 2).

   Phosvitin – It represents 4% of the dry matter of the yolk. α-fosvitin and β-fosvitin were detected in the egg. They contain 8%–10% phosphorus, respectively.

Chemical structure – It is a phosphoglycoprotein consisting of a peptide chain in which 50% of the amino acid residues are serine. They are phosphorylated, which explains their interaction with metal ions (mainly iron).

Biological activity – It is mainly characterized by its antioxidant effect – it inhibits the oxidation of phospholipids (catalyzed by Cu^{2+} and Fe^{2+}). The chelating effect of fosvitine binds to 120 Co^{2+} ions or 113 Mn^{2+} ions. It can also chelate the ferric ion (Fe^{3+}), more precisely the heme iron, in which case the antioxidant capacity is lost.

Livetin – It is a protein compound isolated from egg yolk.

Chemical structure – Through its analytical separatology techniques, three fractions were obtained: α -, β- and γ-livetin. γ-livetin is of particular importance, commonly called Y-YIg immunoglobulin (Yolk Immunoglobulin).

The YIg molecule consists of two light chains and two heavy immunoglobulin chains.

Biological activity – It is known that α-*livetin* is involved in the production of allergies caused in some people, causing respiratory symptoms. With reference to β-*livetin* data are brief and inconclusive. Finally, γ-*livetin* provides passive immunity to the bird. The formation of YIg, in the first phase, takes place in the blood of the bird. From here it reaches the egg yolk. In the second phase, during hatching, YIg passes from the yolk to the embryonic blood, ensuring the immunization of the bird embryo.

3.3.3.2.2   Egg White-Containing Compounds

Egg white is a deposit of various biologically active substances. More important are: ovolysozyme, ovotransferrin, ovomucin, flavoproteins, cystatin, avidin, etc.

Ovolysozyme – Lysozyme is an enzyme (EC 3.2.17) also known as lysozyme "muramidase" and "N-acetylmuramide hydrolase". It has ubiquitous distribution in nature, in all secretions, biological fluids and tissues in human and animal organisms. It is also known to have been isolated from some plants and bacteria.

Chemical structure – Egg lysozyme (for which we use the name ovolizozyme) is a compound with a basic polypeptide structure and character. Chicken egg ovolizozyme contains 129 amino acid residues and has four disulfide bridges (–S–S–). It has a molecular mass of 14.3 kDa.

Biological activity – It resides in the more intense antimicrobial action compared to Gram-positive bacteria. The bacteriostatic and bactericidal effect has been used to preserve various foods and pharmaceuticals in human medicine and veterinary medicine. It is important because offers protection against bacterial and viral inflammatory diseases.

Egg lysozyme has been shown to control "immune imbalance" in autoimmune diseases. It is suitable for use in the production of cheeses because it prevents the growth of the micro-organism *Clostridium tyrobutiricum*. It can also be used to make beer and to make high-quality wines (in these cases the action is to control the growth of lactic acid bacteria). Can be used as a food additive (E 1150).

Ovotransferrin – It is usually denoted by OTf. It is also known as the "conalbumin". It contains 12% of the protein in egg whites.

Chemical structure – OTf is a neutral glycoprotein that ensures the transfer of iron ions (Fe^{3+}) from the chicken oviduct to the egg, where it can contribute to embryonic development. It was isolated from the albuminous fraction of proteins.

Biological activity – It is characterized by antiviral and antimicrobial effects, having therapeutic uses in acute enteritis, especially in children. In the diet, it is used as a nutritional ingredient in certain food supplements, or fortified foods. It is also used in fortifying instant drinks and in sports bars.

Ovomucin – Represents approx. 2%–4% of the proteins present in the egg white and are important for the gelling properties of the egg white.

Chemical structure – It is a sulfated glycoprotein characterized by high molecular weight and the presence of two subunits: α-*ovomucin* – consisting of 2087 amino acid residues, approx. 15% carbohydrates and has a molecular weight of 180–220 kDa; β-*ovomucin* – contains 827 amino acids, 60% carbohydrates and has a molecular weight of 400–720 kDa.

Biological activity – Ovomucin has cytostatic effects on tumor cell cultures. It has also been shown to have immunopotentiating activity characterized by stimulation of lymphocytes in the spleen of mice. It is also considered an antibacterial agent that inhibits the colonization of *Helicobacter pylori* in the stomach. It attenuates hypercholesterolemia in rats and inhibits the absorption of cholesterol. The study of ovomucin has been extended, considering that it is of industrial interest for the production of functional foods.

Ovoflavoprotein – Also known as "egg flavoprotein" or "protein-bound riboflavin" – RofBP (Riboflavin Binding/Carrier protein) is specific to birds. RfBP has been found to be present in both egg whites and egg yolks, with a common gene at its origin.

Chemical structure – It is a phosphoglycoprotein consisting of a globular monomer with a molecular weight of 29,200 Da. The protein component consists of 219 amino acid residues and is acidic. There are 18 cysteine residues in the structure linked by 9 disulfide bridges. The 14% carbohydrate component is fucose, mannose, N-acetylglucosamine and sialic acid.

Biological activity – RfBP has been found to be transported by riboflavin (a water-soluble vitamin) in the embryo to support its growth and development.

Cystatin – It was isolated from egg whites. It has also been found in mammals: mice, rats, dogs and humans. Chemical structure. It is a polypeptide compound that has approx. 120 amino acids and a molecular weight of approximately 12–13 kDa. There are at least two disulfide bonds in the chain.

Biological activity – It is suitable for the prevention of cerebral hemorrhages. Cystatin also has an inhibitory effect on cystein proteases. These enzymes play an important role in the evolution of metastatic processes.

Avidin – It is known that avidin is synthesized in the oviduct of birds, reptiles and amphibians and stored in eggs. It was found in raw eggs. It contains 0.05% of the total protein present in egg whites. It is destroyed by boiling. Its exact role in the egg is unknown.

Chemical structure – It is a compound with a high stable glycoprotein structure. The protein component is homotetrameric and each monomer contains 128 amino acid residues. The carbohydrate component is an oligo-saccharide containing mannose and N-acetylglucosamine residues.

Biological activity – It is not fully elucidated. However, avidin in the oviduct has been postulated to have antibacterial action (observed, especially in the genus *Streptomyces*) in birds. In biochemistry, it is known that there is an interaction between biotin (vitamin B7) and avidin, which prevents the absorption of the vitamin in the intestine.

3.3.4   ZOOCHEMICALS FROM BEE PRODUCTS

3.3.4.1   Synoptical Data

Bee products obtained from beekeeping – a distinct field of zooculture – are important from a food and pharmaceutical point of view. In beekeeping, the species *Apis mellifera mellifera* and *Apis mellifera ligustica* – in Europe and *Apis mellifera scutelata* – in Africa are mainly grown. Data on bees and bee products have been known for over 7,000 years.

The group of bee products includes honey, propolis, milk, royal jelly, apitoxin, wax and pollen. In beekeeping, for these products, the expression "bee hexagon" is used (Hepburn, 1986; Macrae et al., 1992; Horn and Lüllmann, 2006; Samarghandian et al., 2017).

The destination of bee products differs. Some are used as foods, others as medicines (in the prevention and/or therapy of various diseases). It is noted that some food and therapeutic attributes are due to the presence of biologically active substances in these products. In larger quantities, the biologically active substances are found in honey, propolis, royal jelly, apitoxin, etc.

3.3.4.2   Representative Compounds

3.3.4.2.1   Honey

It is a bee product obtained by bees from the nectar of flowers from plants of spontaneous flora and agricultural crops. These are also called "honey plants". In extreme situations, the harvesting of honey products can be done from certain insects of the order of hemiptera and from the branches of some trees.

The honey is stored in honeycombs and mainly feeds the bee population in the hive. The surplus is taken for human food needs. Depending on the origin, there are monofloral honey (e.g., acacia, lime, mint, etc.) and polyfloral.

Chemical composition – Mainly the presence of water, carbohydrate and non-carbohydrate compounds is present. Among the carbohydrate compounds are *oses* (pentose and hexose) and holoside (sucrose, maltose, etc.). Heterosides are also present – compounds that contain, along with a carbohydrate component and a non-carbohydrate component (aglycone). In the composition of Romanian honey: water 17.0%; fructose 39.9%; sucrose 2.3% and other components, in minor quantity were determined. Water-soluble vitamins were also detected – especially from the B complex and vitamin C, fat-soluble vitamins – vitamin A and K, as well as enzymes, mineral compounds (Na, K, P, Mg, Cu, Al, Mn, Fe, Si, etc.).

Some biologically active substances have also been found in honey (Crane, 1984; Horn and Lüllmann, 2006). In Figure 3.15, structural formulas of some biologically active substances isolated from honey are presented. Data are compiled from the literature (Viuda-Martos et al., 2008; Alvarez-Suarez et al., 2014; Mijanur Rahman et al., 2014; Pasupuleti et al., 2017).

Their selection was made according to the group of substances, e.g., flavonoid compounds, phenolic derivatives, etc. From the group of flavonoid compounds were mentioned: *hesperetin* (I) - antioxidant; *naringenin* (II) - antioxidant, neuroprotective,

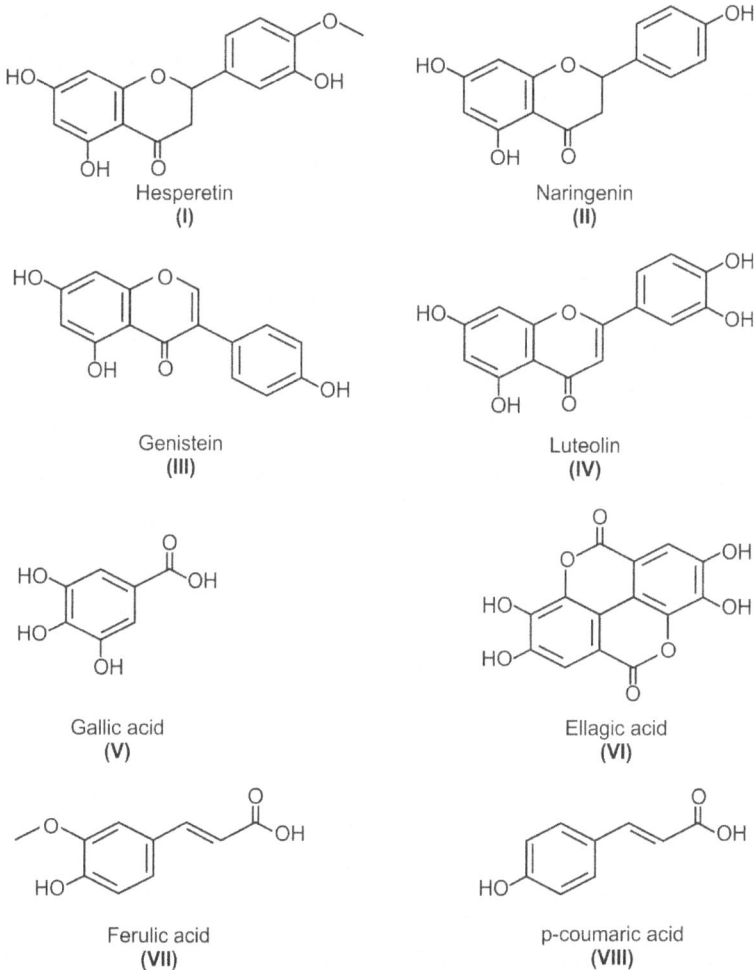

FIGURE 3.15   Flavonoid compounds (I–IV) and phenolic compounds (V–VIII) in honey bee.

genistein (III) - anticancer; *luteolin* (IV) - antioxidant, anti-inflammatory. Some phenolic compounds were: *gallic acid* (V) - antiaxiolytics; *ellagic acid* (VI) - antioxidant, antiproliferative; *ferulic acid* (VII) - antioxidant, anti-inflammatory, neuroprotective; *p-coumaric acid* (VIII) - antigenotic, neuroprotective.

Biological activity – Due to the content of nutrients, but also through the presence of some biologically active substances, honey is important from a physiological and pharmacological point of view (in fact, physiologically active and pharmacologically active).

Flavonoid substances (e.g., apigenin, acasetin, quercetin, galangin, hesperidin, naringein, etc.) have antioxidant, antifungal, antimicrobial, neuroprotective, antitumor and anti-inflammatory action. Substances in the phenolic acid class (e.g., p-coumarin acid, gallic acid, ellagic acid, ferulic acid, etc.) have antigenotoxic, neuroprotective, antianxiolytic, antiproliferative, anti-inflammatory action.

Epidemiological studies on dietary nutrition and the role of antioxidant compounds in decreasing the incidence of cardiovascular disease have been extended in relation to the consumption of bee products, especially honey. Honey has been shown to have antioxidant effects that help prevent disease, including neurovegetative diseases, cancer and diabetes.

Reducing oxidative stress through antioxidant molecules means, in de facto, diminishing the role of reactive species (O^-; $\cdot OH$; H_2O_2 and others), which contribute to the formation of free radicals (Lushchak, 2014). It is known that free radicals affect the quality of food and reduce its nutritional value.

As shown, flavonoids and polyphenols in honey – biologically active substances (in this case physiologically active), intervene in combating oxidative stress. It is mentioned that there are native antioxidant enzymes in the body (genetically encoded), i.e. superoxide dismutase, catalase and glutathione peroxidase. Intensification of antioxidant action of the body can be ensured by the nutritional intake of some compounds, such as tocopherols, ascorbic acid, etc.

It is noted that the "toxic honey" mentioned in history since the Roman Empire has been studied. In the case of toxic honey, we can talk about the presence of toxicologically active substances from certain plants frequented by bees (e.g., *Hyosciamus niger*-henbane, *Datura stramonium*-datura, *Atropa belladonna*-mandrake, *Cannabis sativa*-hemp, etc.).

These sources have been the basis for the intake of toxic compounds such as hyoscyamine, gelsemine, graynotoxin, etc.

3.3.4.2.2   Propolis

Propolis is a resin-specific substance (resin) harvested and stored by bees. Its composition differs, being conditioned by the botanical specificity of the area where beekeeping was practiced (Cardinault et al., 2012; Šturm and Poklar-Ulrih, 2019).

At the origin of propolis are bee-picked resins from buds, young plant branches, tissue damage to some trees and shrubs (Marcucci, 1995; Khalil, 2006). These resins from plant tissues have the role of protecting against the attack of microorganisms, having a repulsive effect on insects.

The mixture of vegetable resins, taken by bees at temperatures of 18°C–20°C, comes into contact with the enzymes in their glandular secretions forming the viscous product, called "propolis". The color of propolis differs, depending on the botanical resource. Green or red can also be found.

Propolis is deposited by bees in the hive, at the assembly areas of the honey frames and the lid. Clogging them with propolis ensures tightness in relation to the external environment (maintaining a constant indoor temperature of about 35°C) and asepsis of the internal environment in the hive.

The generic name of propolis (gr.*pro*-avant; *polis*-city) derives from its role in the protection of the hive. In the early Hellenistic interpretation, propolis was assigned the role of "defender of the city" (in this case "of the hive"). The use of propolis in ethnopharmacology was known 300 years BC. by Egyptians, Persians and Greeks.

Chemical composition – There are different compounds in propolis. The first factor that conditions the composition is represented by the plants present in the geographical area where beekeeping is practiced, so the dependence on the ecosystem. Another factor that influences the composition is the pharyngeal secretions of bees that produce many transformations.

The hydrolysis reactions of heterosides and flavonoids are included in this framework. It is generally estimated that propolis contains 50%–55% resins (resins); 30% fatty acids; 10% essential oils; 5% pollen and 5% various organic and mineral substances. Organic substances include flavonoids, various phenolic derivatives and their esters, and volatile aromatic derivatives. Among the minerals are Fe, Ca, Zn, Cu and Mn. There are also vitamins, e.g., vitamins C and E, and B-complex vitamins (Cardinault et al., 2012; Ahangari et al., 2018; Zullkiflee et al., 2022).

From the group of biologically active substances isolated from honey, propolis and royal jelly are important. The structural formulas of some flavonoid compounds and one aliphatic compound are shown in Figure 3.16.

Important flavonoid compounds present in honey and propolis are: *apigenin* (IX) - antibacterial and anti-inflammatory; *acacetin* (X) - antiallergic, antineoplastic; *galangin* (XI) - antioxidant, antineoplastic; *quercetin* (XII) - antiallergic, antibacterial, anti-inflammatory, antineoplastic; *fisetin* (XIII) - antiallergic, antibacterial, antineoplastic; *chrysin* (XIV) - antibacterial, anti-inflammatory, antineoplastic; *phenethyl ester of caffeic acid* (XV) - antitumor. A compound present in royal jelly was also included: *10-hydroxy-2-decenoic acid* (XVI).

Systematic research on various propolis samples – taken from 29 countries on various continents – has led to the detection of more than 300 different compounds. The main group of compounds and their number are as follows: flavonoids 92, phenols 126, terpenoids 46, various fatty acids 17, alkaloids 21 and other compounds 22.

Table 3.2 lists groups and subgroups of compounds and their derivatives isolated from propolis from various sources on Earth. The information regarding the composition and chemical structure of some substances from the group of phenolic and terpenoid compounds isolated from propolis is also remarkable. Figure 3.17 presents data compiled from the specialized literature mentioned at the beginning of the subchapter. Some biologically active substances in propolis are set out in more detail below, and their biological activities are mentioned. These include phenolic compounds: *2,2-dimethyl-8-prenylchromene* (XVIII) - antimicrobial; *3-prenylcinnamic acid* (XVIII) allylic ester - antimicrobial; and kemferide (XIX) - antitumor. Also tepenoid compounds (terpenoids): *isocupresic acid* (XX) - antifungal; *13C-symphioreticulic acid* (XXI) - antitumor; *farnesol* (XXII) – antifungal are included.

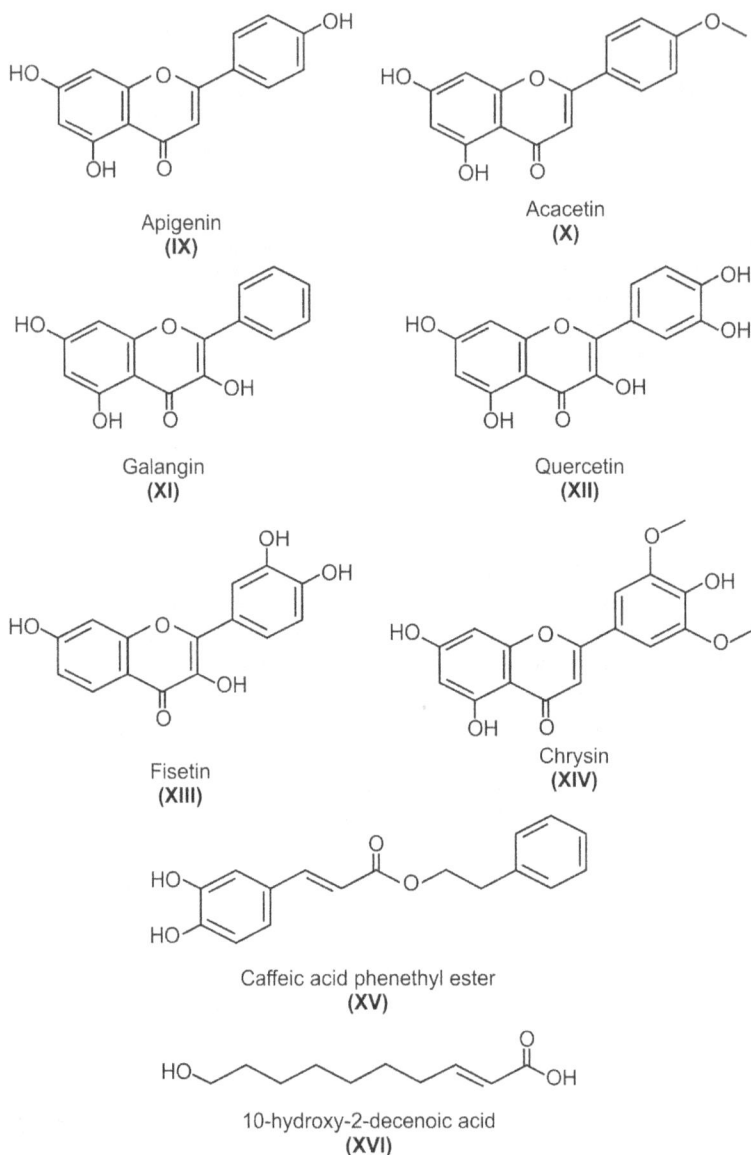

Apigenin
(IX)

Acacetin
(X)

Galangin
(XI)

Quercetin
(XII)

Fisetin
(XIII)

Chrysin
(XIV)

Caffeic acid phenethyl ester
(XV)

10-hydroxy-2-decenoic acid
(XVI)

FIGURE 3.16 Flavonoid compounds in honey and propolis (IX–XV) and an aliphatic compound in propolis and royal jelly (XVI).

Biological activity – Propolis has known pharmacologically active properties for thousands of years. The therapeutic effects attributed to propolis are dependent on its composition, which in turn is conditioned by the existing plant sources available to bees in a limited geographical area where their movement takes place. A more detailed study of the biological (pharmacological) activity in vitro and in

TABLE 3.2

Groups of Chemical Compounds in Propolis

No.	Groups of Compounds	Subgroups of Compounds and Derivatives
1	Flavonoids	Flavans, Flavonones, Flavononols, Flavones, Flavonols, Chalcones, Flavonoid glycosides, Phenylated flavonoids
2	Phenols	Phenolic glycerides, Esterified phenolic acids, Xanthone, Propanoid phenyl derivatives, Glycosylated phenylpropanoids, Phenolic glycosides, Stilbenes, Lignans, Phenolic acids, Other phenolic compounds
3	Terpenoids	Monoterpenes, Sesquiterpenes, Diterpenes, Triterpenes
4	Fatty acids	Fatty acids and derivatives, Glycosylated fatty acids
5	Alkaloids	Alkaloids (e.g., papaverine, thebaine, lobelin, etc.), alkaloid derivatives (e.g., nicotinealdehyde semicarbazone)
6	Other compounds	Compounds with various structures (e.g., ethoxy sulfonate, carbamazepine, 1-butylisoquinoline)

Source: After Šturm and Poklar-Ulrih (2019)

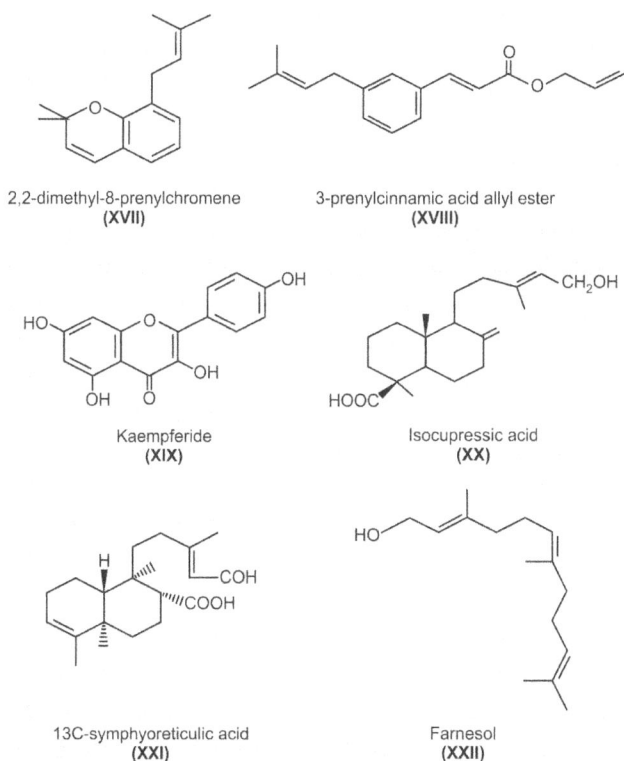

2,2-dimethyl-8-prenylchromene
(XVII)

3-prenylcinnamic acid allyl ester
(XVIII)

Kaempferide
(XIX)

Isocupressic acid
(XX)

13C-symphyoreticulic acid
(XXI)

Farnesol
(XXII)

FIGURE 3.17 Phenolic compounds (XVII–XVIII) and terpendoid compounds (XIX–XXII) from propolis.

vivo has been done in the last decades, finding that there are antimicrobial, antifungal, antiviral, antiparasitic, antioxidant, anti-inflammatory, antiangiogenic activity, etc. More detailed data are presented below.

Antioxidant activity – This is due to the presence of polyphenolic compounds and vitamins C and E. In vivo, propolis significantly reduces the level of lipoperoxidase in various organs (e.g., liver, kidneys, lungs and brain). It also modulates the action of antioxidant enzymes (catalase, superoxide dismutase and glutathione peroxidase).

Anti-inflammatory activity – It is explained by the action on molecules from the immune system (e.g., IL-6) and the inhibition of enzymes involved in inflammatory processes (i.e., cyclooxygenase, lipoxygenase, NADPH oxidase, etc.).

Antiangiogenic activity – It is a remarkable action because reduces angiogenesis in vitro and in vivo by limiting neovascularization (new vessel formation) and inhibiting cell proliferation and migration.

Antimicrobial activity – It has been found to various species of the genera *Staphylococcus, Streptococcus, Proteus, Psudomonas, and Salmonella.* Such microorganisms are involved in otorhinopharyngeal, gastrointestinal, genital, diseases. Experimental research has suggested that propolis, through its various constituent compounds, inhibits the growth of bacteria, by blocking cell division, disorganizing the cytoplasm, inhibiting protein synthesis and inhibiting cell adhesion.

Antifungal activity – It resides in the action on fungi of the genera *Candida, Aspergillus* and *Mycrosporum.*

Antiviral activity – It is characterized by its action on many viruses, e.g. mixoviruses, polioviruses, coronaviruses, rotaviruses and adenoviruses. It has a prophylactic action against the influenza virus (having antineuromidase action).

Antiparasitic activity – There are studies on the action of propolis in trichomoniasis, trypanosomiasis (responsible for sleeping sickness), leishmaniasis and intestinal parasitosis called lambliasis (caused by *Giardia lamblia*).

Immunomodulatory activity – It has been found that in vivo and in vitro induce effects on cells involved in innate and acquired immunity. Propolis has also been found to cause a decrease in the level of pro-inflammatory prostaglandins and leukotrienes and an increase in the level of anti-inflammatory cytokines.

Antitumor activity – Studies have shown that propolis has antiproliferative effects in relation to many tumor cell lines (blood, skin, colon, breast, prostate, liver, etc.).

Other effects – The data on the preventive effect against neutropenia, anemia and thrombocytopenia following chemo- and radiotherapy treatments were also remarkable. It has also been shown to be useful in the treatment of intoxications caused by *xenobiotics*: drugs (e.g., paracetamol) and environmental pollutants (e.g., heavy metals).

3.3.4.2.3   Royal Jelly

Royal jelly (fr.-*gelée royale*; germ.-*Bienenköniginnenfuttersaft*) is a viscous product resulting from the secretion of the hypopharyngeal and mandibular glands of "worker bees".

Chemical composition – There are various substances in the royal gel: 10%–20% carbohydrates; 60%–70% water; 9%–18% proteins and amino acids; 4%–8% lipids; vitamins (thiamine, riboflavin, pyridoxine, niacin, pantothenic acid, biotin and folic acid), steroids, biopterin, neopterin, minerals (trace elements), 4-hydroxybenzoic acid ester (used in cosmetology and pharmaceuticals), enzymes; antibiotics. Among the specific components present in royal jelly are: royalactin, royalsin, jelleines and aspimin (Horn and Lüllmann, 2006; Kunugi and Ali, 2019).

Biological activity – Royal jelly is a "superfood" consumed only by the "queen bee" (usually called "queen"), hence the name "royal gell". This ensures the "biological cycle" of bees in the hive. From the components of royal jelly, royalactin – monomeric glycoprotein with a molecular weight of 57 kDa has special importance, which contributes essentially to the morphological evolution of the larva, with evolutionary development toward the queen (queen). Also important are royalisin and jelly – antimicrobial peptides that increase the immunity of the queen's precursor larva (Kunugi and Ali, 2019).

This product is considered an important therapeutic remedy for humans, appreciated not only in traditional medicine but also in modern medicine. It has many pharmacological effects including antibacterial, antitumor, antiallergic, anti-inflammatory and immunomodulatory effects.

3.3.4.2.4   Apitoxin

Apitoxin, also known as "bee venom", is a product secreted by various bee species and used by them as a means of defense against predators. Apitoxin is a toxicologically active substance known for its complex toxic effects in humans (Meier and White, 1995; Pucca et al., 2019; Wehbe et al., 2019).

Chemical composition – Apitoxin is a complex substance made from a mixture of different compounds. These include proteins and peptides dominated by melittin and apamin; biologically active amines (e.g., histamine and dopamine), enzymes (e.g., phosphorylase A2 and hyaluronidase), etc. It also contains carbohydrates, phospholipids and some volatiles. General data on these compounds are presented in Table 3.3.

A brief description of the major constituents of apitoxin is given below. There are general references to their chemical composition and biological activity.

Melittin – (gr.-μελιτα, *melitta*-bee) is the main constituent of apitoxin. This compound, which is higher in apitoxin, is also known to have cytostatic action.

Chemical composition – highlights a cationic polypeptide structure in which there are 26 amino acid residues. The crude formula is $C_{131}H_{229}N_{39}O_{31}$. Biological activity is defined by the role of enzyme inhibitor of protein kinase C, involved in the functioning of the sodium-potassium pump in the synaptosomal membrane. It is considered a "lysis factor" of the cell membrane.

Apamin – Its composition shows a polypeptide with 18 amino acid residues and two sulfide bonds. It has a molecular weight of approx. 5,200–7,800 Da. Biological

TABLE 3.3
Chemical Compounds Isolated from Apitoxin - General Data

No.	Groups	Main Compounds	Dry Substance (%)
1	Proteins and peptides	Melittins	40.0–50.0
		Secapin	0.5–2.0
		Mast cell degranulation peptide	2.0–3.0
		Procamine	1.0–2.0
		Protease inhibitor	13.0–15.0
2	Biologically active amines	Histamine	0.5–1.6
		Dopamine	0.13–1.0
		Norepinephrine	0.1–0.7
3	Specific enzymes	Phospholipase A2	10.0–20.0
		Hyaluronidase	1.5–2.0
		a-Glucosidase	0.6

activity resides in neurotoxic effects. It blocks calcium-activated potassium channels in the nervous system.

Figure 3.11 shows the chemical structure for α-lipoic acid. It highlights a pentacycle structure (with two sulfur atoms) and a side chain with a carboxylic group. From a chemical point of view, it is called D-6,8-dithiooctanoic acid. In nature, α-lipoic acid is found mainly in acyclic form, but it also exists in cyclic form.

Adolapin – It is a polypeptide of 103 amino acid residues. It has been shown to have anti-inflammatory, antinociceptive (analgesic) and antipyretic action by blocking prostaglandin synthesis and inhibiting cyclooxygenase activity.

Mast cell degranulation peptide – It is also known as "401 peptide". There are 22 amino acid residues in the composition of this peptide, with two disulfide bonds in the chain.

Biological activity – lies in the inhibition of potassium channels in mast cells and the release of histamine. It is also an epileptogenic neurotoxin.

Phospholipase A2 – It is a 128 amino acid polypeptide with four disulfide bonds. It has enzymatic activity, being considered the most toxic in the complex formed with melitin. It is also called "*bee hemolytic factor*" because it cleaves phospholipids from the composition of cell membranes. In vitro it has antitrepanocidal and antibacterial effects, as well as antitumor properties.

Experimentally, in mice, phospholipase A2 was found to decrease hepatotoxicity caused by acetaminophen. Recent studies have shown that this enzyme also has immunoprotective action in various conditions, e.g., asthma, Parkinson's and Alzheimer's diseases.

Hyaluronidase – This compound (with enzymatic specificity) decomposes hyaluronic acid, dilates the capillaries in the tissues causing an increase in blood flow in the affected area and thus allows the diffusion of apitoxin.

Biological activity – Apitoxin is generally used in complementary/alternative medicine due to its pharmacologically active effects. In apitherapy (sometimes called apitoxotherapy), it has applications in rheumatic diseases of the joints and spine (spondylosis, discopathy, low back pain, etc.) due to the effect of peptide 401 and melitin acting on the immune system (it is, in fact, an immunostimulant). It is also used to treat nervous system disorders (depression, insomnia and headaches) and so on.

3.3.4.2.5   Bee Wax

It is produced by "bees". It has the scientific name of "white wax". It is secreted by four pairs of "ceriferous glands" located on the bee's abdomen. After the bees chew this secretion, "adulteration with pollen and propolis" occurs. Thus, in the end, the color gradually becomes yellow, then brown. The wax is extracted from the combs with a beeswax extractor (solar or electric).

Chemical composition – Beeswax consists of a mixture of compounds containing 14% hydrocarbons, 50% esters over (of which 35% monoesters, 14% diesters, and 3% triesters), 12% various fatty acids, long chain alcohols (called fatty alcohols), etc. (Hepburn, 1986).

It is reiterated that cerides (waxes) are included in simple lipids along with glycerides, steroids and etholides. Structurally, cerides are mainly esters of (higher) fatty acids with higher aliphatic monoalcohols (also called fatty alcohols).

These two groups of substances are presented below. The main fatty acids in *cerides* – for which the number of carbon atoms (C_n) and the trivial (usual) and chemical names are mentioned – are the following: C_{14}- myristic acid (n-tetradecanoic); C_{16}- palmitic acid (n-hexadecanoic acid); C_{24}-lignoceric (n-tetracosanoic acid); C_{26}- cerotic acid (n-hexacosanoic acid); C_{30}- melisic acid (n-tricontanoic); C_{18}- oleic acid (cis-9,10-octadecenoic acid) et al. Also, higher fatty monoalcohols have been isolated, more important (initially mentioning C_n) are: C_{16}- cetyl alcohol (n-hexadecanoic); C_{20}- arachyl alcohol (n-eicosanol); $C26$- ceryl alcohol (n-hexacosanol) and C_{30}- melyssil alcohol (n-triacontanol) et al.

Figure 3.18 shows the structural formulas of some wax compounds bees: monoesters – myricyl palmitate (XXIII), myricyl cerotate (XXIV), cetyl lignocerate (XXV) and some fatty acids – lignoceric acid (XXVI), cerotic acid (XXVII).

The substances present in the wax give it the property of being preserved for a long time. It is suitable for physical transformations in heating and filtration – so it can be kept in small pieces.

Biological activity – Beeswax (similar to vegetable waxes) is an edible product with negligible toxicity. It can be used in food as a food additive. In the European Union, the additive is known under no. E901.

Wax can be used to seal (glaze) food (e.g., surface-coated cheeses) in relation to air. In the pharmaceutical field, it can be used in the production of ointments and suppositories (Fratini et al., 2016). In cosmetics, it is used to obtain creams, lotions, ointments, lipsticks, etc. It has been observed that wax has a high affinity for chemical xenobiotics in the environment (e.g., pesticides, industrial pollutants, etc.).

3.3.4.2.6   Bee Pollen

It is presented in the form of microscopic grains (with a size of 2.5–250 µm) collected by bees from the anthers of flower stamens.

Chemical composition – Investigations into the pollen have concluded that it contains approx. 200 substances represented by: 22.7% proteins rich in amino acids (e.g., valine, lysine, cysteine, threonine, histidine, phenylalanine, tryptophan, glutamine, etc.), lipids (e.g., triglycerides, steroids and phospholipids), unsaturated fatty

$H_3C-(CH_2)_{14}-CO + O-CH_2-(CH_2)_{28}-CH$

Residue of palmitic acid | Residue of myricyl alcohol

Myricyl palmitate
(**XXIII**)

$H_3C-(CH_2)_{24}-CO + O-CH_2-(CH_2)_{28}-CH_3$

Residue of cerotic acid | Residue of myricyl alcohol

Myricyl cerotate
(**XXIV**)

$H_3C-(CH_2)_{22}-CO + O-CH_2-(CH_2)_{14}-CH_3$

Residue of lignoceric acid | Residue of cetyl alcohol

Cetyl lignocerate
(**XXV**)

$H_3C-(CH_2)_{22}-COOH$

Lignoceric acid
(**XXVI**)

$H_3C-(CH_2)_{24}-COOH$

Cerotic acid
(**XXVII**)

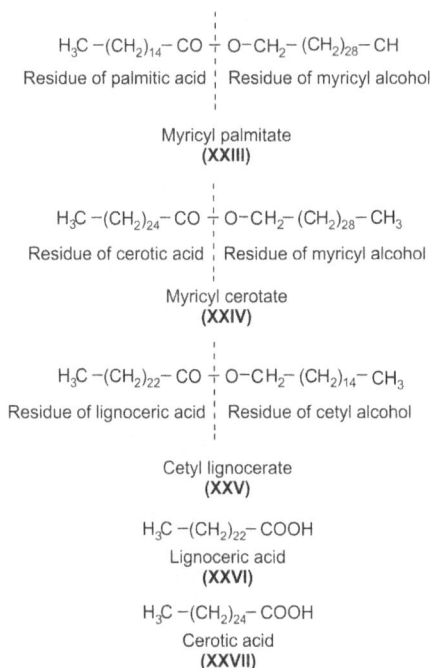

FIGURE 3.18   Monoesters and fatty acids isolated from beeswax.

acids (linoleic and arachidonic), flavonoids, nucleoproteins, vitamins (e.g., fat-soluble A, E, D and water-soluble B_1, B_2, B_6, C, etc.), carbon dioxide, sulfur dioxide and mineral trace elements (Komosinska-Vassev et al., 2015).

Biological activity – Bee pollen is used in apitherapy due to its proven action: antifungal, antimicrobial, antiviral, anti-inflammatory, immunostimulant and local analgesic. Pharmacological studies have also shown a lipid-lowering effect by decreasing total lipids and triglycerides (Bogdanov, 2014).

The anti-inflammatory effect has also been shown to inhibit the activity of cyclooxygenase and lipoxygenase, the enzymes responsible for converting arachidonic acid to compounds such as prostaglandin and leukotrienes.

The antibiotic effect of pollen ethanolic extract was observed in Gram-positive bacteria (e.g., *Staphylococcus aureus*) and Gram-negative bacteria (e.g., *E. coli*, *Klebsiella penumoniae* and *Pseudomonas aeruginosa*) as well as the effect on fungi (e.g., *Candida albicans*). In diseases of the nervous system, bee pollen has been used in conditions of stress, asthenia, depression, drug addiction, etc.

3.4   ZOOCHEMICALS PRESENT IN ANIMAL TISSUES

3.4.1   SYNOPTICAL DATA

With reference to zoochemicals with attributes of biologically active substances (as mentioned in Section 3.3) they were included in two categories, namely: (i) zoochemicals from various *food groups*, i.e. milk and dairy products, meat and viscera, eggs from birds, and bee products; (ii) zoochemicals present in *animal tissues*, i.e., various fatty acids, alkaloids, specific pigments, a.o.

The distribution of zoochemicals in tissues can be considered as a completion of the information because these compounds are also found in foods of animal origin. They were presented separately because in certain situations zoochemicals are obtained as "tissue extracts" necessary in diet (e.g., processing of food supplements, fortification of certain foods). Extracts are used also in the fields of pharmaceuticals and cosmetics. There are also situations in which distinction is required for zoochemicals in the form of alkaloids from various animal tissues. In such cases, it is necessary to exclude food from consumption (e.g., the presence of alkaloids in the tissues of some aquatic animals). In the situations exposed above, we encounter biologically active zoochemicals of physiological, pharmacological, and toxicological interest.

The group of biologically active zoochemicals, which will be discussed in this section, includes: *fatty acids* – presenting general characteristics, types of fatty acids (ω-9, ω-6, ω-3), importance of ω-3 fatty acids with examples and the derivatives of linoleic acid called conjugated linoleic acids (CLA). In the case of *alkaloids*, general data will be discussed about those originated from aquatic animals, from terrestrial animals and bacteria. For *animal pigments*, only general data on melanin, anthraquinone a.o. will be presented. Some of them are being extracted for use in food processing (e.g., anthraquinones extracted from cochineal).

3.4.2 REPRESENTATIVE COMPOUNDS

3.4.2.1 Fatty Acids - Generalities

They are omnipresent compounds found in various lipids, e.g., glycerides, steroids, glycerophospholipids, sphingolipids, etc. (Armstrong, 1989; Şerban and Roşoiu, 1992; Alasalvar et al., 2011; Bane et al., 2014; Gârban, 2018a). Some general data are repeated in order to understand the way of notation (of carbon atoms) specific to fatty acids. The general formula for a saturated fatty acid is exemplified (Figure 3.19):

In the chain with n - carbon atoms, the main carbon atoms denoted by, α, β, ω, involved in the oxidation processes (α-, β- and ω-oxidation) are presented.

Notations of C numbers are important for highlighting the positioning of the double bonds, which influence the nutritional and pharmacological specificity.

3.4.2.1.1 Unsaturated Fatty Acids of Types ω-3, ω-6 and ω-9

The unsaturated fatty acids of types ω-6 and ω-3 are considered *essential fatty acids* because their endogenous biosynthesis is not possible. In the body, the need for unsaturated fatty acids (PUFA), including EPA and DHA, is provided by food intake (e.g., foods with high unsaturated fatty acids and dietary supplements). Also, such compounds are used in pharmaceutical products.

Regarding the ω (omega) positions, a distinction is made between ω-9, ω-6 and ω-3 acids. In Figure 3.20 are presented: *oleic acid* (II) – i.e. octadecamonoenoic; *linoleic* acid (III) – i.e. octadecadienoic and *linolenic acid* (IV) – i.e. octadecatrienoic.

$$H_3C-(CH_2)_{n-4}-\underset{3}{CH_2}-\underset{2}{CH_2}-\underset{1}{COOH}$$

(I)

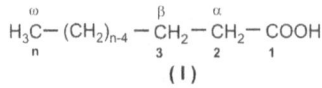

FIGURE 3.19   Fatty acid – general formula.

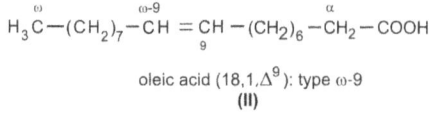

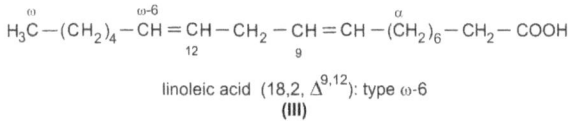

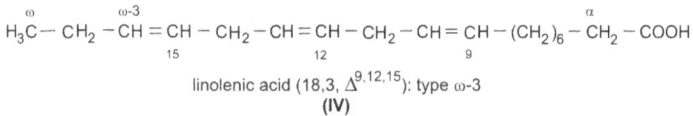

$$H_3C-(CH_2)_7-\overset{\omega\text{-}9}{CH}=\underset{9}{CH}-(CH_2)_6-\overset{\alpha}{CH_2}-COOH$$

oleic acid $(18,1,\Delta^9)$: type ω-9

(II)

$$H_3C-(CH_2)_4-\overset{\omega\text{-}6}{CH}=\underset{12}{CH}-CH_2-CH=\underset{9}{CH}-(CH_2)_6-\overset{\alpha}{CH_2}-COOH$$

linoleic acid $(18,2,\Delta^{9,12})$: type ω-6

(III)

$$H_3C-CH_2-\overset{\omega\text{-}3}{CH}=\underset{15}{CH}-CH_2-CH=\underset{12}{CH}-CH_2-CH=\underset{9}{CH}-(CH_2)_6-\overset{\alpha}{CH_2}-COOH$$

linolenic acid $(18,3,\Delta^{9,12,15})$: type ω-3

(IV)

FIGURE 3.20   The chemical structure of unsaturated fatty acids (ω-9, ω-6, ω-3).

The annotations in numbers in the case of linoleic acid – for example – give the number of C atoms (e.g., 18), the number of double bonds (e.g., 3) and the position of the double bonds (e.g., $\Delta^{9,12,15}$). It is a ω-acid (although it has other double bonds, too).

3.4.2.1.2   Type ω-3 Fatty Acids

Chemical structure – More important, from a nutritional and pharmacological point of view, are ω-3 fatty acids. This group of compounds includes two polyunsaturated fatty acids – PUFA (Poly-Unsaturated-Fatty-Acids). They are: *eicosapentaenoic acid* – EPA (V) with 20 carbon atoms and *docosahexaenoic* acid – DHA (VI) with 22 carbon atoms. The structural formulas of these polyunsaturated acids are shown in Figure 3.21. Omega-3 fatty acids are found in foods of animal origin and are

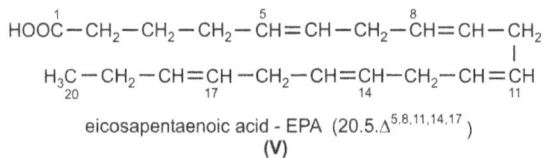

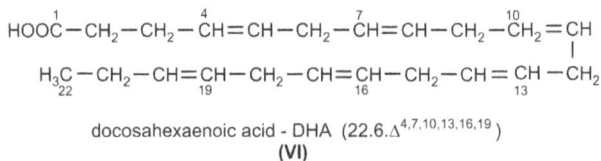

$$\underset{1}{HOOC}-CH_2-CH_2-CH_2-\underset{5}{CH}=CH-CH_2-\underset{8}{CH}=CH-CH_2$$

$$\underset{20}{H_3C}-CH_2-CH=CH-CH_2-CH=CH-CH_2-CH=CH$$

eicosapentaenoic acid - EPA $(20.5.\Delta^{5,8,11,14,17})$

(V)

$$\underset{1}{HOOC}-CH_2-CH_2-\underset{4}{CH}=CH-CH_2-\underset{7}{CH}=CH-CH_2-\underset{10}{CH_2}=CH$$

$$\underset{22}{H_3C}-CH_2-CH=CH-CH_2-CH=CH-CH_2-CH=CH-CH_2$$

docosahexaenoic acid - DHA $(22.6.\Delta^{4,7,10,13,16,19})$

(VI)

FIGURE 3.21   Structural formulas of eicosapentaenoic acid (EPA) and docosahexaenoic acid (DHA).

important for maintaining/protecting health. Along with EPA and DHA, it includes linoleic acid among the fatty acids in the group of fatty acids with increased nutritional/pharmacological importance (Vazhappilly and Chen, 1998; Gârban, 2000).

Biological activity – The beneficial action of unsaturated fatty acids in food causes changes in some hematological parameters, e.g., decreased platelet aggregation, increased coagulation time and vasodilation. In the pathobiochemistry of lipid metabolism, it is known the need for an optimal ratio of unsaturated fatty acids/saturated fatty acids which is approx. 0.8. The increase in this ratio is associated with the hypercholesterolemic effect. Omega-3 fatty acids have more intense anti-atherogenic effects by reducing cholesterol and triacylglycerols.

For this reason, they are used as ingredients in food supplements to optimize parameters specific to clinical chemistry – detectable in medical laboratories (e.g., steroids-cholesterol, glycerides-triglycerides). Also, they can be used in medical clinic, as pharmaceutical products, and administered for prophylactic, metaphylactic and therapeutic purposes. Table 3.4 presents some characteristic data on the natural distribution and benefits of these two main compounds in the PUFA group, namely: EPA and DHA (Cannella and Giusti, 2000; Alasalvar et al., 2011).

Eicosapentaenoic acid (EPA) – is also known by the trivial term *timnodonic acid.* It belongs to the PUFA group with 20 C atoms and 5 double bonds. During metabolic processes in the EPA, prostaglandin-3 which inhibits platelet aggregation, thromboxane-3 and leukotriene-5 from the group of eicosanoid compounds are formed (Burdge et al., 2002; Gârban, 2018a). In the human diet, EPA is found in fish oil from cod liver, herring, mackerel, sardines or edible sea seashells that consume phytoplankton. It can also be found in non-animal sources, e.g. microalgae of the genus *Monodus, Chlorella, Phaeodactylum*, etc.

Decosahexaenoic acid (DHA) – also has the trivial name of *cervonic* acid. This PUFA has 22 C atoms and 5 double bonds. It can be obtained from fish oil or microalgae oil from the genra *Crypthecodinium cohnii* and *Schizochytrium limacinum*.

TABLE 3.4
Biologically Active Fatty Acids: Natural Distribution, Effects

Fatty Acid Types	Natural Distribution	Consumption Benefits
EPA	• fish from cold waters, e.g. salmon, herring, tuna, trout	• anti-inflammatory effects • provides cardiovascular protection • may reduce the risk of cardiac arrhythmias • protects against type 2 diabetes • protects against Alzheimer's disease • protects against cancer
DHA	• cold water fish, e.g., salmon, herring, tuna, trout • eggs • fish oil	• alleviates depressions • relieves schizophrenia disorders • reduces attention deficit • important for brain development and baby eyes

In low amounts, DHA and EPA can be formed in the body through the biosynthesis of α-linoleic acid (this path of biogenesis is better known for its formation in breast milk).

In the body, DHA is present in larger amounts in phospholipids. In cytology, it was detected in the plasma membrane of neurons. It intervenes in the transport of choline, glycine, taurine, the function of K channels, etc. It plays an important role in cognitive processes (children and elderly) and improves postpartum maternal depression. These fatty acids are suitable for use as ingredients in food supplements and pharmaceuticals.

3.4.2.1.3   Conjugated Linoleic Acids

This denomination includes a mixture of positional and geometrical isomers of linoleic acid (octadecadienoic). They are in *trans* fatty acid and *cis* fatty acid form and are characterized by the presence of conjugated double bonds. These compounds are formed, in natural conditions, in the rumen compartment of the stomach in polygastric animals (ruminants), under the action of the microbial enzymes (Aydin, 2005; Kathuria et al., 2019). Nowadays, it is known that such mixture can be formed also in the digestive tract of non-ruminant animals and even in the gastrointestinal tract of humans (Aydin, 2005). The CLA formation is facilitated by the increased temperature.

Chemical structure – The characteristic compounds of CLA derive from the initial form of linoleic acid (see structure III in Figure 3.20) in which there are two double bonds at the C_9–C_{10} and C_{12}–C_{13} positions. From linoleic acid, various conjugated derivatives can be formed (CLA), with the two double bonds in different positions. Since CLA contains double bonds, cis and trans isomers are also distinguished. Three types of CLA (A, B and C models) are presented in Figure 3.22.

Model A CLA has its two double bonds at the C_8–C_9 and C_{10}–C_{11} positions. The first double bond has a cis configuration (C_8–C_9) and the second double bond has a trans configuration (between C_{10}–C_{11}).

Model B CLA has its two double bonds at the positions C_9–C_{10}, and C_{11}–C_{12}. The first is in cis configuration and the second is in trans configuration.

Model C CLA has its two double bonds the C_{10}–C_{11} and C_{12}–C_{13} positions. The first double bond has a trans configuration (between C_{10}–C_{11}) and the second double bond has a cis configuration (between C_{12}–C_{13}).

Biological activity – The major food sources for CLA are ruminant meat and dairy products. Studies revealed some beneficial effects of CLA for human organisms. It was shown that CLA inhibits the development of cancer cells, reduces the risk of cardiovascular disease, strengthens the immune system, develops smooth muscles and improves milk secretion, etc.

3.4.2.2   Alkaloids

3.4.2.2.1   Alkaloids from Aquatic Animals

In nature, there are also animal alkaloids from sea and ocean water, generically called "marine alkaloids" (marine zooalkaloids). Such zoochemicals with alkaloid attributes are formed from a *β*-carboline (Figure 3.23), which has arginine as a precursor.

Linoleic acid (LA) $(18.2.\Delta^{9c,12t})$
(see fig. 3-2)

Conjugated linoleic acid (CLA) $(18.2.\Delta^{8c,10t})$ - model A
(VII)

Conjugated linoleic acid (CLA) $(18.2.\Delta^{9c,11t})$ - model B
(VIII)

Conjugated linoleic acid (CLA) $(18.2.\Delta^{10t,12c})$ - model C
(IX)

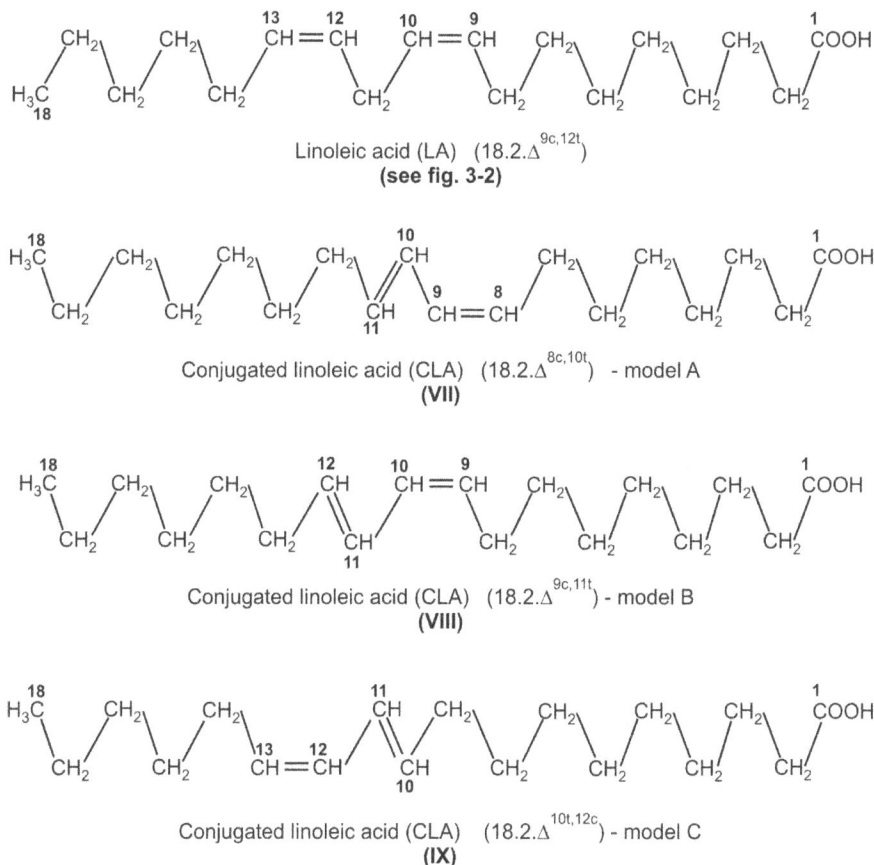

FIGURE 3.22 Linoleic acid (LA), Conjugated linoleic acids (CLA): model A (VII); model B (VIII); model C (IX).

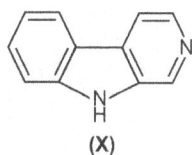

(X)

FIGURE 3.23 Carboline heterocycle.

Among a few alkaloids produced in animal organisms – mollusks (shells, snails) there are zoochemicals that originate from β -carboline. The best known are saxitoxin and tetrodotoxin. These are biologically active compounds with toxicologically active attributes through neurotoxic effects (Banu et al., 1982; Chang et al., 1997; Clark et al., 1999; Bane et al., 2014; Gârban, 2018b).

The (more complex) chemical names of these zoochemicals can be found in various catalogs, treatises that include data compiled by IUPAC. The chemical structures of saxitoxin (XI) and tetrodotoxins (XII) are given in Figure 3.24.

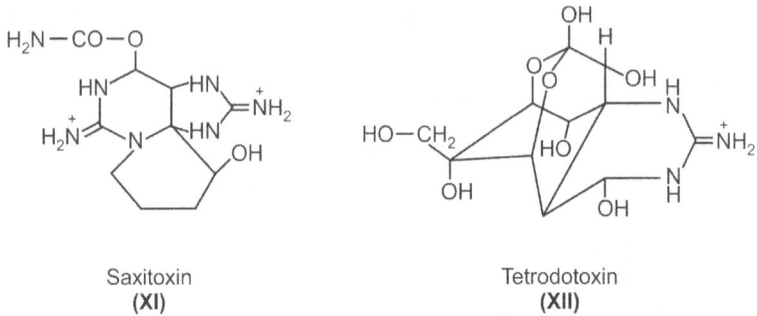

Saxitoxin
(XI)

Tetrodotoxin
(XII)

FIGURE 3.24   Alkaloids of animal origin.

Saxitoxin (STX) – is an alkaloid extracted from the shells of the genus *Saxidomus* present in seawater. Human contamination is caused by the consumption of mollusks (especially mussels) which in turn have consumed toxic algae (Smith et al., 2001).

Chemical structure – It reveals the presence of a tetrahydrogenated purine heterocycle that has many substituents. This neurotoxin is naturally produced by certain species of dinoflagellate marines (e.g., *Alexandrium* sp., *Gymnodinium* sp., etc.) and certain cyanobacteria (e.g., *Anabaeba* sp., *Cylindro spermopsis* sp., etc.). From the marine environment, saxitoxin accumulates, in particular, in bivalve shells of the genus mentioned above.

Biological activity – It resides in neurotoxic effects after the consumption of shells. The mechanism of the neurotoxin action lies in the selective blockade of sodium channels in the neuronal axons and synapses.

In this way, the normal functioning of the nerve cells is disrupted and paralysis occurs. Saxitoxin is reversibly bound to sodium channels. Specifically, binding is done directly to the pores that limit channel proteins in the cell wall. In experimental studies, the toxicity of SXT in mice has been shown to increase the semi-lethal dose (LD_{50}) in i.p. is 10 µg/kg.

In humans, studied under accidental circumstances, STX aerosols were found to show inhalation toxicity at 5 mg/min/m^3. The disease in humans has generally been caused by ingestion (consumption) of contaminated shells. In lethal intoxications, there is a progression of paralytic symptoms leading to respiratory failure, followed by exitus.

Studies in laboratory on animals (guinea pigs) have shown that the lethal effects of STX can be reversed by 4-aminopyridine (Chang et al., 1997). So it is a possible antidote. There are no studies in humans.

Tetrodotoxin (TTX) – is another alkaloid from the group of zoochemicals. It was isolated from fish of the order *Tetraodontiformes*. This order includes: balloon fish, moon fish, etc. In general, they are ocean fish. Later it was also discovered in other marine animals, e.g., the blue octopus, the rough triturus and the moon snails, which accumulate TTX.

Originally it was an infection with symbiotic bacteria of the genus *Pseudoalteromonas, Pseudomonas* and *Vibrio* (Bane et al., 2014).

Chemical structure – It is characterized by the presence of a *quinazoline heterocycl*e. In fact, it is a compound with the structure of amino perhydro quinazolone.

Biological activity – It is characterized by a mechanism of action that lies in blocking the sodium channels in the cell membrane of the nervous system. Thus, the potential action of neurons is inhibited by binding to specific areas of voltage channels (gates). The effects are initially manifested by nausea, vomiting, paralysis of the lips and tongue, and later paralysis of the limbs. Finally, paralysis of the respiratory tract sets in, when exitus becomes imminent.

Aplisinopsin – obtained from marine organisms: sponges of the genus *Aplysinopsis* and corals of the genus *Dendrothyllia* sp. The substance, with specific alkaloid, has antimicrobial and antitumor activity.

3.4.2.2.2 *β Alkaloids from Terrestrial Animals and Bacterium*

Advances in the study of alkaloids have shown that they can be isolated from various animal organisms – integrating into the category of *zoochemicals*. Over time, "zooalkaloids from terrestrial organisms", marine organisms and bacteria have been isolated (Gârban, 2018a). Zooalkaloids have been isolated from terrestrial animals: *castoramine and muscopyridine*. Pyocyanin was isolated from the bacteria.

Muscopyridine – isolated from the male body of the deer species *Moschus moschiferus* – a deer-sized and civet animal (*Viverra zibetha*) – a species of Asian cat. For these animals, the secretion product of the abdominal glands acts as a pheromone produced during the reproduction period. The natural compound extracted from the glands contains muscopyridine (alkaloid specific), muscone (cyclic ketone), testosterone, androsterone, etc. Muscopyridine has odorizing properties and is used in cosmetology (perfumes) and pharmacology.

Castoramine – from the North American beaver (*Castor canadensis*) and from the European beaver (*Castor fiber*);

Pyocyanin – is an alkaloid isolated from the bacterium *Pseudomonas aeruginosa*. This compound is water-soluble, blue-green and has antifungal action.

3.4.2.3 Pigments of Animal Origin

3.4.2.3.1 *Pterin Pigments*

Structurally, pterins are dicyclic compounds with nitrogen heteroatoms. Figure 3.25 presents the main pterin pigments derived from the nucleus called *pteridine* (XIII), which formally consists of a pyrimidine ring and a pyrazine ring. Pterin pigments are found in higher quantities in the wings and skin of insects. These pigments have various colors, e.g., white, yellow, red, purple, blue, sometimes being colorless.

The first isolated compounds (as early as 1895) were two natural colored products: *leucopterin* (XIV) and *xanthopterin* (XV) – see Figure 3.25. They were isolated

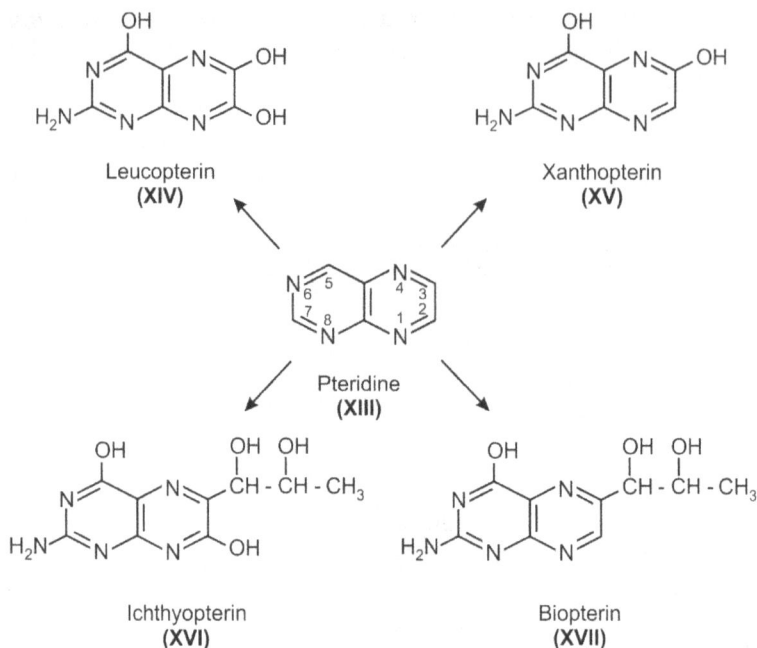

FIGURE 3.25   Pteridine and pterin chromatic derivatives.

from the scales from the wings of butterflies (gr. *ptera*-wing). Later *ichthyopterin* (XVI) and *biopterin* (XVII) were isolated.

Leucopterin – white, isolated from the wings of the white butterfly on cabbage called little bee, i.e. *Pieris brassicae.*

Xantopterin – yellow, on the wings of a yellow butterfly *Gonepteris rhamni* also on the yellow bands on the bodies of bees and wasps. Xanthopterin inhibits the biosynthesis of melanin pigments.

Ichthyopterin – a colorless or blue substance which has been isolated from scales of fish belonging to the genus *Ciprinus* spp.

Biopterin – a light yellow substance that has biological effects similar to vitamin B_2 (riboflavin).

From pterin, alloxazine (XVIII) is derived which is, in fact, a tricyclic condensed compound (Figure 3.26). It is synthetically obtained from o-phenyl-endiamine and alloxane. Alloxazine is the origin of riboflavin (XIX) - see Figure 3.26. Vitamin B_2 (riboflavin) is a yellow substance with a slight green fluorescence.

Yellow compounds, generically called *flavins*, have also been isolated from various animal tissues, e.g., *hepatoflavin* (from the liver). They were also isolated from some animal products, e.g., *ovoflavin* and *lactoflavin*. Alloxazine can form flavoproteins – the yellow color of which are enzymatic compounds whose structure is *flavin mononucleotide* (FMN) and flavin adenine dinucleotide (FAD). These enzymes are carriers of hydrogen in various redox reactions in the body.

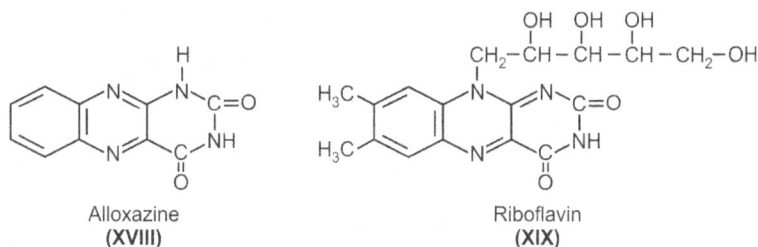

FIGURE 3.26   Alloxazine and riboflavin – structural formulas.

3.4.2.3.2   Melanin Pigments

In animal tissues, *melanin* was isolated from invertebrates (present in the cuticle covering the body of insects), and from vertebrates from hair and skin. The mentioned distribution is also characteristic of man, knowing the fact that the absence of melanin is due to the lack of the *tyrosinase* enzyme, which converts tyrosine up to melanin, and albinism appears. This is a condition with genetic determinism, known in pathobiochemistry (Gregoire, 1971; Feuer and Iglesia, 1985; Gârban, 2018b).

Phenylalanine and tyrosine metabolization is achieved by oxidation reactions, cyclization and polymerization resulting in compounds of the type dihydroxyphenylalanine (DOPA), its quinone, and finally phenolmelanins result in the presence of cysteine.

3.4.2.3.3   Anthraquinone Pigments

In the food field, among the anthraquinone pigments, cochineal extract is used – a carmine pigment. Cochenilla is extracted from the dry body of the female insect of *Dactylopius coccus* species that lives on the *Nopalea coccinellifera* cactus. The dye is actually carminic acid. It is used in the food industry as a coloring agent for a wide variety of products (meat, sausages, drinks, desserts, a.o), as cosmetic dye, as dye in the textile industry, etc.

Regarding the natural compounds of animal origin, it is necessary to mention the fact that they have been less studied in relation to the natural compounds of vegetable origin. Although known for a long time, even since antiquity, investigations into the chemical composition have encountered, and still do, difficulties. These are mainly due to the fact that defining the chemical structure is often difficult.

The first difficulty in analytical/bioanalytical investigation lies in the fact that the biogenesis of these substances often takes place during metabolic processes (especially in the catabolism stage). In such cases, molecular fragments with a biologically active specificity are released. One such example is provided by so-called "encrypted peptides" that are released during food metabolism (see "encrypted peptides" in milk and dairy products).

Another difficulty originates from the fact that the investigation is laborious and requires the use of instrumental analytical methods (based on physicochemical analysis). Although these analyses provide some speed, impediments occur due to the low stability of metabolites. Therefore, certain biological effects can be observed, but the chemical structure of the compounds that induce these effects cannot be rigorously determined. Investigating the composition of biologically active substances also presents difficulties due to their presence in mixtures.

In such situations, the biological activity of the mixture can be ascertained. A concrete example is provided by certain bee products for which the biological effects are well known, but information on the structure has been long awaited, e.g., propolis, apitoxin, etc. Persevering in analytical investigations, with in vivo experiments and conclusive clinical observations, notable information was obtained.

Gradual recognition of the role of biologically active substances in food and their contribution to improving health has been made possible by the scientific information gained. To this end, it is recalled that the effects of natural foods and functional foods containing zoochemicals have been studied, following their physiologically active role (Thomas and Earl, 1994; Prates and Mateus, 2002).

The study of the physiologically active and pharmacologically active character of zoochemicals, in the sense given by Prates and Mateus (2002), also involves the investigation in four thematic areas: (i) clinical trials; (ii) animal studies; (iii) in vitro experimental studies and (iv) epidemiological studies.

The gradual discovery of new aspects of biological activity (from beneficial effects to adverse effects) has continuously interested new groups of researchers and specialized laboratories. For this purpose, the aim was to investigate the chemical structure and the structure-activity relationship of the compounds obtained by extraction in the form of zoochemicals. Obviously, in some cases, the synthesis of products with similar biological activity has been attempted in the laboratory.

REFERENCES

Ahangari Z., Naseri M., Vatandoost F. - Propolis: chemical composition and its applications in endodontics, *Iran. Endodontic J.*, 2018, 13(3), 285–292.

Alasalvar C., Miyashita K., Shahidi F., Wanasundara U. - *Handbook of Seafood Quality, Safety and Health Applications*, John Wiley & Sons, West Sussex, 2011.

Alhasaniah A.H. - L-carnitine: nutrition, pathology, and health benefits, *Saudi J. Biol. Sci.*, 2023, 30, 103555 https://doi.org/10.1016/j.sjbs.2022.103555.

Alvarez-Suarez J.M., Gasparrini M., Forbes-Hernández T.Y., Mazzoni L., Giampieri F. - The composition and biological activity of honey: a focus on Manuka honey. *Foods*, 2014, 3, 420–432.

Anton M., Nau F., Nys Y. - Bioactive egg components and their potential use, pp. 237–244, in *Proceedings of the XIth European Symposium on the Quality of Eggs and Egg Products*, Doorwerth, The Netherlands, 23–26 May, 2005.

Armstrong F.B. - *Biochemistry*, 3rd edition, Oxford University Press, New York-Oxford, 1989.

Aydin R. - Conjugated linoleic acid: chemical structure, sources and biological properties, *Turk. J. Vet. Anim. Sci.*, 2005, 29(2), 189–195.

Bane V., Lehane M., Dikshit M., O'Riordan A., Furey A. - Tetrodotoxin: chemistry, toxicity, source, distribution and detection, *Toxins*, 2014, 6(2), 693–755.

Banu C., Preda N., Vasu S.S. - *Food Products and Their Innocuity* (in Romanian), Editura Tehnică, Bucureşti, 1982.

Barker G.A., Green S., Askew C.D., Green A.A., Walker P.J. - Effect of propionyl-L-carnitine on exercise performance in peripheral arterial disease, *Med. Sci. Sports Exerc.*, 2001, 33(9), 1415–1422.

Bogdanov S. - Pollen: production, nutrition and health: a review, *Bee Product Sci.*, 2014, https://www.bee-hexagon.net.

Burdge G.C., Jones A.E., Wootton S.A. - Eicosapentaenoic and docosapentaenoic acids are the principal products of α-linolenic acid metabolism in young men, *Br. J. Nutr.*, 2002, 88 (4), 355–363.

Bustamante J., Lodge J.K., Marcocci L., Tritschler H.J., Packer L. - Alpha-lipoic acid in liver metabolism and disease. *Free Radical Biol. Med.*, 1998, 24, 1023–1039.

Cannella C., Giusti A.M. - Conjugated linoleic acid - a natural anticarcinogenic substance from animal food, *Italian J. Food Sci.*, 2000, 12, 123–127.

Canty D.J., Zeisel S.H. - Lecithin and choline in human health and disease, *Nutr. Rev.*, 1994, 52, 327–339.

Cardinault N., Cayeux M.-O., Percie du Sert P. - La propolis: origine, composition et propriétés, *Phytothérapie*, 2012, 10, 298–304.

Chang F.C., Spriggs D.L., Benton B.J., Keller S.A., Capacio B. R. - 4-Aminopyridine reverses saxitoxin (STX) - and tetrodotoxin (TTX) -induced cardiorespiratory depression in chronically instrumented guinea pigs, *Fundam. Appl. Toxicol.*, 1997, 38(1), 75–88.

Clare D.A., Swaisgood H.E. - Bioactive milk peptides: a prospectus. *J. Dairy Sci.*, 2000, 83, 1187–1195.

Clark R.F., Williams S.R., Nordt S.P., Manoguerra A.S. - A review of selected seafood poisoning, *Undersea Hyperbaric Med.*, 1999, 26(3), 175–184.

Crane, E. - Bees, honey and pollen as indicators of metals in the environment. *Bee World*, 1984, 65(1), 47–49.

Daria V., Abdallah S., Delgado Z.L. - Contribution à l'étude éthnozoopharmacologiques des Îles Canaries, pp. 205-206, în Médicaments et aliments. Approche ethnopharmacologique, Deuxième colloque européen d'Éthnopharmacologie/Onzième conférence internationale d'Ethnomédicine, Heidelberg 24–26 mars, 1993, Publ. Société Française d'Éthnopharmacologie, Paris, 1996.

Devlin M.T. (Ed.) - *Textbook of Biochemistry with Clinical Correlations*, 7th edition, Wiley-Liss, Hoboken, New Jersey, 2010.

Ensminger A.H., Ensminger M.E., Konlande J.E., Robson J.R.K. - *The Concise Encyclopedia of Foods and Nutrition*, 2nd edition, CRC Press, Boca Raton, 1995.

Feuer G., de la Iglesia F.A. - *Molecular Biochemistry of Human Disease, Vol. I*, CRC Press, Boca Raton, 1985.

Fleming A.- On a remarkable bacteriolytic element found in tissues and secretions. Proc. R. Soc. B., 1922, 93, 306–317.

Fratini F., Cilia G., Turchi B., Felicioli A. - Beeswax: a minireview of its antimicrobial activity and its application in medicine, *Asian Pac. J. Trop. Med.*, 2016, 9(9), 839–843.

Frei B., Kim M.C., Ames B.N. - Ubiquinol-10 is an effective lipid-soluble antioxidant at physiological concentrations, *Proc. Natl. Acad. Sci.*, 1990, 87, 4879–4883.

Galloway S.D.R., Broad E.M. - Oral L-Carnitine supplementation and exercise metabolism. *Mon. Bull. Chem.*, 2005, 136, 1391–1410.

Gârban Z. - *Human Nutrition, Vol. I, Basic Issues* (in Romanian), Editura Didactică şi Pedagogică R.A., Bucureşti, 2000.

Gârban Z. - *Biochemistry: Comprehensive Treatise, Vol. II. Biochemical Effectors* (in Romanian), 5th edition, Publishing House of the Romanian Academy, Bucureşti, 2018a.

Gârban Z. - *Quo Vadis Food Xenobiochemistry*, 3rd edition, Publishing House of the Romanian Academy, Bucharest, 2018b.

Goldberg I. (Ed.) - *Functional Foods - Designer Foods, Pharmafoods, Nutraceuticals*, Chapman & Hall, New York, 1994.

Golinelli P.L, Del Aguila E.M., Flosi Paschoalin V.M., Silva J.T., Conte -Junior C.A. - Functional aspect of colostrum and whey proteins in human milk, *J. Hum. Nutr. Food Sci.*, 2014, 2(3), 1035–1044.

Gomes M.B., Negrato C.A. - Alpha-lipoic acid as a pleiotropic compound with potential therapeutic use in diabetes and other chronic diseases. *Diabetol. Metab. Syndr.*, 2014, 6 80. https://doi.org/10.1186/1758-5996-6-80.

Gregoire P.E. - *Biochimie Pathologique*, Presses Academiques Europeennes, Bruxelles-Libraire Maloine SA, Paris, 1971.

Guyton A.C., Hall J.E. - *Textbook of Medical Physiology*, 11th edition, Elsevier-Saunders, Philadelphia, 2006.

Hepburn H.R. - *Honeybee and Wax*, Springer Verlag, Berlin-Heidelberg, 1986.

Horn H., Lüllmann C. - *Das Grosse honigbuch: Entstehung, Gewinnung, Gesundheit und Vermarktung*, 3. Auflage, Kosmos Verlag GmbH and Co, Stuttgart, 2006.

Huopalahti R., López-Fandiño R., Anton M., Schade R. (Eds.) - *Bioactive Egg Compounds*, Springer-Verlag, Berlin-Heidelberg, 2007.

Kanwar J.R., Kanwar R.K., Sun X., Punj V., Matta H., Somasundaram M., Parratt A., Puri M., Sehgal R. - Molecular and biotechnological advances in milkproteins in relation to human health, *Curr. Protein Peptide Sci.*, 2009, 10(4), 308–338.

Kathuria D., Gautam S., Sharma S., Sharma K.D. - Animal based bioactives for health and wellness. *Food Nutr. J.*, 2019, 4, 203. https://doi.org/10.29011/2575-7091.100103.

Khalil M.L. - Biological activity of bee propolis in health and disease, *Asian Pac. J. Cancer Prev.*, 2006, 7, 22–31.

Komosinska-Vassev K., Olczyk P., Kafmierczak J., Mencner L., Olczyk K. - Bee pollen: chemical composition and therapeutic application, *Evidence Based Complementary and Altern. Med.*, 2015, https://doi.org/10.1155/2015/297425.

Kunugi H., Ali A.M. - Royal Jelly and its components promote healthy aging and longevity: from animal models to humans, *Int. J. Mol. Sci.*, 2019, 20, 4662; https://doi.org/10.3390/ijms20194662.

Lee B.J., Huang Y.C., Chen S.J., Lin P.T. - Effects of coenzyme Q10 supplementation on inflammatory markers (high-sensitivity C-reactive protein, interleukin-6, and homocysteine) in patients with coronary artery disease. *Nutrition*, 2012, 28, 767–772.

Leszek J., Inglot D.A., Janusz M., Liszowski J., Krukowska K. - Colostrinin: a Proline-Rich Polypeptide (PRP) complex isolated from ovine colostrum for treatment of Alzheimer's disease. A double-blind, placebo-controlled study, *Arch. Immunol. Ther. Exp.*, 1999, 47, 377–385.

Lushchak V.I. - Free radicals, reactive oxygen species, oxidative stress and its classification, *Chem. Biol. Interact.*, 2014, 224, 164–175.

Macrae R., Robinson R.K., Sadler M.J. - *Encyclopedia of Food Science and Nutrition*, Academic Press, London, 1992.

Madureira A.R., Pereira C.I., Gomes A.M.P., Pintado M.E., Xavier Malcata F.- Bovine whey proteins - Overview on their main biological properties. *Food Res. Int.*, 2007, 40(10), 1197–211. https://doi.org/10.1016/j.foodres.2007.07.005.

Maher J. - Zoonutrient supplementation via functional food formulas, *Dyn. Chiropractic*, 2007, 25(22), 1–4.

Marcucci M.C. - Propolis: chemical composition, biological properties and therapeutic activity. *Apidologie*, 1995, 26, 83–99.

Martelli A., Testai L., Colletti A., Cicero A.F.G. - Coenzyme Q10: clinical applications in cardiovascular diseases, *Antioxidants*, 2020, 9, 341 https://doi.org/10.3390/antiox9040341.

Meier J., White J. - *Clinical Toxicology of Animal Venoms and Poisons*, CRC Press, Inc., Boca Raton, 1995.

Mijanur Rahman M., Gan S. H., Khalil M.I. - Neurological effects of honey: current and future prospects, *Evidence Based Complementary Altern. Med.*, 2014, https://doi.org/10.1155/2014/958721.

Motohashi N., Gallagher R., Anuradha V., Gollapudi R. - Coenzyme Q10 (Ubiquinone): It's implication in improving the life style of the elderly, *Med. Clin. Rev.*, 2017, 3, 10.

Möller P.N., Scholz-Ahrens E.K., Schrezenmeir J. - Bioactive peptides and proteins from foods: indication for health effects, *Eur. J. Nutr.*, 2008, 47(4), 171–182.

Orlando G., Rusconi C. - Oral L-carnitine in the treatment of chronic cardiac ischemia in elderly patients, *Clin. Trial J.*, 1986, 23, 338–344.

Orsi N. - The antimicrobial activity of lactoferrin: current status and perspectives, *Biometals*, 2004, 17, 189–196.

Overvad K., Diamant B., Holmi L., Holmer G., Mortensen S.A., Stender S. - Coenzyme Q10 in health and disease. *Eur. J. Clin. Nutr.*, 1999, 53, 764–770.

Packer L., Witt E.H., Tritschler H.J. - Alpha-Lipoic acid as a biological antioxidant, *Free Rad. Biol. Med.*, 1995, 19, 227–250.

Pallotti F., Bergamini C., Lamperti C., Fato R. - The roles of coenzyme Q in disease: direct and indirect involvement in cellular functions, *Int. J. Mol. Sci.*, 2022, 23, 128 https://doi.org/10.3390/ijms23010128.

Pasupuleti V.R, Sammugam L., Ramesh N., Gan S.H. - Honey, propolis, and royal jelly: a comprehensive review of their biological actions and health benefits, *Oxid. Med. Cell. Longevity*, 2017, https://doi.org/10.1155/2017/1259510.

Paulson D.J., Shug A.L. - Tissue specific depletion of L-carnitine in rat heart and skeletal muscle by D-carnitine, *Life Sci.*, 1981, 28, 2931–2938.

Peterson J., The editors of Seafood Business - *Seafood Handbook: The Comprehensive Guide to Sourcing, Buying and Preparation*, 2nd edition, John Wiley & Sons Inc., New York-Hoboken-New Jersey, 2009.

Prates J.A.M., Mateus M.R.P.C. - Functional foods from animal sources and their physiologically-active components, *Revue Méd. Vét.*, 2002, 153, 3, 155–160.

Pucca B.M., Cerni A.F., Oliveira S.I., Jenkins P.T., Argemí L., Sørensen V.C., Ahmadi S., Barbosa E.J., Laustsen H.A. - Bee updated: current knowledge on bee venom and bee envenoming therapy, *Front. Immunol.*, 2019, 10 https://doi.org/10.3389/fimmu.2019.02090.

Reed L.J., Gunsalus I.C., Schnakenberg G.H.F., Soper Q.F., Harold E., Boaz E.H., Kern S.F., Thomas V, Parke T.V. - Isolation, characterization and structure of α-lipoic acid, *J. Am. Chem. Soc.*, 1953, 75(6), 1267–1270.

Réhault-Godbert S., Guyot N., Nys Y. - The golden egg: nutritional value, bioactivities, and emerging benefits for human health, *Nutrients*, 2019, 11, 684 https://doi.org/10.3390/nu11030684.

Rizzon P., Biasco G., Dibase M. - High doses of L-carnitine in acute myocardial infarction: metabolic and antiarrhythmic effects, *Eur. Heart J.*, 1989, 10, 502–508.

Sachan D.S., Cha Y.S. - Acetyl-carnitine inhibits alcohol dehydrogenase, *Biochem. Biophys. Res. Comm.*, 1994, 203, 1496–1501.

Samarghandian S., Farkhondeh T., Samini F. - Honey and health: a review of recent clinical research, *Pharm. Res.*, 2017, 9, 121–127 https://doi.org/10.4103/0974-8490.204647.

Smith E.A., Grant F., Ferguson C.M.J., Gallacher S. - Biotransformations of paralytic shellfish toxins by bacteria isolated from bivalve molluscs, *Appl. Environ. Microbiol.*, 2001, 67 (5), 2345–2353.

Strecker A.-Über eizige neue Bestandtheile der Schweinegalle. *Ann Chem Pharm*. 1862, 183, 964–965.

Šturm L., Poklar-Ulrih N. - Advances in the propolis chemical composition between 2013–2018: a review, *eFood*, 2019, 1(1), 24–37.

Şerban M., Roşoiu N. - *Bioactive Substances in Marine Organisms* (in Romanian), Publishing House of the Romanian Academy, Bucureşti, 1992.

Thomas P.R., Earl R. (Eds.) - *Opportunities in the Nutrition and Food Sciences - Research Challengers and the Next Generation of Investigators*, National Academy Press, Washington, DC, 1994.

Ulvi H., Aygul R., Demir R. - Effect of L-carnitine on diabetic neuropathy and ventricular dispersion in patients with diabetes mellitus. *Turkish J. Med. Sci.*, 2010, 40, 169–175.

Vazhappilly R., Chen F. - Eicosapentaenoic acid and docosahexaenoic acid production potential of microalgae and their heterotrophic growth, *J. Am. Oil Chem. Soc.*, 1998, 75(3), 393–397.

Viuda-Martos M., Ruiz-Navajas Y., Fernández-López J., Pérez-Álvarez J.A. - Functional properties of honey, propolis, and royal jelly, *J. Food Sci.*, 2008, 73(9), R117–R124.

Ward R.E., Bruce J. - Zoonutrients and health, *Food Technol.*, 2003, 57, 33–39.

Wehbe R., Frangieh J., Rima M., El Obeid D., Sabatier J.-M., Fajloun Z. - Bee venom: overview of main compounds and citoactivities for therapeutic interests, *Molecules*, 2019, 24, 2997; https://doi.org/10.3390/molecules2416299.

West J.B. (Ed.) - *Best and Taylor's Physiological Basis of Medical Practice*, 12th edition, Lippincott Williams and Wilkins, Philadelphia, 1991.

Zdrojewicz Z., Herman M., Starostecka E. - Hen's egg as a source of valuable biologically active substances, *Postepy Higieny i Medycyny Doswiadczalnej (Adv. Hygiene Exper. Med.)*, 2016, 70, 751–759.

Zhu X., Zeisel S.H. - Choline and phosphatidylcholine, pp. 392–396, in *Encyclopedia of Human Nutrition*, 2nd edition (Caballero B., Ed.), Elsevier, Amsterdam, 2005.

Zullkiflee N., Taha H., Usman A. - Propolis: its role and efficacy in human health and diseases, *Molecules*, 2022, 27, 6120 https://doi.org/10.3390/molecules27186120.

Yamamura Y., Folkers K., Ito Y. (Eds.) - *Biomedical and Clinical Aspects of Coenzyme Q, Vol. 2*, Elsevier/North-Holland Biomedical Press, Amsterdam, 1980.

4 Prebiotics

4.1 INTRODUCTION

A number of compounds from the class of carbohydrates with an oligosaccharide structure are known as prebiotics. These compounds stimulate in a beneficial way the intestinal flora (primarily from the colon) and at the same time suppress the development of undesirable bacteria such as proteolytic bacteria (e.g., *Clostridium* sp., *Pseudomonas*). Prebiotics participate in biochemical interactions specific to the metabolic processes at the level of the colon, involving the activity of microorganism, influencing thus the physiological activity at the level of the digestive tract.

The concept *prebiotics* was introduced in 1995 by Gibson and Roberfroid. According to these authors, prebiotic was considered a "non-digestible food ingredients that beneficially affect the *host organism* by selectively stimulating the growth and/or activity of a bacterium or a small number of bacteria in the colon and thus improving the host health" (Gibson and Roberfroid, 2008).

In general, the prebiotic compounds are considered to be food ingredients that are not digested in the stomach and small intestine, which is why they are called *non-digestible*. Once in the large intestine, these substances, with a predominantly oligosaccharide composition, are subjected to digestion under the action of "specific bacteria" referred to generically as "probiotics".

The presence of prebiotics facilitates digestion in the large intestine and provides the potential substrate for enzymes in the class of hydrolases that are also involved in fermentation processes produced by intestinal bacteria (Ensminger et al., 1995; Guyton and Hall, 2006; Yannai, 2004; Roberfroid, 2007).

Prebiotics are a selective substrate for a small number of bacterial strains beneficial to the body, which are resident in the colon. These microorganisms constitute the so-called *microbiota*. Mainly, as mentioned, prebiotics are chemical compounds with oligosaccharide structure being of nutritional and pharmaceutical interest. Thus, due to their physicochemical properties and physiological role, they are used in the food and pharmaceutical industry – being ingredients, stabilizers, excipients or precursors in the synthesis of chemical derivatives for therapeutic products (Oku, 1999; Charalampopoulos and Rastall, 2009; Gârban, 2014; Wan et al., 2020).

4.2 CHEMICAL COMPOSITION AND NATURAL DISTRIBUTION

Prebiotics are chemical compounds belonging to the class of carbohydrates. Carbohydrate compounds are also known as "oses", "saccharides" or "sugars".

Prebiotics present, especially in "dietary fiber", belong to various groups of oligosaccharides, such as oligofructose and inulin (Gibson et al., 1995), trans-galactosylated disaccharides (Ito et al., 1993) and others. Mainly, prebiotics are represented by different carbohydrate compounds dominated by oligosaccharides. In this regard,

DOI: 10.1201/9781032702520-4

there are exemplified: fructo-oligosaccharides; lactosucrose; polydextrose, etc. Such compounds can result in heat processing of food (Schrezenmeir and de Vrese, 2001; Panesar et al., 2013).

In biochemistry, the chains of oligosaccharides, according to classical biochemistry, are considered as compounds containing 2–8 monosaccharides. In nutrition, with regard to prebiotics, reference is sometimes made to "oligosaccharides" that do not comply with this "canon".

It is an option for some nutritionists who seek to simplify the explanations regarding prebiotics: chemical structure, biological-activity, structure-activity relationship, etc.

Obviously, in this approach are pursued aspects related to metabolism and, evidently, to the biodegradative processes in which they participate.

To establish the chemical composition of prebiotics, it is currently recommended to use physico-chemical analytical methods (usually known as "instrumental analysis").

Regarding the natural distribution of prebiotics, it is to be mentioned that they can be found, in low concentrations, in terrestrial plant-derived foods, such as cereal grains (barley, wheat, oats and rye), vegetables (sugar beet, garlic, leeks, chicory, onion, artichoke, soybean, etc.), fruits (asparagus, banana, tomato, etc.), blue agave plant, etc. Also, prebiotics are present (in small quantity) in aquatic plant-derived foods, e.g. seaweeds and microalgae. Other sources of prebiotics are represented by terrestrial animal-derived foods, like honey, cow's milk, a.o. as well as from aquatic animals, e.g. obtainment of chitin and its derivatives from zooplankton cuticles (Sabater-Molina et al., 2009; Varzakas et al., 2018; Cherry et al., 2019; Lopez-Santamarina et al., 2020; Kaur et al., 2021). The existence of prebiotics that do not belong to oligosaccharides should also be mentioned. Among these are the flavanol derivatives in cocoa (Tzounis et al., 2011), which have been found to stimulate the lactic acid bacteria.

Because of their low concentration in foods, some prebiotics are industrially produced by using lactose, sucrose, and starch as raw materials (Al-Sheraji et al., 2013).

4.3   CLASSIFICATION AND NOMENCLATURE

4.3.1   Classification

For the classification of prebiotics used as food ingredients, certain distinct criteria were accepted: (i) *chemical criteria* – which take into account the specifics of the composition; (ii) *physiological and microbiological criteria* – based on the evaluation of metabolic aspects in relation to the microbiota.

4.3.1.1   Chemical Criteria

The evaluation based on these criteria implies the knowledge of the composition of the oligosaccharides and obviously of the constituent monosaccharides. According to these criteria, the following are distinguished:

- fructo-oligosaccharides;
- galacto-oligosaccharides;
- isomalto-oligosaccharides;
- xylo-oligosaccharides, etc.

These classification criteria are used in this chapter.

4.3.1.2 Physiological and Microbiological Criteria
According to Roberfroid (2007), the defining criteria for this group of compounds are:

- resistance to physiological processes occurring in the gastrointestinal tract;
- fermentation under the action of food microbiota;
- selective stimulation of the growth and/or activity of a limited number of beneficial bacteria.

4.3.2 NOMENCLATURE

In the case of prebiotics, to understand their specificity and, subsequently, the interactive particularities, it is important to establish the nomenclature in relation to the composition. In this sense, the constituent monosaccharides are considered as a priority: D-fructofuranose; β-D-galactopyranose; β-D-glucopyranose; α-D-xylopyranose.

In a broader context – characteristic of prebiotics – they are also called glycane-specific oligosaccharides, such as hexosans and pentosans. Hexosans include: glucans, fructans, mannans and galactans, and pentosans include arabans and xylans.

The more rigorous names of prebiotic compounds refer, as appropriate, to the monomer and oligomer or a name that refers to different constituents, e.g., lactulose containing galactose and fructose; xylo-oligosaccharose containing xylose, etc.

4.4 PREBIOTICS VERSUS PROBIOTICS

An emphasis on the biochemical and physiological role of prebiotics becomes possible in the case of the related approach of the prebiotic-probiotic relationship. For this purpose, some general data are presented.

4.4.1 PREBIOTICS

In terms of chemical composition, prebiotics are oligosaccharide compounds. They are found in various foods of plant origin, e.g., various grains, lentils, bananas, tomatoes, spinach, garlic, chicory, etc. A presentation of the main groups of prebiotics is given in Table 4.1.

Prebiotics are not broken down by enzymes in the gastrointestinal tract (from the oral cavity to the large intestine). Thus, they reach the large intestine where, under the action of microorganisms (bacteria), they are subjected to fermentation processes.

TABLE 4.1
The Main Prebiotics

No.	Specification
1	Fructo-oligosaccharides
2	Galacto-oligosaccharides
3	Isomalto-oligosaccharides
4	Xylo-oligosaccharides
5	Lactosucrose
6	Lactulose
7	Raffinose

The presence of dietary fibers and their contribution to the intestinal transit and the development of a flora suitable for digestion in the colon led to certain ingredients with prebiotic attributes to be added to foods.

4.4.2 PROBIOTICS

Probiotics - are microorganisms characterized by beneficial metabolic effects that are present in human and animal foods, intervening in the modification of the *colon microbiota*.

The term *probiotic* (lat. *pro*-for; gr. *bios*-life) was first used by Lilly and Stillwell (1965) to characterize "substances produced by one microorganism to stimulate the growth of another". It should be noted that this term contrasts with the term *anti-biotic* (lat. *anti*-against; gr. *bios*-life). Regarding the antithetic specificity of these terms, it is understandable why the substances/products integrated into these classes of compounds can be included in the field of biologically active substances.

Information about probiotics is based on studies undertaken in the early twentieth century by Metchnikoff and developed over time with advances in food and nutrition science. Probiotics have been discussed in the last two decades in an extensive context of beneficial effects for animals and humans (Parker, 1974; Fuller, 1989; Schrezenmeir and de Vrese, 2001).

The problem of probiotics was known in Roman antiquity. The historian Plinius Secundus mentioned the recommendation, since 76 BC, regarding the consumption of fermented milk in gastroenteritis (cited by Bottazzi, 1983). Over time, research in the field of microbiology, nutrition, and food science (a term used today for fields of food study) has attested to the beneficial role of probiotics in human nutrition. It is now accepted that probiotics are "a preparation or product that contains various live microorganisms (in sufficient numbers) to colonize the host organism providing beneficial effects on it" (Wood, 1992).

In the general sense, probiotics are considered "microorganisms that, when administered in adequate amounts, provide health benefits" (Gibson and Roberfroid, 2008). It is a process of colonization of the "indigenous microflora", which populates the gastrointestinal tract represented by a complex mixture of bacteria present under normal conditions in the intestine. Probiotics have been studied and recommendations have been developed by WHO and FAO expert groups.

TABLE 4.2
The Main Probiotics

Genus	Species
Lactobacillus	L. acidophilus
	L. delbrueckii
	(ssp. bulgaricus)
	L. brevis
	L. casei
	L. fermentum
	L. lactis
	L. rhamnosus
Bifidobacterium	B. adolescentis
	B. bifidum
	B. breve
	B. thermophilum
Streptococcus	S. thermophilus
	S. salivarius
	(ssp. thermophilus)
Saccharomyces	S. cerevisiae
	S. boulardii
Bacillus	B. cereus
	B. coagulans

Characteristics of probiotics – Probiotics allowed for use must be well taxonomically characterized (genus, species). Table 4.2 shows the main probiotics, mentioning their genus and species. The effect of certain combinations of microorganisms must also be known on probiotics (Fuller, 1989). In particular, investigations into the determination of antibiotic resistance are recommended; effects on metabolic processes (e.g., lactic acid formation; bile salt conjugation action); side effects, epidemiological aspects, etc.

Experimental animal studies and human observations have been performed for probiotics. Thus, beneficial effects have been highlighted in the case of allergies, diarrhea, lactose intolerance, hypercholesterolemia, *Helicobacter pylori* infection, inflammation, ulcerative colitis, eczema, etc.

The term microbiota refers to the species of microorganisms, designated in the older specialized treatises as *microflora*. So, the microbiota is represented by microorganisms that are sustainably adapted inside the living organism.

In current terminology, there is also the term *microbiome* which refers to the genome of living microorganisms in the human body, animal or plant (except for pathological conditions). At present, knowledge about the evolution of the microbiome is still deficient. The microbiome is the expression of the ecological conditions signifying the environment (temperature, pH, the presence of hormones, the absence of light, the type of mucous tissue in the organs, etc.) in which the microbial community in the intestine of the host organism is located.

These environmental conditions may change over time. In general, a co-evolution of the microbiota with the host organism is discussed.

The concept of microbiota brings to mind the notion of "microbial community", in relation to microbial biodiversity and microbial ecology. There is, in fact, a lasting interaction between the various microorganisms in the intestine and between them and the host organism. In this context, it can distinguish sustainable functional interactions that progress from consensualism to symbiosis.

It is also mentioned that there are non-dairy products, such as sauerkraut, fermented cereals or other plant-based foods that contain viable probiotic-specific microorganisms (e.g., *Lactobacillus plantarum*).

Thus, an extension of the definition of probiotics that were limited to dairy products only (Schrezenmeir and deVrese, 2001) is possible. For information, it is mentioned that there is also a *plant microbiota*. A characteristic example is the rhizosphere microbiota which is known in plant physiology and biochemistry. This is specific to vascular plants.

4.4.3 Synbiotics

Synbiotics are defined as products of food interest that contain both prebiotics and probiotics. Prefixation by "syn" is justified by the fact that the two ingredients act synergistically (Sekhon and Jairath, 2012). It is known that probiotics are more active in the small intestine, and prebiotics actually intervene only in the large intestine.

A careful, verified combination of them becomes beneficial. In this context, the nature of the ingredients and their amount is important for their suitability in the "probiotic/prebiotic combination".

The term prebiotics – characterizes a group of chemical compounds and concerns biochemistry, and the term probiotics defines a category of microorganisms and is of interest to microbiology.

Their association in the digestive environment facilitates co-participation in physiological processes. The terminological relation reveals the existence of the prefixes "pre" (lat. before) and respectively "pro" (lat. for), which (simplifying notions) means lexical attributes.

There are currently conclusive examples of such experimentally confirmed dual prebiotic-probiotic combinations. Table 4.3 shows the main associations known for the efficiency of the physiological and biochemical effects produced in the colon.

TABLE 4.3
Prebiotic - Probiotic Combinations

Prebiotic Composition	Bacterial Genus (Species)
Fructo-oligosaccharides (inulin)	*Bifidobacterium* sp.
	Lactobacillus rhamnosus
Galacto-oligosaccharides	*Bifidobacterium* sp.
Xylo-oligosaccharides	*Bifidobacterium lactis*

Synbiotics are now used to process food and even to obtain "food supplements", which are considered associations of food ingredients with a beneficial effect on digestion.

In obtaining these associations, in fact, the synergism of the biochemical interactions specific to the metabolic processes is taken into account. Prebiotics are considered, as shown, as non-digestible food ingredients. These may be natural, extracted and/or synthetic chemical compounds. In terms of chemical composition, the vast majority are oligosaccharides.

4.5   PREBIOTICS AND NUTRITION

From a nutritional and biochemical point of view, prebiotics are not digested in the stomach and small intestine. Their metabolism occurs only in the large intestine and, more precisely, in the colon. At this level, the metabolic processes that affect prebiotics, i.e. non-digestible oligosaccharides, are triggered by microorganisms. These microorganisms belong to the group of bacteria beneficial to digestion. Understanding the peculiarities of prebiotics digestion, absorption and metabolism can be facilitated by a brief knowledge of anatomical data and physiological aspects related to the digestion and absorption of nutrients (see Chapter 1).

The absorption of nutrients is achieved mainly in the intestine. In absorption, from the point of view of physiology and biochemistry, there are two phases: *intraluminal* which affects the transit of nutrients through the intestinal wall and *intracellular* which concerns the metabolism of nutrients inside the cell.

It is mentioned that the metabolization of prebiotics has similarities in the body of humans and mammals. Particular aspects of digestion and absorption are highlighted in the large intestine where probiotics participate in physiological processes and their specific biochemical interactions (for details, see Fuller, 1989; Gârban, 2000; Gibson and Roberfroid, 2008; Sekhon and Jairath, 2012).

4.6   GROUPS OF PREBIOTIC COMPOUNDS

Various groups of prebiotics include compounds belonging to the class of carbohydrates (oses), more precisely to the subclass of polysaccharides. This subclass includes: glycans - polysaccharides without amino carbohydrates and glycosaminoglycans (GAG) - polysaccharides with amino carbohydrates (Gârban, 2014). The more detailed classification of glycans is shown in Figure 4.1. It comprises, as a whole, homo- and heteroglycans.

Prebiotics include only some of the mentioned compounds, namely: fructans – oligosaccharides, in which the carbohydrate is fructose (D-Fru_f); galactans – oligosaccharides with identical monosaccharide monomer represented by galactose (b-D-Gal_p); isomalto-oligo saccharide – with the specific monomer α-D-glucopyranose (1-D-Glc_p), etc.

Glycans

- Homoglycans (Homopolysaccharides)
  - Hexosans
    - glucans
    - fructans
    - mannans
    - galactans
  - Pentosans
    - arabans
    - xylans

- Heteroglycans (Heteropolysaccharides)
  - consisting of different monosaccharides
  - consisting of uronic acids (polyuronides)

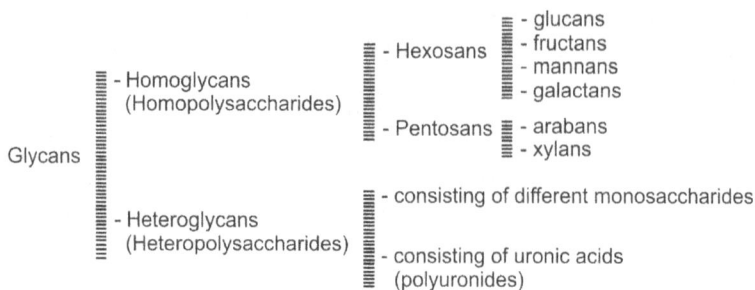

FIGURE 4.1   Classification of glycans.

4.6.1   Specific Structure and Biological Activity

4.6.1.1   Monosaccharide Precursor Compounds

Investigations into prebiotics have led to the identification of new compounds whose structure differs not only in the nature of carbohydrate monomers (monoglycerides) but also in the number of *mers (repeating units),* i.e. the degree of polymerization - denoted by n.

The composition of prebiotic-specific oligosaccharide includes various monosaccharides, comprising: D-fructofuranose; β-D-galacto-pyranose; β-D-glucopyranose, etc. These compounds can be considered as precursors of the oligosaccharide in the prebiotic constitution. Figure 4.2 presents the chemical formulas of some monosaccharide compounds present in prebiotics (Gârban, 2018).

Monosaccharides from the constitution of oligosaccharides of prebiotic have the α or β anomeric types. The cyclization forms of monosaccharide can be: furanosic (f) or pyranosic (p). In the structural formulas of the prebiotics presented in this chapter, only the positions of the hydroxyl groups (OH) marked by vertical lines (existing custom in biochemistry) were mentioned for monosaccharides. Oligosaccharide prebiotics include: fructo-oligosaccharides (inulins), galacto-oligosaccharides, iso-malto-oligosaccharides, xylo-oligosaccharides, etc.

Also included in the group of prebiotics are sometimes atypical disaccharide and trisaccharide, which are specific to non-digestible food ingredients. Such compounds are lactulose (disaccharide), lactosucrose (trisaccharide) and the like. These compounds exist in nature and are continuously formed. The degree of polymerization (n), in most cases, is difficult to specify. For these reasons, the presentation of rigorous chemical structures is more difficult.

4.6.1.2   Fructo-Oligosaccharides

Fructo-oligosaccharides are natural hexosan-type polysaccharides resulting from polycondensation of D-fructofuranose (D-Fru$_f$). In nature, they are synthesized by many plants at the level of the roots (rhizomes). These compounds belong to a subclass of substances called *fructans* or *fructosans* (Apolinário et al., 2014).

In the case of fructans, there are in fact two different types with a structure:

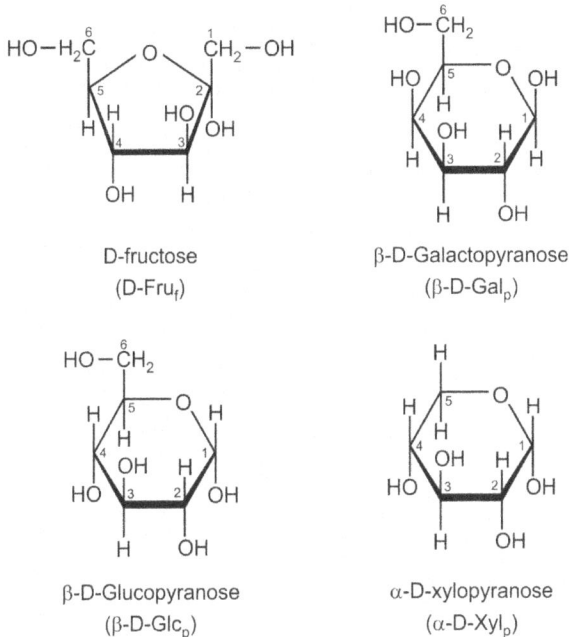

FIGURE 4.2   Carbohydrate monomers present in the composition of prebiotics.

a. of *inulin type* - with β (2--> 1) bonds which are synthesized in plant cells;
b. of *levan type* - with β (2--> 6) bonds which are formed in bacterial cells.

Inulin was discovered in 1804 by the German researcher Valentin Rose (cited by Flückiger and Tschirch, 1885). It was extracted in a hot aqueous medium and considered a "specific substance" existing in the roots of the plant *Inula helenium* (sea grass). In the (older) biochemistry treatises it is also found under the names *of heletin, alatin*, etc.

The classic treatises on Biochemistry/Organic Chemistry describe poly- or oligosaccharide fructose and have the general name of *inulin*. The degree of polymerization (n) is different. The n values for inulin presented in the classical literature are 2–60.

For oligosaccharides the value n is of the order of 2–8 monomers. When discussing prebiotics from the fructose group, the great variation in the degree of polymerization is sometimes ignored. In general, in the case of these compounds, when treating the problem of prebiotics, the expression "inulin" is used to indicate that they differ in the degree of polymerization and (possibly) in the existence of another monosaccharide residue (e.g., glucose).

Chemical structure – Inulins are polysaccharides consisting mainly of D-fructo-furanose (D-Fru$_f$) units. Fructose groups are in most cases related to positions β (2--> 1). Figure 4.3 shows the molecular structure of two fructo- oligosaccharides detected in nature: *polyfructoside (I) inulin* – formed by polycondensation of

or

β - D - Fru$_f$ -(1→2) [β - D - Fru$_f$]$_{n-2}$ (1→2) - β - D - Fru$_f$

(I)

or

α - D - Glc$_p$ -(1→2)- β - D - Fru$_f$ - [β - D - Fru$_f$]$_{n-2}$ (1→2)- β - D - Fru$_f$

(II)

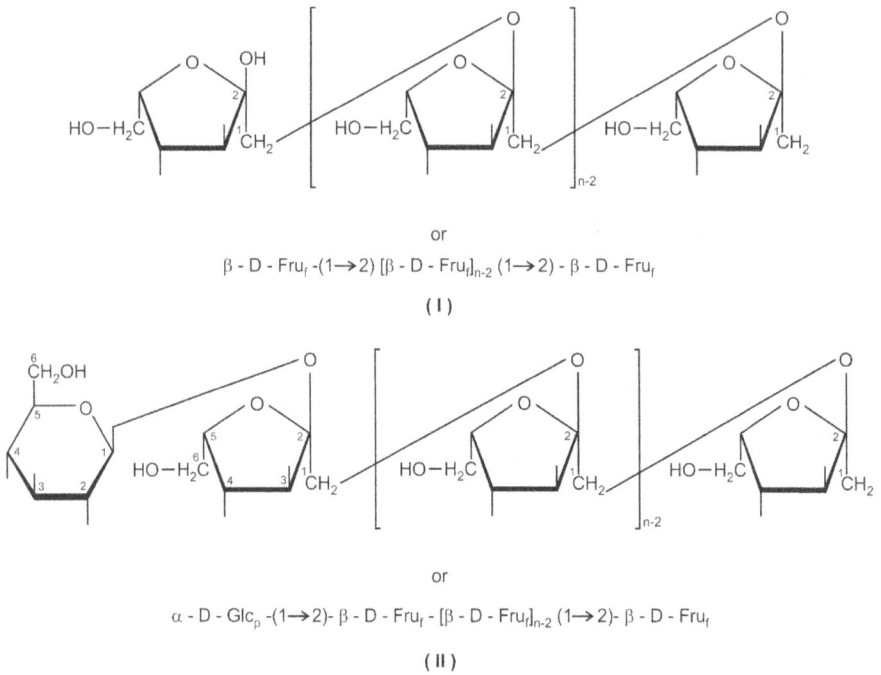

FIGURE 4.3   Fructo-oligosaccharides – chemical structure.

β-D-fructofuranose which is a homopolymer; polyfructoside inulin – monosaccha-rides (II) – which contains a β-D-Glu$_p$ residue, a compound which is a heteropolymer.

There are numerous hydroxyl groups in the inulin that can contribute to the forma-tion of intra- and intermolecular bonds. X-ray diffraction studies have shown that in the crystalline phase inulin intermolecularly interacts, engaging nearby chains, and in the gel phase, it forms three-dimensional colloidal particle structures. Hydrogen bonds between inulin particles contribute to gel stabilization (André et al., 1996).

Biological activity – Biodegradation of inulin occurs in the large intestine (in the colon) by the intervention of microorganisms of the genus *Bifidobacterium* sp. These bacteria prefer to use compounds derived from the biodegradation of inulin as an energy source (Boeckner et al., 2001). It is even mentioned that there is a bifi-dogenic effect that lies in the selective increase in the number of Bifidobacteria in the colon (Tuohy et al., 2001; Langlands et al., 2004). The administration of inulin in the diet also leads to an increase in the size and diversification of Bifidobacterium species without changing the amount of blood lipids. Inulin in food stimulates the absorption of Ca and Mg (Scholz-Ahrens et al., 2007). The presence of inulin in the digestive tract was also associated with the formation of a more stable microbiota.

Research on the role of inulin compounds and microorganisms in the genus *Bifidobacterium* has extended to carcinogenesis problems (Rowland et al., 1998). Such studies often look at the specificity of the action of symbiotics in metabolic processes.

4.6.1.3   Galacto-Oligosaccharides

The oligo- and polysaccharide compounds in which the hexose monomer is α-D-galactopyranose are called galactans or galactosans. The group of galacto-oligosaccharides includes prebiotics which are formed from lactose under the action of the enzymes glycoside hydrolase and galactoside transferase. In food, the main source of lactose is cow's milk (it is known that about 50% of the carbohydrates in milk are lactose). Whole lactose in wheat is also important for food needs. Milk oligosaccharide compounds differ greatly. Thus, a study of human milk (Miller and McVeagh, 1999) shows that there may be more than 130 such combinations in which monomers differ greatly in detection: D-glucose, D-galactose, L-fucose, N-acetylglucosamine, sialic acid.

Chemical structure – The composition of a galacto-oligosaccharide (III) contains mainly residues of D-galactopyranose ($D\text{-}Gal_p$). The chemical bonds between the $D\text{-}Gal_p$ molecules are made in positions 1--> 4. The general chemical structure of a galacto-oligosaccharide is shown in Figure 4.4.

During the decomposition of lactose – a process also called transgalactosylation – under the action of β-galactosilase (lactase) other compounds may be formed which also contain a residue of D-glucopyranose (Rabiu et al., 2001; Roberfroid, 2007).

Biological activity – Galacto-oligosaccharides interact well with probiotic bacterial microorganisms of the genus *Bifidobacterium*. The effect on the immune system of the breastfeed infants (human milk) is well known. The immune effect is also exerted in adults by the fact that inflammation in the colon is attenuated and there is an increase in the amount of short-chain fatty acids (e.g., butyric acid, propionic acid). Butyrate is known to have an anticancer effect, being important in lipid metabolism in epithelial cells, and propionate has an anti-inflammatory effect. Galacto-oligosaccharides are also involved in stimulating the growth of bacteria in the genera *Bifidobacterium* and *Lactobacillus*. This is important in ensuring the digestive transit of food and the reduction of bacteria in the digestive tract (Shoaf et al., 2006). The explanation lies in the fact that these prebiotics become the substrate for the fermentation processes produced by the mostly anaerobic bacteria present in the colon.

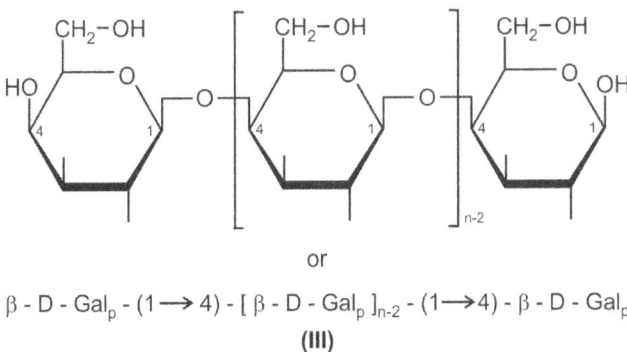

or

$\beta \text{ - D - } Gal_p \text{ - } (1 \longrightarrow 4) \text{ - } [\, \beta \text{ - D - } Gal_p \,]_{n-2} \text{ - } (1 \longrightarrow 4) \text{ - } \beta \text{ - D - } Gal_p$

(III)

FIGURE 4.4   Galacto-oligosaccharide.

4.6.1.4   Isomalto-Oligosaccharides

The carbohydrate monomer in the isomalto-oligosaccharide constitution is α-D-glucopyranose (α-D-Glc$_p$). The prebiotic compounds in this group have a low energy (caloric) intake and are considered non-digestible sweeteners. They are important in the development of the flora in the colon, especially the genera *Bifidobacterium* and *Lactobacillus*.

Chemical structure – In the composition of prebiotics from the isomalto-oligosaccharide group, the constituent monomer is α-D-Glc$_p$. A typical compound is isomalto-oligosaccharide (IV) in which there are a number of n-monomers (Figure 4.5). This is the chemical structure.

The monomers in the composition of isomaltotriose and isomaltotetros have bonds $\alpha(1\text{-} \to 6)$. In them are involved the glycosidic hydroxyl from position α from C$_1$ of the first molecule and the alcoholic hydroxyl from position C$_6$ of the next molecule and so on. There are also connections $\alpha(1 \to 4)$ in the case of panose. There are other representative prebiotics from the isomalto-oligosaccharide group: *isomaltotriose* (V); *panose* (VI) and *isomaltotetrose* (VII).

Their abbreviated chemical structures are shown in Figure 4.6. Enzymatically, their synthesis starts with starch or isomaltose oligomers (Vetere et al., 2000). There are two stages: (i) hydrolysis of starch by α-amylase and β-amylase; (ii) transglucosylation of the glucoside groups of maltase by the intervention of α-D-glucosidase. Finally, in the case of starch synthesis, the yield is low – approx. 40% of the starch is converted to isomalto-oligosaccharide. Very high yields can be obtained for laboratory synthesis when yeasts are used to remove digestible carbohydrates (i.e. glucose, maltose and maltotriose).

Biological activity – Isomalto-oligosaccharides support mainly the activity of lactic acid bacteria and ensure the production of butyrate. These prebiotics are used as ingredients for "functional foods". In the diet, they beneficially stimulate the microbiota of the large intestine, especially microorganisms of the genera *Bifidobacterium and Lactobacillus*.

In general, the degree of polymerization of isomalto-oligosaccharide is 3–5 for compounds of food interest. The recommended amount for consumption is 1.5 g/kg body weight.

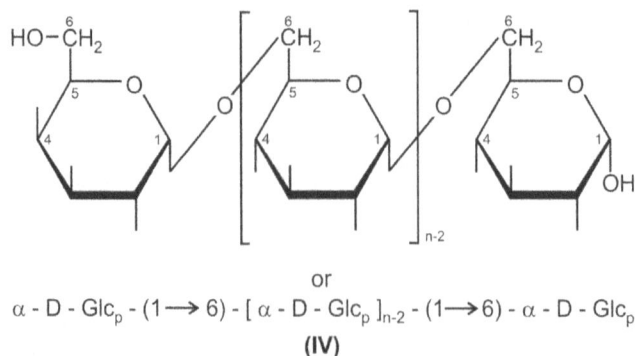

or

α - D - Glc$_p$ - $(1 \longrightarrow 6)$ - [α - D - Glc$_p$]$_{n-2}$ - $(1 \longrightarrow 6)$ - α - D - Glc$_p$

(IV)

FIGURE 4.5   Isomalto-oligosaccharide.

or

α - D - Glc$_p$ - (1 → 6) - α - D - Glc$_p$ - (1→6) - α - D - Glc$_p$

(V)

or

α - D - Glc$_p$ - (1 → 6) - α - D - Glc$_p$ - (1→4) - α - D - Glc$_p$

(VI)

or

α - D - Glc$_p$ - (1 → 6) - α - D - Glc$_p$ - (1→6) - α - D - Glc$_p$ - (1→6) - α - D - Glc$_p$

(VII)

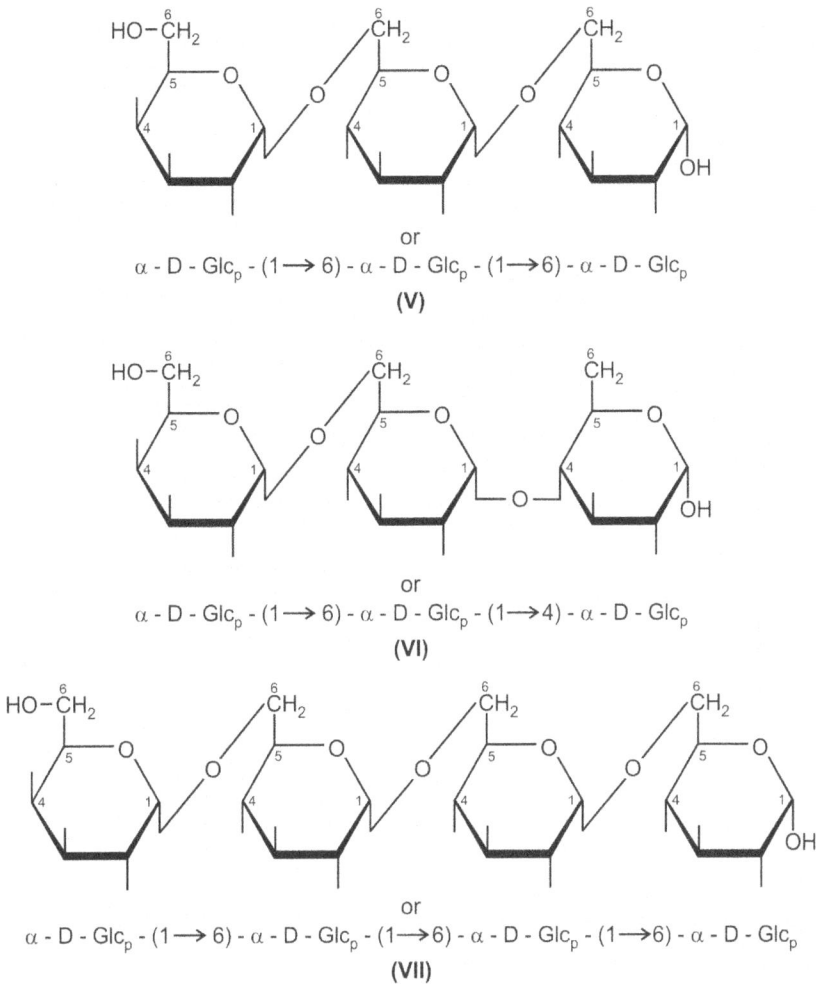

FIGURE 4.6   Isomalto-oligosaccharides.

Higher concentrations (exceeding 40 g/day) may cause digestive disorders with characteristic symptoms: bloating, flatulence and diarrhea. Use in food processing has been approved by EFSA.

For the production of these prebiotics on an industrial/commercial scale, starch from cereals, rice and potatoes, is used. The effects on the human body include digestion, decreased glycemic index and prevention of tooth decay (mainly caused by the formation of insoluble glucan on the surface of the teeth – the so-called dental plaque). In terms of biochemistry, the interaction of isomalto-oligosaccharide prebiotics with microorganisms in the colon increases the amount of immunoglobulins A (IgA) in that intestinal segment and stimulates the production of interferon γ at intraepithelial lymphocytes level. In hepatocellular carcinoma to mice, prebiotics of the isomalto-oligosaccharide group have been found to inhibit cell proliferation and induce apoptosis (Xiao et al., 2011). It also inhibits cell migration and adhesion.

4.6.1.5 Xylo-Oligosaccharides

In the structure of poly- and oligosaccharides in the xylan subclass, the monomer is a pentosic compound with pyranose cyclization called β-D-xylopyranose. Xylo-oligosaccharide compounds are present in vegetables, cereals (bran), milk and honey.

Chemical structure – Xylo-oligosaccharides consist of monomers of β-*D-xylopyranose* (β-D-Xil$_p$) which is released during enzymatic hydrolysis. Figure 4.7 presents the composition of a xylo-oligosaccharide (VIII) showing the general chemical structure and the abbreviated formula.

Biological activity – These oligosaccharides are necessary for probiotics – microorganisms of the genera *Lactobacillus* sp. and *Lactococcus* sp. In the case of xylo-oligosaccharides from oat bran, it has been found that the fermentation produced by these microorganisms is preferred to obtain compounds of an acidic nature (Kontula et al., 1998).

In the case of xylo-oligosaccharides introduced into pure cultures of probiotic microorganisms of the species *Bifidobacterium lactis*, other beneficial microorganisms have developed and metabolic processes have been activated. From a metabolic point of view, fermentation processes are more efficient in the case of wheat bran.

The physiological effects of xylo-oligosaccharides are characterized by the prevention of the adhesion of pathogenic bacteria to the intestinal epithelium, which has been observed in vitro and in vivo. These effects are found in humans and animals (Ebersbach et al., 2012). From a physiological and biochemical point of view, it has been found that they intervene in lowering blood sugar and serum cholesterol levels.

4.6.1.6 Lactosucrose

Lactosucrose is a prebiotic found in the digestive tract of humans and animals. It is a non-digestible compound that (in the large intestine) interacts with microorganisms of the genus *Bifidobacterium*, stimulating their activity.

Chemical structure – The structure of this trisaccharide compound is β-D-galactopyranose (β-D-Gal$_p$), α-D-glucopyranose (α-D-Glu$_p$), β-D-fructofuranose (β-D-Fru$_f$)

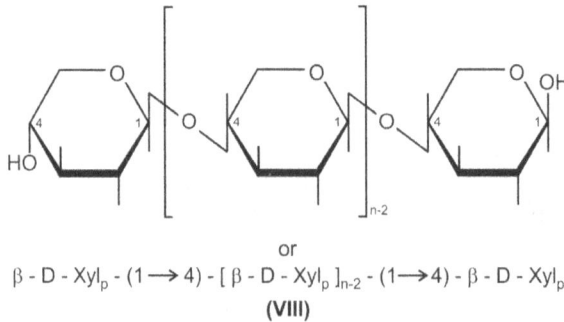

or

$$\beta - D - Xyl_p - (1 \longrightarrow 4) - [\, \beta - D - Xyl_p\,]_{n-2} - (1 \longrightarrow 4) - \beta - D - Xyl_p$$

(VIII)

FIGURE 4.7   Xylo-oligosaccharides.

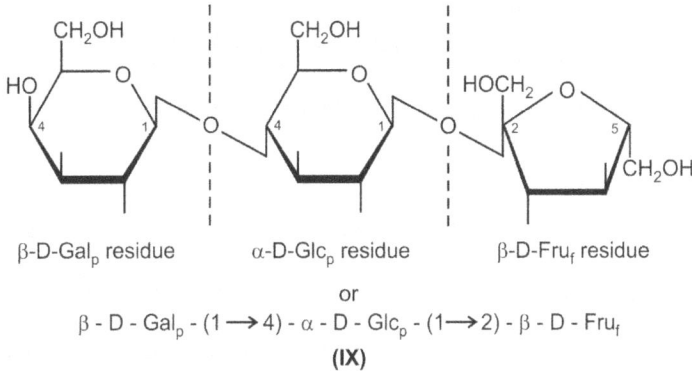

β-D-Gal$_p$ residue α-D-Glc$_p$ residue β-D-Fru$_f$ residue

or

β - D - Gal$_p$ - (1 $\longrightarrow$ 4) - α - D - Glc$_p$ - (1$\longrightarrow$2) - β - D - Fru$_f$

(IX)

FIGURE 4.8   Lactosucrose.

and linked in positions 1--> 4 and 1--> 2, respectively. The structure of lactosucrose (IX) is shown in Figure 4.8.

It is formed in the body during metabolic processes. Currently, lactosucrose is obtained by bioconversion-based synthesis (Choi et al., 2004) starting from lactose and sucrose in the presence of enzymes produced by selected microorganisms (e.g., *Paenibacillus polymyxa*).

Biological activity – Lactosucrose in the body stimulates the development of beneficial bacteria (especially *Bifidobacterium*) and reduces the population of anaerobic bacteria. In terms of biochemistry (metabolic) lactosucrose causes a decrease in the concentration of ammonia (so it is related to ammoniogenesis). Experimentally, it has been found in humans (Teramoto et al., 2006) that this prebiotic reduces the amount of short-chain fatty acids and increases the absorption of calcium in the intestine. In relation to lipid metabolism, it has been found in animal experiments (Mizote et al., 2009) that lactosucrose reduces the accumulation of lipids in the body by depressing (decreasing) their intestinal absorption, mainly of serum triglycerides.

4.6.1.7   Lactulose

It is a disaccharide that has the attribute of a prebiotic compound being non-digestible. The inclusion in the group of fructans is not rigorous for the consideration that in vitro it has been found to exhibit some resistance to enzyme treatment, which reduces prebiotic efficacy (Gibson and Roberfroid, 2008).

Chemical structure – Structurally, lactulose is the disaccharide of β-D-galactopyranose (β-D-Gal$_p$) and β-D-fructofuranose (β-D-Fru$_f$) with binding in positions 1--> 4. According to IUPAC, the full name is: *4-O-D-galactopyranosyl-β-D-fructofuranose*. The usual name is galactosyl-β (1--> 4) fructose. Figure 4.9 presents the structural formula of lactulose (X). This is, in fact, the main disaccharide obtained by "transglycosylation".

Biological activity – The metabolism of lactulose results in short-chain organic acids in the large intestine: acetic acid, lactic acid. They react with ammonia resulting from metabolic processes and thus reduce the level of ammonia.

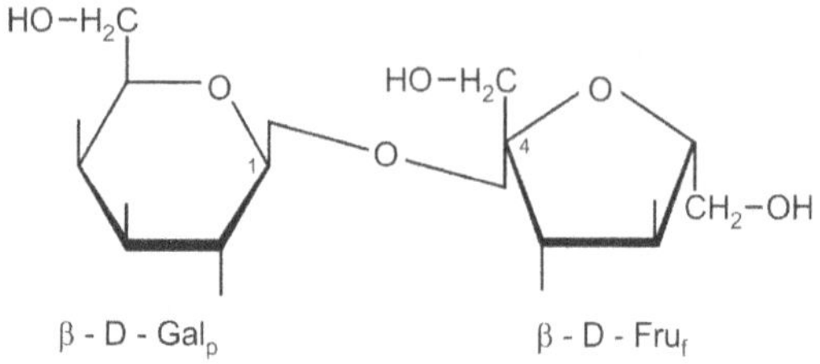

β - D - Gal$_p$ β - D - Fru$_f$

or

β - D - Gal$_p$ - (1 $\longrightarrow$ 4) β - D - Fru$_f$

(X)

FIGURE 4.9   Lactulose.

For this reason, it is used in the prophylaxis and treatment of hyperammonemia in liver diseases. In nutrition, this prebiotic can be obtained from lactose and fructose by enzymatic synthesis in the presence of β-galactosidase. This thermostable enzyme is obtained from *Sulfolobus solfataricus* (Kim et al., 2006).

In pharmacology, lactulose is used as a laxative and osmosis regulator (recommended for hepatic encephalopathy). It is also suitable for therapeutic use in cirrhosis because it limits the absorption of ammonia resulting from metabolic processes. For the above reasons, it is recommended in diets (foods containing lactulose) and in pharmacotherapy (lactulose syrups).

4.6.1.8 Polydextrose

It is known as a non-digestible polymer. For this reason, it is used as a substitute for carbohydrates (a favorite of sucrose) in food (Fava et al., 2007; Gibson and Roberfroid, 2008). It also has the advantage of ensuring an increased intestinal transit.

Chemical structure – In polydextrose, there are α-D-glucopyranose (α-D-Glc$_p$) residues with binding in positions 1--> 4 and 1--> 6 (Figure 4.10).

The existence of these bonds is at the origin of the branching of specific macromolecular chains that form the prebiotic called polydextrose (XI).

Biological activity – Polydextrose in the body causes changes in the intestinal flora. It increases the populations of the genera *Lactobacillus* and *Bifidobacterium*, and decreases the number of microorganisms of the genus Bacteroides which are anaerobic Gram-negative bacteria (e.g., *B. fragilis*, *B. vulgatus*, etc.).

Fava et al. (2007) noted that the effect of polydextrose on the intestinal flora is more intense in the distal colon. This observation attests to the increase in the efficiency, conditioned by time, of the prebiotic-probiotic interaction. In lipid metabolism, polydextrose is involved in the significant reduction of apolipoproteins, especially in the classes (Apo A, Apo B, Apo C, etc.), Apo A-I and Apo A-II types (Saku et al., 1991).

Polydextrose in food has also been found to interfere with mineral metabolism, increasing the level of calcium absorption in the small intestine. The effects of polydextrose are also attributed to its presence in the diet, conferring feeling full, thus limiting the consumption of foods with energetic attributes.

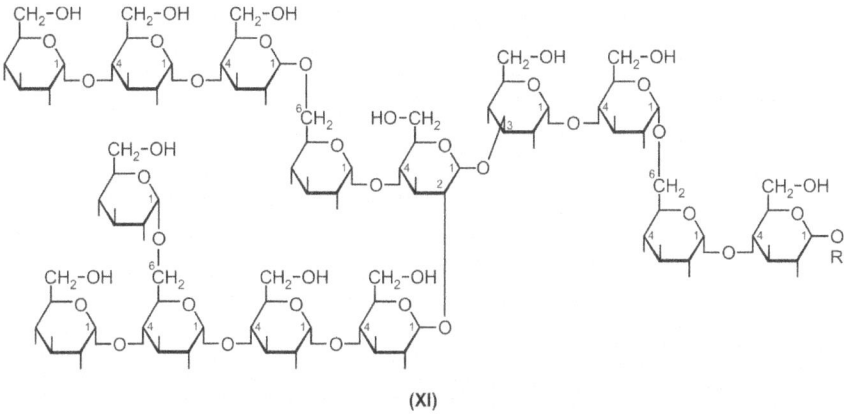

(XI)

FIGURE 4.10   Polydextrose.

4.7   PREBIOTICS IN FOODS AND PHARMACEUTICAL PRODUCTS

4.7.1   PREBIOTICS IN FOOD PRODUCTS

In relation to prebiotics, mainly oligosaccharides, and probiotics, represented by bacteria beneficial to the body, hold the attention of some particular aspects of biochemical and physiological interest. For the metabolization of oligosaccharides, it is necessary for their integration into the bacterial cell in the presence of the enzyme system phosphoenolpyruvate phosphotransferases (Maze et al., 2007). Transmembrane transport involves cell membrane proteins, and ATP is used as an energy source. In general, nutrition issues are discussed in connection with prebiotics and probiotics (Ito et al., 1993; Sanz et al., 2009).

The prebiotic specificity of oligosaccharides makes it possible to be used in the food industry as ingredients in food supplements or as addition to various foods – resulting in fortified foods (Varzakas et al., 2018). Being water-soluble but non-digestible compounds, some prebiotics are used as bulking agents (e.g., polydextrose -E1200) in the food industry, especially for low-fiber food products. It is to be mentioned that the organoleptic properties of the final foods are not influenced. The use of this food additive is not allowed in food supplements for infants and young children.

Prebiotics present in humans' food have beneficial consequences. Foods rich in prebiotics n improve digestion and metabolism, reduce the intestinal pH, regulate bowel movement, improve calcium absorption (thus maintaining bone density), enhance immune system, increase the level of IgG and assure a healthy microbiome a.o.

The problem of prebiotics is also concerned with feeding animals with reference to nutritional and economic aspects.

4.7.2 Prebiotics in Pharmaceutic Products

In the pharmaceutical industry, some prebiotics are used as a stabilizer for protein-based drugs, as auxiliary therapeutic agents for certain diseases (e.g., constipation, diabetes). Other prebiotics (e.g., polydextrose) can be used as excipient for medicines, as granulation aid, humectant, binder and viscosity-increasing agent, coating agent, etc.

Studies revealed that prebiotics have prophylactic action, e.g., reduce the level of serum cholesterol (lowering the risk of atherosclerosis), lower the risk of colorectal cancer (by inducing apoptosis), enhance the immune system (preventing infectious disease), increase the production of hormones which suppress appetite and thus prevent obesity and consecutively atherosclerosis, prevent inflammation. It is claimed that they can prevent necrotizing enterocolitis, can decrease the population of harmful bacteria by *Lactobacilli, Bifidobacteria*, etc. Prebiotics are also used in metaphylaxy; it help to restore the intestinal balance of useful bacteria during/after treatment with antibiotics, after acute diarrhea, a.o. (Appanna, 2018; Davani-Davari et al., 2019; Wan et al., 2020; Kaur et al., 2021).

4.8 OBTAINMENT OF PREBIOTIC COMPOUNDS – GENERALITIES

Oligosaccharide derivatives produced by metabolism lead to the formation of specific monosaccharide compounds, e.g., galactose, xylose, or even their alcoholic derivatives, e.g. xylitol. The case of compounds with such applications is exemplified.

> *D(+) Galactose* – During metabolization, it is converted into glucose if liver function is normal. Residual metabolites are excreted in the urinary tract. For this reason, the amount of galactose in the urine (galactosuria) is a "test" in assessing liver function.
>
> *D(+) Xylose* – It results from metabolic processes. For this reason, it is suitable for use for diagnostic purposes as a "test" to assess resorption in the small intestine.
>
> *Xylitol* – It is a derivative of xylose. It tastes sweet. It is a sweetener in the diet of diabetics. It is not degraded by enzymes in the oral cavity. It is used in the manufacture of chewing gum.

In food processing, the aim is to obtain prebiotics that are suitable for use as specific "non-nutritive" ingredients. For this purpose, for example, prebiotics from the fructo-oligosaccharides group can be obtained by specific methods based on controlled enzymatic hydrolysis. The obtainment of fructo-oligosaccharides and galacto-oligosaccharides is exemplified.

4.8.1 Obtainment of Fructo-Oligosaccharides

To obtain fructo-oligosaccharides prebiotics, specific technological processes can start from sucrose or from monoglucopoly-fructose inulin (Figure 4.11). In both cases, specific methods based on "controlled enzymatic hydrolysis" are used.

$$D\text{-}Glu_p\text{-}(1 \longrightarrow 2)\text{-}D\text{-}Fru_f$$

Sucrose

Fructofuranosidase

Fructo-oligosaccharides

$$D\text{-}Glu_p\text{-}(1 \longrightarrow 2)\text{-}[\beta\text{-}D\text{-}Fru_f]_{n-1}\text{-}(1 \longrightarrow 2)\beta\text{-}D\text{-}Fru_f$$
monogluco-oligofructose inulin
$n = 2 - 4$

$$D\text{-}Fru_f\,(1 \longrightarrow 2)\text{-}[\text{-}D\text{-}Fru_f]_{n-1}\text{-}(1 \longrightarrow 2)\,\beta\text{-}D\text{-}Fru_f$$
oligofructose inulin
$n = 1 - 9$

$$D\text{-}Glu_p\text{-}(1 \longrightarrow 2)\text{-}[\beta\text{-}D\text{-}Fru_f]_{n-1}\text{-}(1 \longrightarrow 2)\,\beta\text{-}D\text{-}Fru_f$$
monogluco-oligofructose inulin
$n = 2 - 9$

Inulinase

$$Glu_p\text{-}(1 \longrightarrow 2)\text{-}[\beta\text{-}D\text{-}Fru_f]_n$$

monogluco-polyfructose inulin
$(n = 2 - 65)$

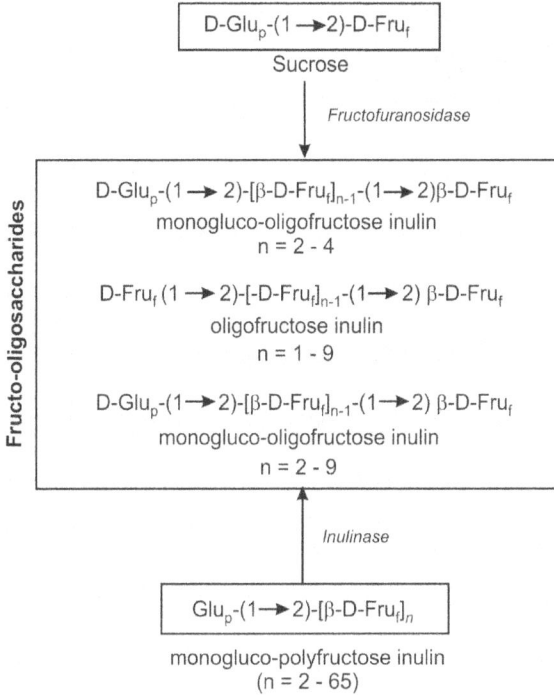

FIGURE 4.11 Obtaining fructo-oligosaccharides from sucrose and monoglucopoly-fuctose inulin.

a. *Obtaining from sucrose* – Processing takes place in the presence of the enzyme fructofuro-nosidase. There is a "transfructosylation reaction" which results in: monogluco-oligofructose inulin (with a chain of 4–6 monosaccharide residues); monogluco-oligofructose inulin (with a chain of 4–11 monosaccharide residues); oligofructose inulin (with 3–9 fructose residues).

b. *Obtaining from monoglucopoly-fructose inulin* – This compound – as we call it – is a polysaccharide that belongs to the class of fructans. Processing takes place in the presence of inulinase. In the structure of the primary compound (which has become a substrate) it is a carbohydrate residue (gluco-pyranose).

The polyfructose catenary sequence has 3–66 fructose residues (fructo-furanose). The resulting compounds are oligofructose derivatives with a composition similar to the derivatives obtained by transfructosylation of sucrose. For this reason, they are shown in Figure 4.11. The differences lie in the specifics of processing technologies (which go beyond the scope of this volume.

4.8.2 OBTAINMENT OF GALACTO-OLIGOSACCHARIDES

Prebiotics with a galacto-oligo-carbohydrate structure can be obtained by synthesis from lactose-disaccharide in the composition of which is galactose and glucose. The reaction takes place in the presence of the β-galactosidase (lactase) enzyme – Figure 4.12.

The action of lactase produces the so-called "trans-glycosylation reaction". Following a "controlled enzymatic hydrolysis" reaction, various galacto-oligosaccharides can be synthesized.

Of these, digalacto-monoglucose and trigalacto-monoglucose are suitable for use as prebiotics. Prebiotics from the galacto-oligosaccharide group are of interest to the food and pharmaceutical industries. The chemical synthesis of galacto-oligosaccharides is carried out starting from lactose. The reaction is kinetically controlled, with competition between transferase and hydrolase-specific enzymes present in this reaction.

Transferase is an enzyme called transgalactosylase, which generates complex mixtures of di- and oligosaccharide. Hydrolysis is produced by the enzyme glycosyl hydrolase, which leads to the formation of D-galactose and D-glucose. Kinetically, the lactose hydrolysis reaction is favored. In the synthesis of galacto-oligosaccharide, the enzyme β-galactosidase of bacterial, fungal or vegetable origin is used. In this case, an increased rate can be obtained in the catalysis of the transglycosylation reaction in relation to the hydrolysis reaction.

For this reason, the physico-chemical properties and the physiological specificity vary from one batch to another (in the synthesis processes). In the mixtures obtained, oligogalacto-carbohydrates predominate (approx. 60%), but there are glucose, galactose and lactose residues. It was also found that in case of a high degree of polymerization of galacto-oligosaccharides used in food and drug processing, the solubility, osmolarity, viscosity, degree of sweetening of the products obtained can change.

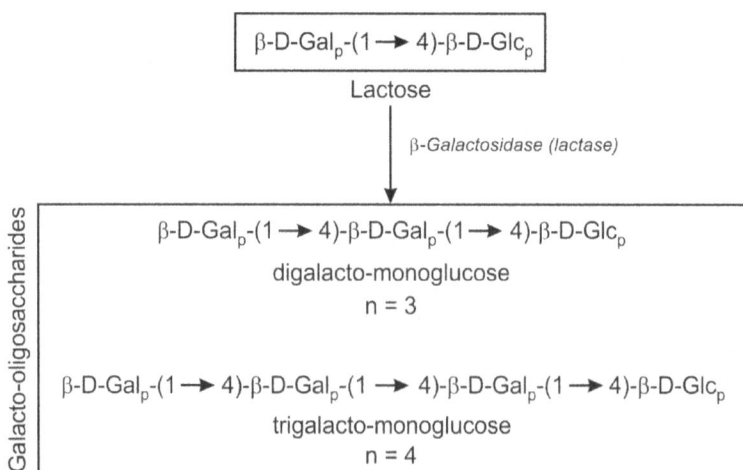

FIGURE 4.12   Obtaining galacto-oligosaccharides from lactose.

For the investigation of prebiotic components, physico-chemical analytical methods by separation (so-called "analytical separatology") are suitable, which also allow the investigation of the chemical structure of a component or mixture of components (Sanz et al., 2009).

Different methods can be used in analytical separatology of prebiotics: thin layer chromatography (TLC); gas chromatography (GC); high performance liquid chromatography (HPLC); capillary electrophoresis (CE), etc. In the structural analysis of prebiotics, mass spectroscopy (MS); nuclear magnetic resonance (NMR); pulsed amperometric detection (PAD); laser-induced fluorescence (LIF), etc. are used. In modern procedures, the so-called "tandem methods" are sometimes used efficiently, e.g., HPLC-MS; CE-LIF et al.

In the analytical practice for the accreditation of a certain method, the reproducibility of the data is important. Thus, it can be validated for use in profile laboratories. The application of these methods for the quantification of prebiotic oligosaccharides is important for food and pharmaceuticals. For example, the "CE-LIF tandem analytical method" is suitable for qualitative and quantitative determinations in the case of galacto-oligosaccharides. The data obtained may be predictive of the physiological effects in the gastrointestinal tract.

REFERENCES

Al-Sheraji S., Ismail A., Manap M., Mustafa S., Yusof R., Hassan F. - Prebiotics as functional foods: a review, *J. Funct. Foods*, 2013, 5, 1542–1553. https://doi.org/10.1016/j.jff.2013.08.009.

André I., Mazeau K., Tvaroska I., Putaux J.-L., Winter W.T., Taravel F.R., Chanzy H. - Molecular and crystal structures of inulin from electron diffraction data, *Macromolecules*, 1996, 29(13), 4626–4635.

Apolinário A.C., de Lima Damasceno B.P., de Macêdo Beltrão N.E. - Inulin-type fructans: a review on different aspects of biochemical and pharmaceutical technology, *Carbohydr. Polym.*, 2014, 101, 368–378.

Appanna V.D. - Dysbiosis, probiotics, and prebiotics: in diseases and health, pp. 81–122, in *Human Microbes - The Power Within*, Spinger Nature, Singapore Pte Ltd., 2018. https://doi.org/10.1007/978-981-10-7684-8_3.

Boeckner L.S., Schnepf M.I., Tungland B.C. - Inulin: a review of nutritional and health implications, *Adv. Food Nutr. Res.*, 2001, 43, 1–63. https://doi.org/10.1016/s1043-4526(01)43002-6.

Bottazzi V. - Food and feed production with microorganisms, *Biotechnology*, 1983, 5, 315–363.

Charalampopoulos D., Rastall A.R. (Eds.) - *Prebiotics and Probiotics Science and Technology*, Springer Science, NY, 2009.

Cherry P., Yadav S., Strain C.R., Allsopp P.J., McSorley E.M., Ross R.P., Stanton C. - Prebiotics from seaweeds: an ocean of opportunity? *Marine Drugs*. 2019, 17(6), 327. https://doi.org/10.3390/md17060327.

Choi H.J., Kim C.S., Jung H.C., Oh D.K. - Lactosucrose bioconversion from lactose and sucrose by whole cells of Panibacillus polymyxa harboring levansucrase activity, *Biotechnol. Prog.*, 2004, 20, 1876–1879.

Davani-Davari D., Negahdaripour M., Karimzadeh I., Seifan M., Mohkam M., Masoumi S.J., Berenjian A., Ghasemi Y. - Prebiotics: definition, types, sources, mechanisms, and clinical applications, *Foods*, 2019, 8(3), 92. https://doi.org/10.3390/foods8030092.

Ebersbach T., Andersen J.B., Bergstrom A., Hutkins R.W., Licht T.R. - Xylo-oligosaccharides inhibit pathogen adhesion to enterocytes in vitro, *Res. Microbiol.*, 2012, 163, 22–27.

Ensminger N.M., Ensminger A., Konlande E.I., Robson J.R. - *The Concise Encyclopedia of Food and Nutrition*, CRC Press, Boca Raton, FL, 1995.

Fava F., Makivuokko H., Siljander-Rasi H., Putaala H., Tiihonen K., Stowell J., Tuohy K., Gibson G., Rautonen N. - Effect of polydextrose on intestinal microbes and immune functions in pigs, *Br. J. Nutr.*, 2007, 98, 123–133.

Flückiger F.A., Tschirch A. - *Grundlagen der Pharmacognosie*, Zweite Auflage, Verlag von Julius Springer, Berlin, 1885.

Fuller R. - Probiotics in man and animals, *J. Appl. Bact.*, 1989, 66(5), 365–378.

Gârban Z. - *Human Nutrition. Vol. I* (in Romanian), Editura Eurobit, Timişoara, 2000.

Gârban Z. - *Biochemistry: Comprehensive Treatise, Vol. I. Basics of Biochemistry* (in Romanian), 5th edition, Publishing House of the Romanian Academy, Bucureşti, 2014.

Gârban Z. - *Biochemistry: Comprehensive Treatise, Vol. II. Biochemical Effectors* (in Romanian), 5th edition, Publishing House of the Romanian Academy, Bucureşti, 2018.

Gibson G.R., Beatty E.R., Wang X., Cummings J.H. - Selective stimulation of bifidobacteria in the human colon by oligofructose and inulin, *Gastroenterology*, 1995, 108, 975–982.

Gibson G.R., Roberfroid M.B. - Dietary modulation of the colonic microbiota: Introducing the concepts of probiotics, *J. Nutr.*, 1995, 125, 1401–1412.

Gibson G.R., Roberfroid M.B. - *Handbook of Prebiotics*. CRC Press, Boca Raton FL, 2008.

Guyton A.C., Hall J.E. - *Textbook of Medical Physiology*, 11th edition, Elsevier-Saunders, Philadelphia, 2006.

Ito M., Kimura M., Deguchi Y., Miyamori-Watabe A., Yajima T., Kan T. - Effect of trans-galactosylated disaccharides on the human intestinal microflora and their metabolism, *J. Nutr. Sci. Vitaminol. (Tokyo)*, 1993, 39, 279–288.

Kaur A.P., Bhardwaj S., Dhanjal D.S., Nepovimova E., Cruz-Martins N., Kuča K., Chopra C., Singh R., Kumar H., Şen F. - Plant prebiotics and their role in the amelioration of diseases, *Biomolecules*. 2021, 11(3), 440. https://doi.org/10.3390/biom11030440.

Kim Y.-S., Park C.-S., Oh D.-K. - Lactulose production from lactose and fructose by a thermostable ß-galactosidase from *Sulfolobus solfataricus*, *Enzyme Microb. Technol.*, 2006, 39(4), 903–908.

Kontula P., von Wright A., Mattila-Sandholm T. - Oat bran beta-gluco- and xylo-oligosaccharides as fermentatve substrates for lactic acid bacteria, *Int. J. Food Microbiol.*, 1998, 45, 163–169.

Langlands S.J., Hopkins M.J., Coleman N., Cummings J.H. - Prebiotic carbohydrates modify the mucosa associated microflora of the human large bowel, *Gut*, 2004, 53, 1610–1616.

Lilly D.M., Stillwell R.H. - Probiotics. Growth promoting factors produced by micro-organism, *Science*, 1965,147, 747–748.

Lopez-Santamarina A., Mondragon A.C., Lamas A., Miranda J.M., Franco C.M., Cepeda A. - Animal-origin prebiotics based on chitin: an alternative for the future? A critical review, *Foods*, 2020, 9(6), 782. https://doi.org/10.3390/foods9060782.

Maze A., O'Connell-Motherway M., Fitzgerald G.F., Deutscher J., van Sinderen D. - Identification and characterization of a fructose phosphotransferase system in Bifidobacterium breve UCC2003, *Appl. Environ. Microbiol.*, 2007, 73, 545–553.

Miller J.B., McVeagh P. - Human milk oligosaccharides: 130 reasons to breast-feed, *Br. J. Nutr.*, 1999, 82, 333–335.

Mizote A., Taniguchi Y., Takei Y., Koya-Miyata S., Kohno K., Iwaki K., Kurose M., Oku K., Chaen H., Fukuda S. - Lactosucrose inhibits body fat accumulation in rats by decreasing intestinal lipid absorbtion, *Biosci. Biotechnol. Biochem.*, 2009, 73, 582–587.

Oku T. - Special physiological functions of newly developed oligosaccharides, pp. 202–218, in *Functional Foods. Designer foods, Pharmafoods, Nutraceuticals* (Goldberg I., Ed.), Chapman and Hall Food Science Book, Aspen Publ., Inc., Gaithersburg, Maryland, 1999.

Panesar P.S., Kumari S., Panesar R. - Biotechnological approaches for the production of prebiotics and their potential applications, *Crit. Rev. Biotechnol.* 2013, 33, 345–364. https://doi.org/10.3109/07388551.2012.709482.

Parker R.B. - Probiotics, the other half of the antibiotic story, *Am. Nutr. Health*, 1974, 29, 4–8.

Rabiu B.A., Jay A.J., Gibson G.R., Rastall R.A. - Synthesis and fermentation properties of novel galacto-oligosaccharides by beta-galactosidases from Bifidobacterium species, *Appl. Environ. Microbiol.*, 2001, 67, 2526–2530.

Roberfroid M. - Prebiotics: the concept revisited, *J. Nutr.*, 2007, 137, 830S–837S.

Rowland I.R., Rumney C.J., Coutts J.T., Lievense L.C. - Effect of *Bifidobacterium longum* and inulin on gut bacterial metabolism and carcinogen-induced aberrant crypt foci in rats, *Carcinogenesis*, 1998, 19, 281–285.

Sabater-Molina M., Larqué E., Torrella F., Zamora S. - Dietary fructooligosaccharides and potential benefits on health, *J. Physiol. Biochem.* 2009, 65, 315–328.

Saku K., Yoshinaga K., Okura Y., Ying H., Harada R., Arakawa K. - Effects of polydextrose on serum lipids, lipoproteins and apolipoproteins in healthy subjects, *Clin. Ther.*, 1991, 13, 254–258.

Sanz M.L., Ruiz-Matute A.I., Corzo N., Martinez-Castro I. - Analysis of prebiotic oligosaccharides, pp. 465–534, in *Prebiotics and Probiotics Science and Technology* (Charalampopoulos D., Rastall A.R., Eds.), Springer Science, NY, 2009.

Scholz-Ahrens K.E., Ade P., Marten B., Weber P., Timm W., Acyl Y., Glüer C.C., Schrezenmeir J. - Prebiotics, probiotics and synbiotics affect mineral absorbtion, bone mineral content, and bone structure, *J. Nutr.*, 2007, 137, 838S–846S.

Schrezenmeir J., de Vrese M. - Probiotics, prebiotics, and synbiotics - approaching a definition, *Am. J. Clin. Nutr.*, 2001, 73(2suppl), 361S–364S. https://doi.org/10.1093/ajcn/73.2.361s.

Sekhon B.S., Jairath S. - Prebiotics, probiotics and synbiotics: an overview, *J. Pharm. Educ. Res.*, 2012, 1(2), 13–36.

Shoaf K., Mulvey G.L., Armstrong G.D., Hutkins R.W. - Prebiotic galactooligosaccharides reduce adherence of enteropathogenic *Escherichia coli* to tissue culture cells, *Infect. Immun.*, 2006, 74, 6920–6928.

Teramoto F., Rokutan K., Sugano Y., Oku K., Kishino E., Fujita K., Hara K., Kishi K., Fukunaga M., Morita T. - Long-term administration of 4G-ß-D-galactosyl-sucrose (lactosucrose) enhances intestinal calcium absorption in young women: a randomized, placebo-controlled 96-wk study, *J. Nutr. Sci. Vitaminol. (Tokyo)*, 2006, 52, 337–346.

Tuohy K.M., Kolida S., Lustenberger A.M., Gibson G.R. - The prebiotic effects of biscuits containing partially hydrolysed guar gum and fructo-oligosaccharides - a human volunteer study, *Br. J. Nutr.*, 2001, 86, 341–348.

Tzounis X., Rodriguez-Mateos A., Vulevic J., Gibson G.R., Kwik-Uribe C., Spencer J.P. - Prebiotic evaluation of cocoa-derived flavanols in healthy humans by using a randomized, controlled, double-blind, crossover intervention study, *Am. J. Clin. Nutr.*, 2011, 93(1), 62–72. https://doi.org/10.3945/ajcn.110.000075.

Varzakas T., Kandylis P., Dimitrellou D., Salamoura C., Zakynthinos G., Proestos C. - Innovative and fortified food: probiotics, prebiotics, GMOs, and superfood, pp. 67–129, in *Preparation and Processing of Religious and Cultural Foods* (Ali E. , Nizar N., Eds.), Elsevier, London-Amsterdam, 2018.

Vetere A., Gamini A., Campa C., Paoletti S. - Regio-specific transglycolytic synthesis and structural characterization of 6-O-alpha-glucopyranosyl-glucopyranose (isomaltose), *Biochem. Biophys. Res. Commun.*, 2000, 274, 99–104.

Wan X., Guo H., Liang Y., Zhou C., Liu Z., Li K., Niu F., Zhai X., Wang L. - The physiological functions and pharmaceutical applications of inulin: a review, *Carbohydr. Polym.* 2020, 246, 116589. https://doi.org/10.1016/j.carbpol.2020.116589.

Wood B.J.B. (Ed.) - *The Lactic Acid Bacteria. Volume 1. Lactic Acid Bacteria in Health and Disease*, Elsevier Applied Science Publishers, London-New York, 1992.

Xiao C.L., Tao Z.H., Guo L., Li W.W., Wan J.L., Sun H.-C., Wang L., Tang Z.-Y., Fan J., Wu W.-Z. - Isomalto oligosaccharide sulfate inhibits tumor growth and metastasis of hepato-cellular carcinoma in nude mice, *BMC Cancer*, 2011, 11, 150.

Yannai S. - *Dictionary of Food Compounds*, CRC Press, Boca Raton FL, 2004.

5 Plant Growth Bioregulators

5.1 INTRODUCTION

Compounds from the class of plant growth bioregulators (PGBs) have attributes of biologically active substances, the reason for which they are considered "biochemical effectors". Such compounds participate in specific biochemical interactions of metabolic processes, contributing to their acceleration or deceleration (Went, 1926; Goodwin and Mercer, 1983; Arteca, 1996; Taiz and Zeiger, 1998; Heller et al., 2004).

In chemistry/biochemistry, the general name of plant growth regulators (PGR) is known for certain organic chemical compounds that intervene in the regulation of physiological processes specific to plant organisms. With reference to the substances mentioned above – considering their biochemical and agrobiological importance – we agree to distinguish between "bioregulators" and "regulators".

Thus, the term *plant growth bioregulators* (PGB) will be used for natural chemical compounds resulting from biosynthesis reactions occurring in plant organisms. They were isolated in the form of plant extracts.

In parallel with this terminology, PGR will be used for chemical compounds obtained by organic chemical synthesis (Gârban, 2018). They are used in experiments carried out in plant physiology, biochemistry and plant molecular biology. It also has applicative importance in agrobiology, being of interest to certain agricultural technologies.

These compounds, in small quantities, intervene in the morphogenesis processes involving growth, differentiation and development (e.g., germination, sprouting, flowering, fertilization and ripening). The biosynthesis of PGB occurs in their meristematic tissue (there is no glandular system similar to that of animals). Initially, it was thought that these compounds do not have protein receptors. Currently, it is known that there are specific protein receptors that take over bioregulators/regulators.

Growth bioregulators in plants also ensure the response to abiotic stress (e.g., heat or cold; drought or flooding) and biotic stress (e.g., the presence of herbivores and the action of pathogens).

In general, it is considered that a PGB must:

(i) be an endogenous substance – i.e. synthesized in the plant body; (ii) act in small doses, of the order of micromoles (μM); (iii) constitute an information vector led to a target cell, sensitive to its action, influencing its functioning (Harnborne, 1989; Buchanan et al., 2000; Heller et al., 2004). The above mentions distinguish between a PGR and a trophic substance.

In the case of growth bioregulators, their action can highlight synergistic or antagonistic effects. Also, they can give specific interactions with various chemical compounds (bioconstituents from plants) of protein (e.g., nucleic acids), carbohydrates, vitamins, etc. nature.

DOI: 10.1201/9781032702520-5

Through this mode of action, PGBs can be considered biologically active substances that reach plant tissues, which have the property of influencing biochemical interactions and physiological processes.

It was observed that PGBs – beyond the interventions characterized by activation/inhibition of some internal metabolic processes – also highlighted other less known external functions.

Thus, it was found that bioregulators from the plant kingdom can act as "inter-communication substances" for different plant individuals. In this framework, it was found that a stressed tree emits a bioregulator that informs another tree about an existing stress situation. The stimulus released by the action of a bioregulator can increase the biosynthesis of a "defensive molecule" in receiving plants (e.g., activating the biosynthesis of tannin molecules).

In such situations in plant physiology, comparisons are made between the "defensive molecules" of PGBs and the "stress hormones" in comparative animal physiology and in human physiology.

Also, "defensive molecules" of bioregulators were detected in plants that have been in water shortage for a long time, as well as in plants injured after the attack of some predators. Moreover, it was noted that there are situations in which the presence of predators on injured plants attracts "predators of predators".

Such observations on bioregulators also allow a comparison with "animal pheromones". Although they are of scientific interest for plant physiology, plant molecular biology and phytopathology, these aspects have not been studied in depth. The explanation, possibly, lies in the fact that this topic is of interest for fundamental scientific research, but has less economic importance.

The presentation of the problems related to PGBs requires the contextual approach of minimal information on plant morphology and physiology. Given the fact that there is no subchapter on plant morpho-physiological aspects, references necessary for understanding are made within the sub-chapters that deal distinctly with issues regarding PGBs.

5.2   CHEMICAL COMPOSITION AND NATURAL DISTRIBUTION

From the point of view of chemical structure, plant growth bioregulators/PGRs are biologically active substances with heterogeneous structure. In their constitution, there are heterocyclic nuclei, e.g., indole, gibban, purine, etc. or linear chains, e.g., ethylene.

Depending on the biological effects, the PGBs are classified as growth promoters (accelerating effect), e.g. *auxins, gibberellins, cytokinins*, respectively growth inhibitors (decelerating effect), e.g. *ethylene, abscisic acid.*

Later, other PGBs were also isolated, including *brassinosteroids, strigolactones, jasmonates* and *salicylates.*

Along with these compounds, "inhibiting" substances and "retarding" substances are currently discussed.

In the vegetable kingdom, PGBs are found in various tissues, ensuring the development of metabolic processes by intervening in growth kinetics, morphogenesis control, tropisms (e.g., phototropism, geotropism), photoperiodicity, etc.

Also, under the action of PGBs, the metabolic characteristics can be observed during the formation of fruits and seeds. In plant physiology, latent life, dormancy, periods of active life in plants, etc. are also studied.

In general, PGBs intervene in the regulation of cellular activities, the development of vegetative and reproductive tissues, responses to stress, etc. Details on the general distribution will be presented in each subchapter in which one or another compound from the PGBs category is discussed.

5.3   BRIEF HISTORY

Observations on some aspects of plant biochemistry and plant physiology that followed phototropism, as well as the stimulating effects of their growth, led to the hypothesis of the presence of biologically active substances with attributes of PGBs. They accelerate or decelerate the biochemical interactions in plant physiology, thus distinguishing two categories of compounds, i.e. with promotor, respectively, inhibitor effect.

Auxins, gibberellins and cytochinins were included in the category of promotors (more studied substances). From a historical point of view in the first decades of the twentieth century, important discoveries were made regarding these substances. Thus, for auxins, the contributions made by Went (1926) and Kögl et al. (1934) are cited, and for gibberellins, the studies undertaken by Kurosawa (1926). Regarding cytokinins, the studies carried out by Haberlandt (1913) and later by van Overbeek (1941) in cell cultures are remembered. The first compound isolated and studied was *kinetin* (Miller et al., 1955), later *zeatin* was also isolated (Letham, 1963).

As to the category of inhibitors, there are also some historically specific milestones. Regarding ethylene, observations on the bioregulatory effects by Gane (1934) are known. For abscisic acid, isolated by Eagles and Wareing (1963), the role in abscission of leaves and fruits was noted. More details are given in another section.

Progressively, from a historical point of view, other compounds were also isolated, e.g., brassinosteroids, strigolactones, jasmonates, etc.

Some historically specific details on the PGBs are presented in the sub-chapters related to their chemical structure and biological activity.

Historically, it is known that in 1902, Starling and Bayliss discovered *secretin* – a compound with *interactive tissue attributes*. They defined the substance with such property *hormone*, currently used term in comparative animal physiology and human physiology. Indirectly, the idea of feedback in the case of hormones is suggested.

Studying the transmission ways of phototropic stimulus in vegetal physiology, Paál (1914), used the expression "plant growth regulators". In the following decades, Went and Thimann (1937) studying the stimulating substances for the development of reproductive tissues in plants take over – from comparative animal physiology and human physiology – for the expression "plant growth regulators" – the term "phytohormones".

It is necessary to mention that there is a difference between *hormones* in animals and humans and "plant growth regulators". In the case of hormones, the existence of a specific feedback can be noted. For example, there is such an effect between pituitary hormones and hormones from other tissues: e.g., thyroid-stimulating hormone (TSH) in relation to thyroid hormones (i.e., thyroxin and triiodothyronine);

follicle-stimulating hormone (FSH) in relation to female gonadal hormones (i.e., estrogens and progesterone).

PGRs (sometimes also called "plant hormones" or "phytohormones") act in the vicinity of the "production zone" (tissue biosynthesis) or in a different area of the plant considered the "reception zone" for this requiring migration from the site of biosynthesis to other tissues (Davis, 1995). However, this does not prove the existence of a feedback.

For the reasons mentioned above, in this volume, the term PGBs will be used.

5.4   CLASSES OF CHEMICAL COMPOUNDS

The deepening of studies regarding PGBs led to the demanding evaluation of their biological activity. As previously mentioned, depending on the chemical structure, the classification of as previously mentioned, depending on the chemical structure, the classification of PGBs into five "groups" (sometimes also called "families") was accepted. According to their biological effects, these compounds can be included in two categories: growth promoters: (i) *auxins;* (ii) *gibberellins* and growth inhibitors: (iii) *cytokinins*; (iv) *ethylene* and (v) *abscisic acid* (Guleria et al., 2021). A brief presentation of their properties follows.

Auxins – intervene in the stimulation of growth through cell elongation and in the regulation of cell division and differentiation. They are also involved in the manifestation of geotropism and phototropism, the stimulation of rhizogenesis (with the formation of adventitious roots), as well as in the regulation of abscission.

Gibberellins – initiate the mobilization of the compounds present in the seeds during germination by activating the biosynthesis of enzymes; produce the elongation of "stems"; and stimulate the development of pollen tube growth in biennial plants. It mobilizes cellular bioconstituents that participate in the formation of new cells – by modulating chromosomal transcription. In cereal seeds (rice, wheat, corn, etc.) around the endosperm, there is a layer of aleurone cells. Gibberellins reach the layer of aleurone cells responsible for the production of metabolic enzymes. Gibberellins also reverse stunted growth in plants.

Cytokinins – stimulate cell division, regulate cell differentiation, produce the growth of leaf cells and inhibit leaf senescence.

*Ethylen*e – disrupts cell elongation, disrupts geotropic responses, accelerates leaf senescence and fruit ripening processes and stimulates abscission.

Abscisic acid – produces a general inhibitory effect on cell growth, regulates budding status in plants, regulates the abscission of leaves, flowers and fruits and regulates the functioning of tomatoes in stress situations (e.g., thermal stress, water stress – in plants).

Apart from these main groups of PGBs, specific effects of secondary groups of biologically active substances were also studied: brassinosteroids, strigolactones, jasmonic acid, salicylic acid, etc.

5.4.1 AUXINS

Auxins (gr. *auxein*-growth) are considered biologically active substances from the group of "plant growth bioregulators", which have been detected mainly in the growth organs: buds, the root system (the so-called rhizosphere), pollen, the leaf apparatus (especially in the tips of young leaves) in immature seeds. In the case of the human body, auxins have been isolated from urine. Free auxins (approx. 30% of the total) that move more easily in the tissues and auxins bound to protein supports that are released by the action of proteolytic enzymes were detected in the plant tissues of young plants. Of these, free auxins can be extracted with organic solvents. In plant organisms, auxins are usually in the form of complex mixtures containing growth-regulating substances with stimulatory and inhibitory effects (Went, 1926, 1928; Cohen and Bandurski, 1982; Neamțu, 1983).

5.4.1.1 Chemical Structure

The representative compound for this group of PGBs is *heteroauxin*, which has the chemical name of *3-indolylacetic acid* (I) – Figure 5.1. It was isolated from higher and lower plants (Kögl et al., 1934).

There are several derivatives of 3-indolylacetic acid that contain the indole nucleus in the molecule and are known in the literature under the generic name of auxins. Such compounds have been isolated from cereals (e.g., corn contains heteroauxins approx. 0.5 mg/100 g grains), from the leaves of various plants, from pollen, etc.

From an evolutionary, historical point of view, it is known that in 1933 the group of researchers led by Kögl isolated from human urine a biologically active substance with an acidic character that was called auxin A – auxentriolic acid. The same group of researchers contributed to the isolation of a substance called auxin B – auxeno-lonic acid – from plant tissues (maize germs brought into the oil phase).

Auxin A, under the action of dehydrating agents, turns into auxin B, by removing the hydroxyl groups from the α and β positions, followed by the formation of a ketone group at the carbon atom from the position. Auxin B is stable to light, but is quickly destroyed in acidic or alkaline environments. Figure 5.2 shows the structural formulas of auxins A and B.

In the presence of ultraviolet radiation, auxins lose their biologically active properties. In this situation, for example, from auxin A is formed a lactone, through the cyclization between the carboxylic group in position α and the hydroxyl group in position δ. Auxin A lumilactone, a biologically inactive compound, is thus formed.

Later, another biologically active substance was isolated, also from plant tissues, which highlighted about 50% of the activity of auxin A. This substance, called *heteroauxin*, is *3-indolylacetic acid*.

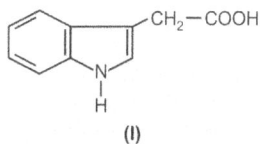

(I)

FIGURE 5.1   Structure of 3-indolylacetic acid.

Auxin A
(auxentriolic acid)

Auxin B
(auxenolonic acid)

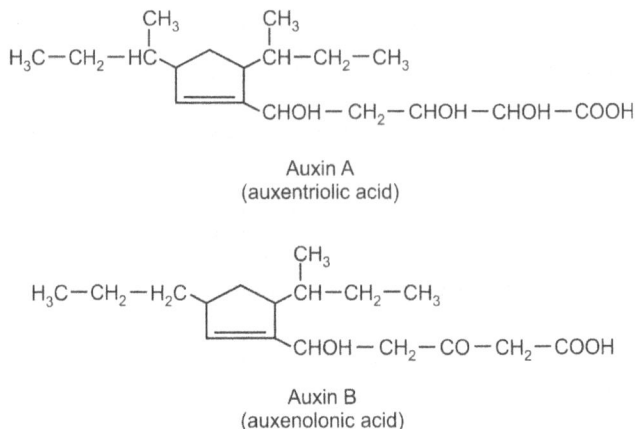

FIGURE 5.2   Structural formulas of auxin A and auxin B.

Initially, 3-indolylacetic acid was isolated from fungi, yeasts, wheat and corn grains, but also from wheat roots. Then it was also extracted from leguminous plants, cabbage, soybeans, spinach, tomatoes, etc.

Biosynthesis of 3-indolylacetic acid – It was highlighted in higher plants, but also in bacteria and fungi. It was also noticed in higher plants. The biosynthesis process starts from tryptophan (Figure 5.3) and goes through an intermediate stage of decarboxylation and the formation of tryptamine or oxidative deamination with the formation of β-indolyl pyruvic acid. In the next step, β-indolyl acetaldehyde and finally 3-indolyl acetic acid are formed. Research on auxins, looking for substances that have chemical structures close to heteroauxin, i.e. *3-indolylacetic acid* (I) and the property of a stimulate plant growth led to the discovery of various chemical derivatives (Pennazio, 2002).

The biogenesis of auxin and other natural derivatives takes place preferentially in young leaves. From the leaves, the physiologically active substances migrate to the growth organs (stems, buds, shoots, etc.).

The migration takes place in the form of *proauxins* – compounds in which the auxin is linked to specific proteins. In the presence of proteolytic enzymes, proauxins are released from proteins and biologically active substances (auxins) and enter into biochemical interactions in the plant tissue at which proteolysis occurs (Simon and Petrášek, 2011).

Figure 5.4 presents various substances isolated from plant tissues. Such compounds are: *indolyl-acetaldehyde* (II) – isolated from cauliflower; *indolyl-acetonitrile* (III) – isolated from white and red cabbage, from tomatoes, from grapes; the *ethyl ester of indolyl acetic acid* (IV) extracted from unripe corn kernels, etc.

Also, various compounds with a carboxylic acid structure were isolated by extraction, such as: *3-indolyl pyruvic acid* (V) – extracted from corn kernels; *3-indolyl carboxylic acid* (VI) – extracted from tomato and cauliflower and *3-indolyl propionic acid* (VII) – extracted from cauliflower. Considering the interest in compounds from this class – characterized by plant growth stimulating properties – new compounds were obtained by organic chemical synthesis: 3-indolyl butyric acid; 1-naphthyl acetic acid; 4-chloro-3-indolyl acetic acid, etc.

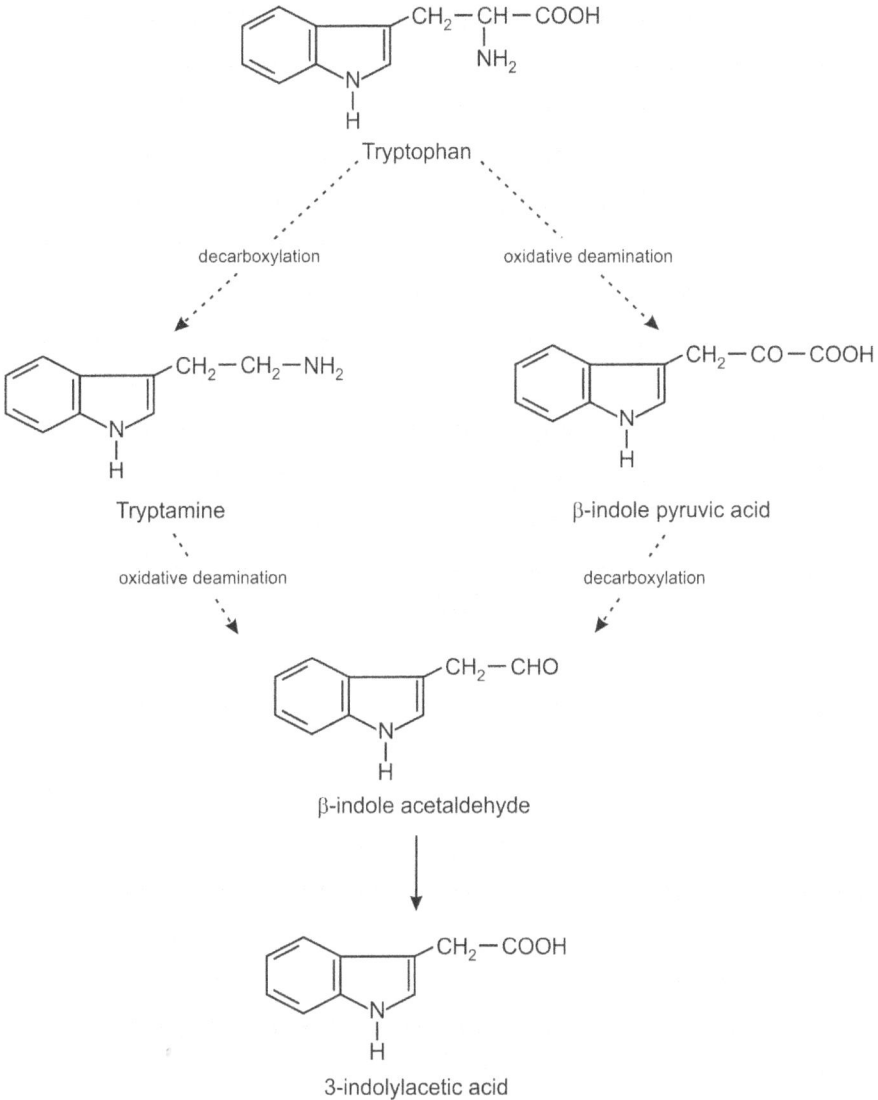

FIGURE 5.3 Biosynthesis of 3-indolylacetic acid from tryptophan.

Such compounds, i.e. PGRs are of interest for agrobiology, being possible to use in reduced concentrations (provided in various normative acts) as a growth stimulator for agricultural plants.

5.4.1.2 Biological Activity

In plant organisms, the biological activity of auxins is characterized by increasing the permeability of cell membranes, facilitating the absorption of water and biomineral compounds. In this way, the intervention in the processes of growth and development is ensured. In this way, the elongation of the cell membranes is produced,

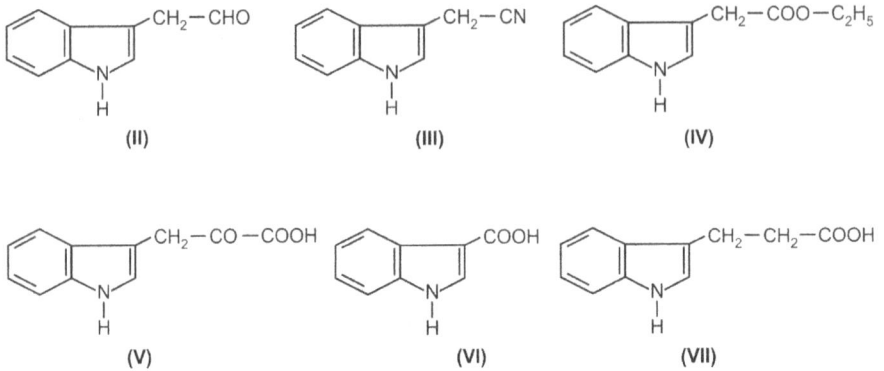

FIGURE 5.4   Natural auxins – structural formulas.

the stimulation of the deposition of new substances, resulting in the thickening of the membranes influencing their permeability and elasticity. In low concentrations, auxins stimulate cell division, and in high concentrations, their elongation occurs.

The absorption of water and ions leads to a decrease in cytoplasmic viscosity and an increase in cell volume. Table 5.1 presents the chemical names of the discussed auxins.

In order to intervene in the metabolic processes, the *chemical compounds* from the auxin group must meet several structural conditions: (i) there must be a cycle with a double bond in the molecule; (ii) there must be, grafted onto the cycle, a side chain in which there is a carboxylic group; (iii) it is also possible to find a carboxylic derivative which during the interactions can generate other carbonyl groups (e.g., aldehydic –CHO, nitrile –CN, ester –CO–O–R); (iv) there must be at least one carbon atom between the ring and the carboxylic group (Cohen and Bandurski, 1982).

In carbohydrate metabolism, auxins increase the amount of reducing carbohydrates, accelerate the hydrolytic processes specific to the starch in the leaves (so that when they ripen, they no longer contain starch).

In the case of lipids, the inhibition of the hydrolytic action of lipases can be distinguished at the level of the stems. This fact leads to a slight increase in lipids in these organs. Protein metabolism under the action of auxins reveals the decrease in the

TABLE 5.1
Chemical Names of Auxins and Derivatives

Chemical Name	Formula
3-Indolylacetic acid	I
3-Indolyl-acetaldehyde	II
3-Indolyl-acetonitrile	III
Ethyl ester of indolylacetic acid	IV
3-Indolylpyruvic acid	V
3-Indolylcarboxylic acid	VI
3-Indolylpropionic acid	VII

amount of amino acids in the leaves and the increase in the stems. Larger amounts of aspartic acid, lysine, valine, leucine, arginine, etc. were found in these organs. In the metabolism of nucleic acids, auxins intervene by stimulating the biosynthesis of DNA through the action of the biosynthesis of some enzymes from the class of oxidoreductases, e.g., peroxidase, alcohol dehydrogenase, phosphorylase, etc. Also, auxins intervene in the splitting of ATP, thus increasing the amount of ADP and inorganic phosphorus (Pi). Thus, there is an increase in the amounts of inorganic phosphates in the stems and roots, followed by a decrease in the leaves.

Also, auxins intervene in the biosynthesis of some water-soluble vitamins from group B (riboflavin and nicotinic acid) and vitamin C, as well as a fat-soluble vitamin – vitamin K.

In plant physiology, it is known that auxins can accelerate or delay the process of seed germination. They intervene on stems and roots, determining their elongation. In the case of the presence of high concentrations of sine, it can determine the appearance of tissue formations with the tumor aspect discussed in molecular biology (Davies, 1995).

5.4.2 Gibberllins

Gibberellins represent a second group of PGRs (phytohormones), which was discovered in 1926 by Kurosawa – a Japanese phytopathologist who was studying the effects produced by the fungus *Gibberella fujikuroi* on young rice plants. It was known that this fungus produces the disease *bakanae* whose effects reside in the intensification of plant growth with the formation of elongated and thin stems, large (narrow and thin) lighter colored leaves and early flowering.

The specific phytopathological observations made by Kurosawa in the case of the "bakanae" disease were based on the finding that the pathological effect is also produced by the free filtrate of cells from the respective fungus.

Gibberellins were initially isolated from fungal sources by Yabuta (1935). In the years to come, the research on a *Gibberella* filtrate led to the isolation of a compound in crystalline form called gibberellin A. This substance was later obtained through chemical synthesis (applied at an industrial level).

In the following years, four compounds with similar structures were obtained from *Gibberella fujikuroi* culture media: gibberellic acid, gibberellin A_1, gibberellin A_2 and gibberellin A_4. Later, changing the culture conditions of the fungus, gibberellins A_7 and A_9 were isolated. Extensive studies on other plants led to the discovery of new gibberellins in beans, peas and pumpkin. From beans, for example, gibberellins A_5, A_6 and A_8 were isolated. Gibberellin compounds have also been isolated from wheat, lettuce (seeds), brown and green algae, coconuts and tree moss, etc. (Mander, 1992; Ifrim, 1997).

For the usual notation of gibberellic compounds, the abbreviations GA_1, GA_2, GA_3, etc. are used. More intense growth stimulating action was highlighted in the case of GA_3. Currently, several dozen gibberellin derivatives are known, generally denoted by the abbreviations: GA_1, GA_2, ... GA_n.

5.4.2.1 Chemical Structure

Structurally, gibberellins are considered to formally originate from the tricyclic hydrocarbon called *perhydrofluorene*. The tetracyclic hydrocarbon called gibberellane (gibban nucleus) is formed from it. Figure 5.5 shows their structural formulas.

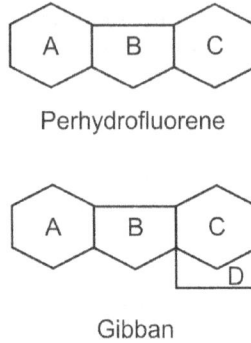

Perhydrofluorene

Gibban

FIGURE 5.5   Hydrocarbons specific to gibberellins.

The chemical structure of gibberellins is best highlighted by two distinct tetracyclic compounds, both acidic in nature that differ in the number of carbon atoms – the C_{19} compound which also has a lactone bond and the C_{20} compound (Figure 5.6).

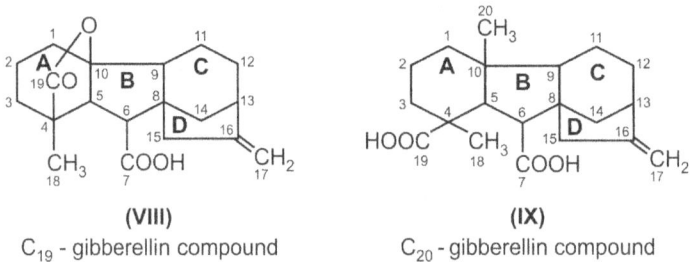

(VIII)
C_{19} - gibberellin compound

(IX)
C_{20} - gibberellin compound

FIGURE 5.6   Structural formulas of gibberellic compounds C_{19} and C_{20}.

Thus, we distinguish the *gibberellin-C_{19}* compound with the chemical name *10,19-norgiberel-16-ene-7-monoenoic acid lactone* and the *gibberellin-C_{20}* compound with the chemical name *gibberell-16-ene-7,19-dioic acid*. In specialized treatises, the C_{19} compound is called gibberellin GA_9, and the C_{20} compound is called gibberellin GA_{12}

Biosynthesis of gibberellins – In higher plants, gibberellins are synthesized starting from mevalonic acid (Figure 5.7), in the chloroplasts of young leaves where metabolic activity is more intense (Sponsel, 1995). A quantitatively reduced synthesis also occurs in seeds (especially during germination), not only in stems but also in flower tissues.

The mechanism of gibberellin biosynthesis involves the formation of an intermediate compound geranyl-geranyl-pyrophosphate, which is a cyclable diterpene. Biosynthesis starts from metabolites that lead to mevalonic acid, from which isopentyl

Mevalonic acid   Isopentenyl pyrophosphate   Geranylgeranyl pyrophosphate

Kaurene   Gibberellin A$_4$

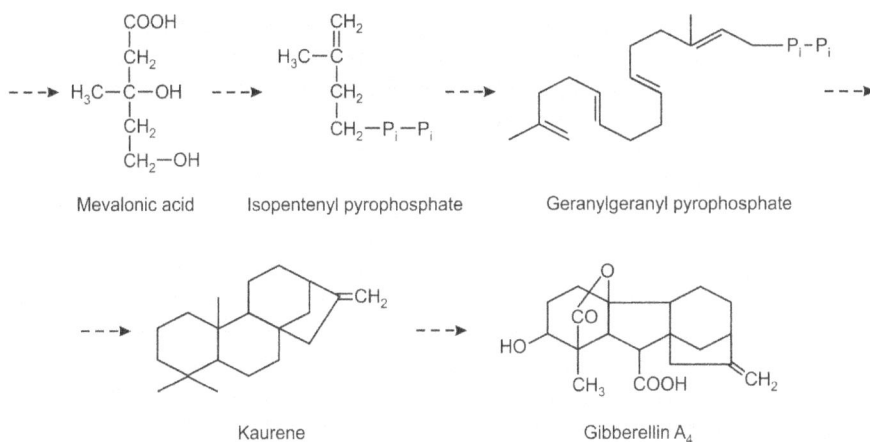

FIGURE 5.7 Biosynthesis of gibberellins with diterpenic intermediate (explanations in text).

pyrophosphate is formed, and from it *geranyl-geranyl-pyrophosphate*. After cyclization, kaurene is formed (dotted lines symbolize the presence of a sequence of intermediate reactions). Finally gibberellin A$_4$ is formed. The detailed mechanism with
the participating enzymes is not fully elucidated.

The biosynthesis of gibberellins is activated by red radiation from the visible
spectrum. Figures 5.8 show the chemical structures of some natural gibberellins, i.e.,

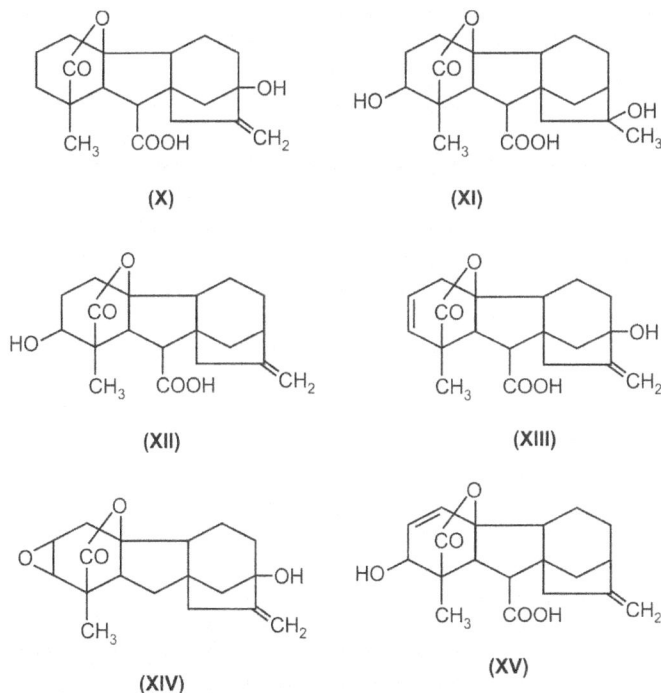

(X)   (XI)

(XII)   (XIII)

(XIV)   (XV)

FIGURE 5.8 Natural gibberellins – structural formulas.

TABLE 5.2
Name of Gibberellins

Chemical Name	Formula
Gibberellic compound C$_{19}$	VIII
Gibberellic compound C$_{20}$	IX
Gibberellin A$_1$	X
Gibberellin A$_2$	XI
Gibberellin A$_3$	XII
Gibberellin A$_4$	XIII
Gibberellin A$_5$	XIV
Gibberellin A$_6$	XV

GA$_1$(X), GA$_2$(XI), GA$_3$(XII), GA$_4$(XIII) and GA$_5$(XIV). They reveal the structural differences mentioned above.

The chemical structure of various gibberellin derivatives is differentiated by the presence of methyl (–CH$_3$), methylene (=CH$_2$), some hydroxyl (–OH) and carboxylic (–COOH) groups. Also, in the structure of numerous compounds, there is a lactone cycle. Table 5.2 presents the trivial (usual) chemical names of the discussed gibberellins (VIII–XIV).

Gibberellic compounds are biologically active substances existing in the vegetable kingdom, which have optically active properties.

For information purposes, it is mentioned that gibberellins synthesized in plant tissues move at a speed of about 5 mm/hour, being "directed", in relation to metabolic processes.

5.4.2.2 Biological Activity

In plant physiology and biochemistry, it is considered that the biological activity of gibberellins is dependent on the structural particularities, the presence of the lactone group and the hydroxyl groups, respectively. The position of the various functional groups mentioned above is also important.

Gibberellins intervene in the production of cell elongation, the increase in cell volume by stimulating the process of cell division.

It was found that they have an important role in increasing the permeability of cell membranes followed by the intensification of the germinative processes of the seeds, the activation of ribosome formation.

From a biochemical point of view, gibberellins intensify the biosynthesis of nucleic acids and specific enzymes. In the case of carbohydrates, it was noted that during germination, gibberellins intervene in the decrease of starch concentration in seeds by intensifying hydrolysis and the release of glucose used in the metabolic processes that accompany growth.

Overall, gibberellins intervene by elongating the stems, increasing the leaf mass, etc. The modification of the general habitat of the plants is actually achieved by the growth of all the organs involved in the growth process. In the case of the administration of gibberellins, the phenomenon of dwarfism (reduced height) is avoided in plants.

Furthermore, gibberellins are believed to be involved in the activation of auxin biosynthesis. The explanation lies in the intervention of gibberellins by stimulating the transformation of tryptophan into tryptamine with the formation of auxins. They also intervene in the biosynthesis of vitamin C and some alkaloids.

Gibberellins also contribute to the activation of enzyme biosynthesis (e.g., amylases, proteases and nucleases) and thereby in carbohydrate and protein metabolism. Intervening in protein metabolism (also affecting nucleic acids), gibberellins delay aging (senescence) in plants.

The effect of gibberellin A_3 highlighted the intensification of the photosynthesis process by fixing carbon dioxide. It also intervenes in the phosphorylation process contributing to the accumulation of phosphoric groups in chloroplasts. However, morphologically, an opening of the green color of the leaves is observed, explained by the decrease in their chlorophyll content with the growth of the whole plant.

In large quantities, the use of gibberellins leads to early flowering, the elongation of floral organs, the acceleration of flower formation and the increase of their weight. There are also inhibitory effects characterized by the reduction of stem size.

Treating plants with gibberellins can produce physiological changes in the development of the ovary followed by the production of seedless fruits – a phenomenon called parthenocarpy. This phenomenon has been observed in tomatoes, grapes, certain varieties of apples, pears, etc.

5.4.3 CYTOKININS

The group of cytokinins includes biologically active substances from the plant kingdom that intervene in cell division. In biology, cell division is the fundamental process by which genetic information is transmitted from one cell to another (within the organism) and from a parent organism to a subsidiary organism (within the species). These problems interest molecular genetics both in the vegetable and animal kingdoms. In plant biology, cell division refers to meristematic tissues and thus the growth processes specific to plants.

Along with cell growth in the body under normal conditions, i.e., *in vivo*, in biology, the growth of cells in sterile cultures, called *in vitro* growth, was also studied. Thus, it was observed that in this case the cell cycle can be influenced (regulated) by various chemical compounds.

The study of the influence of bioregulators (and subsequently of PGRs) was inspired by the existing interrelationship between *Agrobacterium tumefaciens* (a prokaryotic organism – bacteria) and the host plant (a eukaryotic organism). An infection with this bacterium produced a rapid growth of a tumor formation in the host organism.

The growth of cells in sterile cultures is of interest for the study of biochemical interactions in vitro. Cell proliferation under certain conditions was called "callus growth".

From a historical point of view, it is known that the observations on the growth of plant tissues in "tissue cultures" carried out at the beginning of the twentieth century by Gottlieb Haberlandt, led to the concept of "totipotentiality". This was based on the mention of the fact that "theoretically all plant cells have the ability to form whole plants". Some data were published after about a decade (Haberlandt, 1913).

According to the concept of "cellular omnipotence", it is considered that differentiable plant cells can "de-differentiate" or, in some situations, "re-differentiate" with or without the functions of the initial cells.

The discovery of cytokinins – involved in the stimulation of cell division was made much later. In this regard, it is mentioned that in plant cell cultures, Overbeek (1941) obtained formations in the form of "callus" (with undifferentiated cells) in the presence of "milk from coconut" extracted from the fruits of the plant and introduced into the culture medium.

Later, *zeatin* was also discovered, a compound isolated from unripe corn (*Zea mays*) grains (Letham, 1963). It was recorded that 7 mg of crystalline substance was obtained from 60 kg of corn kernels.

Among the cytokinins, the first discovered was *kinetin* (6-furfuryladenine). The name kinetin derives from the fact that it intervenes in the process of cytokinesis – encountered during cell division (Miller et al., 1955). In the specialized literature, the chemical compounds from the group of cytokinins were also mentioned under the names of "phytokinins", "kinetins", "cytokinetins", etc.

5.4.3.1   Chemical Structure

The compounds included in the group of cytokinins, structurally speaking, are derivatives of the purine nucleus. More precisely, it is the amino derivative – the nucleobase called *adenine* (6-aminopurine) with a side chain to the amino group ($-NH_2$) in the C_6 position. The first isolated compound belonging to this group was 6-furfuryladenine commonly called *kinetin*. Figure 5.9 shows the structural formulas of adenine (XVI), kinetin (XVII).

Later, kinetin was also obtained through chemical synthesis in the laboratory.

Subsequent research also led to the obtaining of other biologically active substances that can be included in this group. Among the natural cytokinins isolated over time, the best known is *zeatin* (4-hydroxy-3-methylbutenyl-adenine), which can be presented in the form of geometric isomers: *cis-zeatin* (XVIII) and *trans-zeatin* (XIX). By hydrogenating them, dihydrozeatin (XX) can be formed – Figure 5.10.

From a stereochemical point of view, in the case of zeatin, it was found that *trans-zeatin* – obtained by chemical synthesis – the methyl group ($-CH_3$) in position 3 is in the trans position in relation to the methylene group ($=CH_2$) in position 4 of the side chain.

Along with the mentioned natural compounds, they also isolated other natural cytokinins such as: *isopentenyl-adenine* (XXI), *isopentenyl-adenine-riboside* (XXII) and *trans-zeatin riboside* (XXIII) – Figure 5.10. Among the natural cytokinins in the

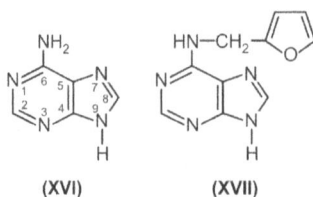

(XVI)   (XVII)

FIGURE 5.9   Adenine and kinetin – structural formulas.

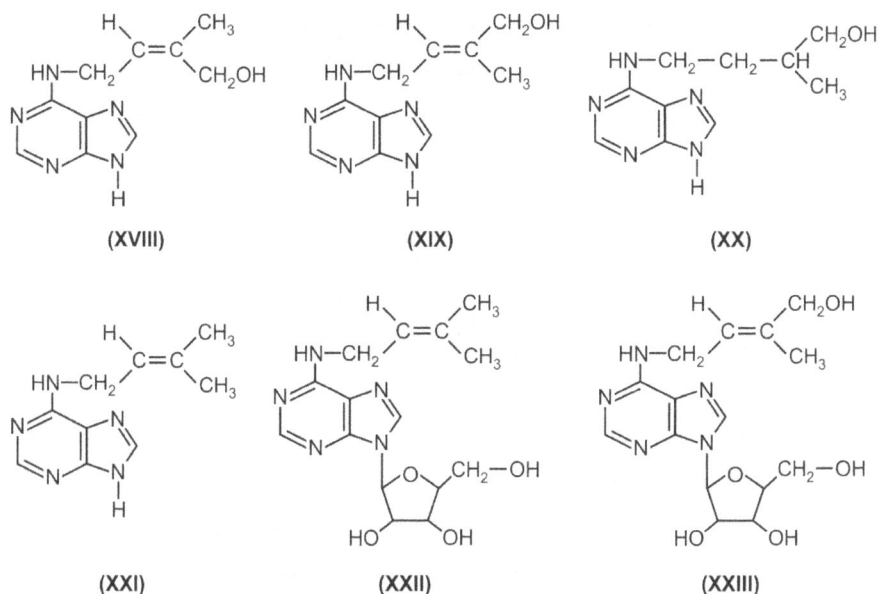

FIGURE 5.10 Natural cytokinins – structural formulas.

plant organism, trans-zeatin is more frequently encountered. Table 5.3 shows the chemical names of cytokinins.

Biosynthesis of cytokinins – In plant tissues, cytokinins are synthesized from bio-degradation compounds of nucleic acids (RNA and DNA). From them comes the purine nucleus, more precisely the adenine residue.

A path of biosynthesis, which is not fully elucidated, starts from ribo-adenosine monophosphate (rAMP) and 2-isopentenylpyrophosphate (for the side chain) and reaches isopentenyl adenine ribotide. Then, under the action of cytokinin synthase, trans-zeatin riboside triphosphate is reached. Two intermediate compounds are formed from trans-zeatin riboside triphosphate (Figure 5.11), from which trans-zeatin riboside and isopentenyl adenine riboside are formed. Finally, isopentenyl adenine and trans-zeatin are formed from these.

TABLE 5.3
Chemical Names of Cytokinins

Chemical Name	Formula
Adenine	XVI
Kinetin	XVII
Cis-zeatin	XVIII
Trans-zeatin	XIX
Dihydrozeatin	XX
Isopentenyl-adenine	XXI
Isopentenyl-adenine-riboside	XXII
Trans-zeatin riboside	XXIII

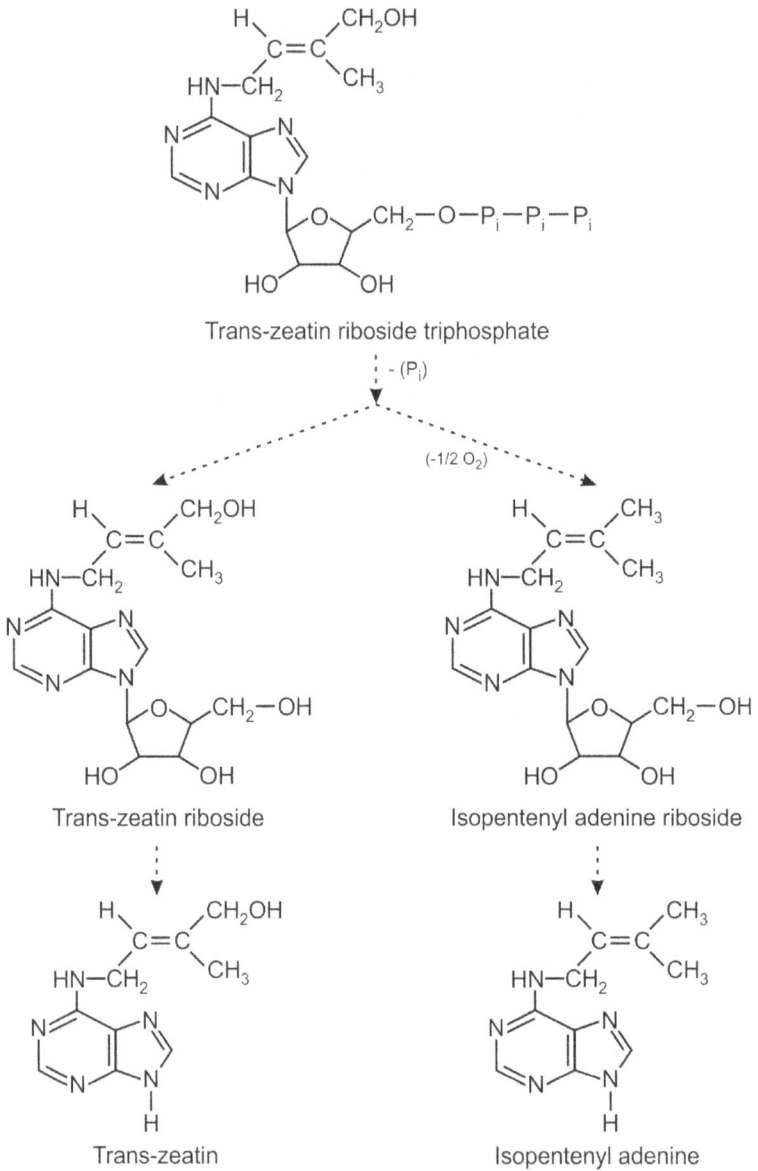

FIGURE 5.11   Biosynthesis of some cytokinins (details in text).

It was found that in plants, biosynthesis occurs at the level of roots. From here, through the vascular tissues, it crosses the stem to other organs of the plant. Small amounts are also synthesized in some plant tissues where they act as a bioregulator. Data on cytokinin biosynthesis have been published by McGaw (1995).

Plant cytokinins can be in free form or conjugated with glucose (via the glycosidic hydroxyl). In the case of zeatin, the conjugation reaction occurs through gluco-conjugation, in which the hydroxyl group of zeatin interacts with the glycosidic hydroxyl of glucose (Glu). Highlighting its glycosidic hydroxyl, the glucose formula will be written $HO-C_5H_{11}O_5$. In the case of glycosyl conjugation, the glycosidase enzyme intervenes, respectively cis-glycosidase and trans-glycosidase (Figure 5.12). Biodegradation of cytokinins is achieved through an oxidative process. For example, under the action of the cytokinin-oxidase enzyme, 3-methyl-2-butenal and adenine are released from isopentenyl adenine.

Native (natural) cytokinins can intervene in the modification of nucleobases in RNA and DNA. In fact, cis-zeatin (the most active form) can be found in many tRNA molecules in living plant cells. The way in which free cytokinins interact with nucleoside monophosphates (NMPs) in nucleotide polymers is not elucidated. The biosynthesis of cytokinins in plants occurs initially in the primary roots. Then it is produced in the apical areas of plant buds, in the primordium areas of leaves and fruits, as well as in seeds.

In general, in plant physiology, the action of cytokinins is discussed in correlation with that of auxins and gibberellins. Thus, the maintenance of cytokinin homeostasis can be better understood (see http://plantphys.info/plant_physiology/cytokinin.shtml, 2012).

In this context, it is mentioned that there is a certain balance between the production of auxins and cytokinins. This balance is ensured by specific genes for the biosynthesis of auxins and cytokinins and the production in DNA of some areas responsible for gene recombination processes. The process is more complex aiming at the production of the "T-DNA system" at the level of plasmid DNA (sometimes referred to only as Ti-plasmid) from *Agrobacterium tumefaciens* (for details, see Gârban, 2009)

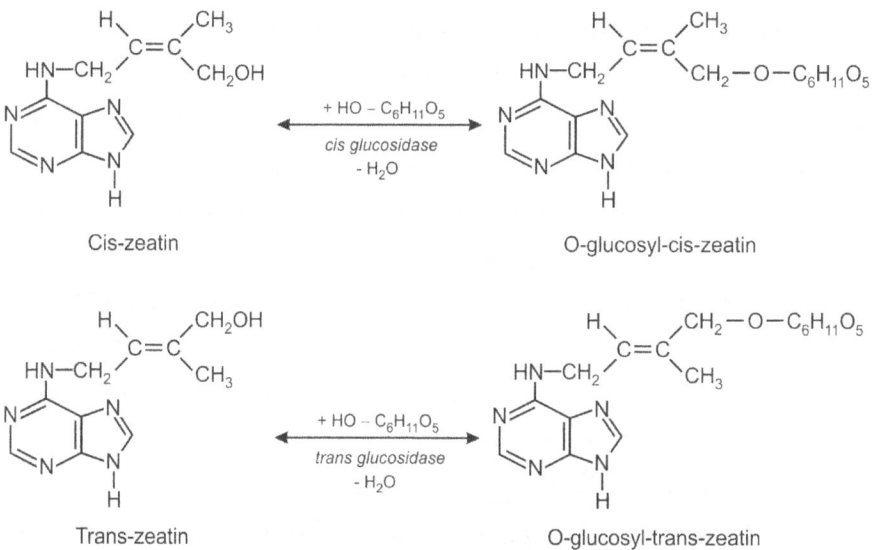

FIGURE 5.12 Conjugation reactions in zeatins.

5.4.3.2 Biological Activity

Cytokinins are PGBs with the role of activators of cell division, but they also intervene in the processes of cell growth and differentiation.

For these reasons, the biological activity of cytokinins is better studied on the tissues that intervene in plant growth: in roots (especially) where they intervene in the stimulation of meristem differentiation, but also in the embryo, the formation of buds, leaves, flowers and fruits. These observations show that the action of stimulating growth is manifested from the period of germination (embryo) to the maturation of the plant (fruits).

The effects of cytokinins, studied experimentally, showed that seeds (after long storage) lose their ability to germinate. The seeds regain this capacity by treating with cytokinin solutions.

Studies on the action of cytokinins on various organs and tissues have highlighted the fact that the effects are more intense in the initial stages of plant growth in relation to the maturation stages. These are distinct peculiarities of plant chronobiology.

Thus, for example, it was found that during the formation of potato tubers, cytokinins stimulate the differentiation process of vegetative buds.

In other species of vegetable plants (such as cabbage, beets, radishes and parsley), cytokinins have a formative role in relation to leaves and stems. Avoiding aging processes is achieved by activating the transport of water and nutrients.

Cytokinins are involved in the biosynthesis of proteins, especially nucleic acids (integrating into the chain of messenger RNA) and chlorophyll. By increasing the amount of chlorophyll, the photosynthesis process is activated. In this way, the action of cytokinins intervenes, as a whole, in the processes of morphogenesis. In excess of water, cytokinins prevent plant aging by retarding leaf senescence.

In plant physiology, it is known that cytokinins increase the resistance of plants to thermal stress (low or high temperatures), resistance to the action of radiation, resistance to infections with viruses or fungi.

5.4.4 ETHYLENE

Ethylene is a PGB with inhibitor effects, discovered by Gane in 1934, and characterized by intervention in the physiological processes involved in plant senescence, e.g., withering of flowers, abscission of leaves and floral petals.

In plant tissues, ethylene binds to protein receptors located in the endoplasmic reticulum membrane of the cell. Its binding occurs at the N-terminal extremity of the protein chain. It is considered that ethylene is synthesized in plant tissues during their development. Tissue distribution presents no difficulties due to the fact that being in a gaseous compound, ethylene dissolves and diffuses easily through the cell membrane.

5.4.4.1 Chemical Structure

Although it is an organic chemical compound known by the state of gaseous aggregation, ethylene (Figure 5.13) is synthesized in plant tissues, for which it has the role of growth bioregulator. Diffusion in tissues is done predilect in the form of a pre-precursor in biosynthesis, known as 1-aminocyclopropane-1-carboxylic acid (AAC).

$$H_2C = CH_2$$

FIGURE 5.13 Ethylene – structural formula.

In superior plants, ethylene biosynthesis has *L-methionine* as the starting compound (Figure 5.14).

This, in the presence of 5′-methylthioribose-1-phosphate under the action of the enzyme *S-adenosylmethionine synthetase* (EC 2.5.1.6) and ribo adenosine triphosphate (rATP) leads to the formation of an intermediate compound called S-adenosylmethionine. Next, under the action of the enzyme *1-aminocyclopropane-1-carboxylate synthase* (EC 4.4.1.14), a non-protein amino acid called *1-aminocyclopropane-1-carboxylic acid* (ACC) is formed.

This takes place under the action of a specific gene family. The production of ethylene (CH_2-CH_2) – in a final step is catalyzed by the enzyme *ACC-oxidase* (EC 1.14.17.4).

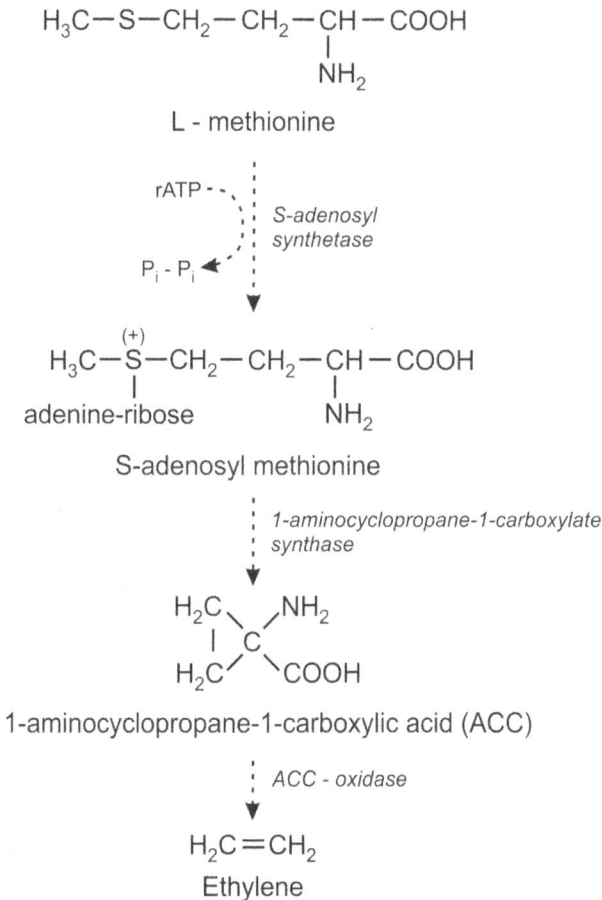

$$H_3C - S - CH_2 - CH_2 - \underset{\underset{NH_2}{|}}{CH} - COOH$$

L - methionine

rATP ⟶

S-adenosyl
synthetase

P_i - P_i ⟵

$$H_3C - \overset{(+)}{S} - CH_2 - CH_2 - \underset{\underset{NH_2}{|}}{CH} - COOH$$

adenine-ribose

S-adenosyl methionine

1-aminocyclopropane-1-carboxylate
synthase

$$\begin{array}{c} H_2C \diagdown \quad \diagup NH_2 \\ | \quad C \\ H_2C \diagup \quad \diagdown COOH \end{array}$$

1-aminocyclopropane-1-carboxylic acid (ACC)

ACC - oxidase

$$H_2C = CH_2$$
Ethylene

FIGURE 5.14 Biosynthesis of ethylene from L-methionine.

In anaerobic conditions, the formation of ethylene is suppressed by Fe^{2+} as a co-factor and by ascorbic acid as a co-substrate, and they activate the enzyme ACC-oxidase. The processes are encoded by a small family of genes detected in plants.

5.4.4.2 Biological Activity

The effect of ethylene at a certain distance from the place of plant tissue biosynthesis is achieved by the release of its precursor – ACC in the plant vascular tissue through which it reaches the place of action. Biotic and abiotic stress can trigger ethylene synthesis in tissues.

The biological activity of ethylene has led researchers to consider this compound released in plant tissues as a PGB produced in quantities that can reach 500 nL g^{-1} h^{-1}, but which is active at very low concentrations, from 10 to 100 nL/L (Abeles et al., 1992; Chaves and de Mello-Farias, 2006).

Ethylene is involved in fruit ripening. Thus, transformations occur during maturation, the taste, aroma, color change (yellow, orange and red colors are highlighted). At the origin of these changes are biochemical reactions that take place successively (in cell biology, the expression "in cascade" is also used).

In fact, a set of genes is activated, which determines the synthesis of a specific mRNA. During translation, proteins are synthesized, including enzymes such as amylase, pectinases, kinases and hydrolases. Thus, amylase transforms starch into disaccharides; pectinases intervene in pectin metabolism; kinases intervene in the transformations of acidic compounds into neutral compounds; hydrolases intervene in the conversion of various organic compounds, e.g., chlorophylls, anthocyanins and odorous substances (collected in vesicles in cells). These changes accompany – in fact – the transition from unripe to ripe fruit.

At the level of plant cell biology, changes in some cellular organelles are discussed. This is how the transition from chloroplasts to chromoplasts occurs. Various glycosides also accumulate in the cell vacuoles.

5.4.5 ABSCISIC ACID

Abscisic acid is the best-known bioregulator of plant growth having inhibitor effects. It was isolated by Eagles and Wareing (1963) from buds of woody species (*Acer pseudoplatanus*, *Betula alba* and *Salix viminalis*) and was called "dormin". In the coming years, it was also isolated from cotton (fruit) by Addicott et al. (1968) being called *abscisin II*. Subsequent studies carried out on these compounds led to the idea that they are identical and were named *abscisic acid*. Gradually it was isolated from numerous plants. As its name suggests, this PGB intervenes in plant leaf abscission. It also mediates the signals received from the environment as a response of plants to thermal and water effects (abiotic stress) and to the action of pathogenic agents (biotic stress).

5.4.5.1 Chemical Structure

Structurally, abscisic acid is an isoprenoid compound with three isoprene residues. This fact led to the idea that such compounds are at its origin. The structural formula is given in Figure 5.15.

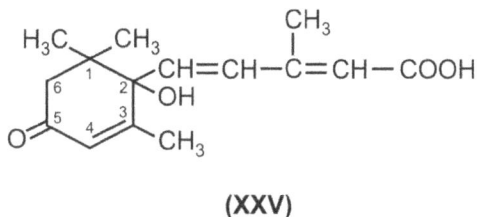

$$\text{(XXV)}$$

FIGURE 5.15   Abscisic acid – structural formula.

The chemical name is *1-hydroxy-2,6,6-trimethyl-4-oxo-2-cyclohexen-1-yl)-3-methyl-2,4-pentadienoic acid.* For the structural formula, there are two ways of writing (both common): the first – represents the formula with the constituent chemical elements; second – a schematic representation suggesting the isoprenoid nature of the compound.

Regarding abscisic acid, there are different opinions regarding its biosynthesis. A first hypothesis claims that it may come from mevalonic acid. Another hypothesis, closer to reality, claims that abscisic acid comes from xantoxin (Walton and Li, 1995; Nambara and Marion-Poll, 2005).

Biosynthesis of abscisic acid – Starting from the above considerations, it can be noted that *violaxanthin* (a carotenoid compound from plants) is transformed into *xantoxin* by photooxidation. Next, this leads to abscisic acid by dehydrogenation (via abscisic aldehyde) and then by oxidation (Figure 5.16).

Abscisic acid is synthesized in the plastids of various plant organs, e.g., roots, flowers and leaves. Also, synthesis occurs in green fruits at the beginning of the winter period, as well as in mature seeds during the dormancy period (winter).

Abscisic acid mobilizes relatively fast in plant organs, being gradually translocated from the roots to the fruits. Thus, a response is produced to the abiotic stress conditions in the environment, e.g., thermal stress, water stress and saline stress.

5.4.5.2   Biological Activity

In the biological activity of abscisic acid, it is mentioned its participation in metabolic reactions that highlight: (i) antiperspirant effects – which induces the closing of the stomata ("pores") at the level of the leaves, reducing thus eliminating water; (ii) inhibiting cell growth and thus inhibiting the premature germination of seeds in the fruit – from a biochemical point of view, the hydrolytic action of amylase on starch is stopped; (iii) decrease in the number of enzymes involved in photosynthesis; (iv) regulating the resting state of dormant buds in the first part of winter and active in the second (when this acid decomposes); (v) increasing the resistance of plants to drought (water stress) and frost (thermal stress), etc.

Abscisic acid is considered the most important natural inhibitor. Its biological activity is antagonistic in relation to plant growth stimulants: indolylacetic acid, gibberellins and cytokinins.

Abscisic acid has also been detected in various genera of fungi (e.g., *Corcospora sp.*, *Botrytis sp.*) It is considered to have a pathological role by suppressing the immune response.

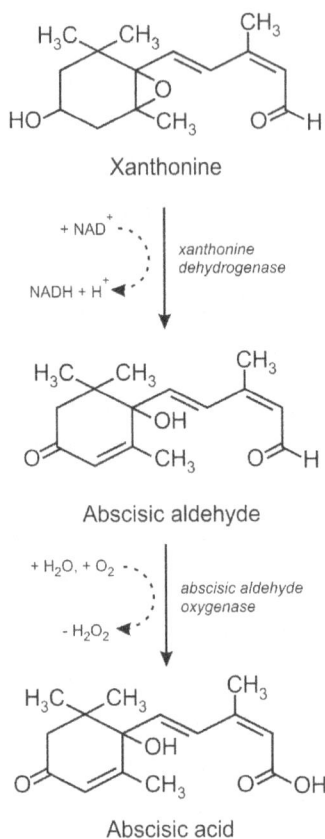

FIGURE 5.16   Biosynthesis of abscisic acid from xantoxin.

In animals, abscisic acid was discovered in metazoa, sponges, including humans. The biological role in animals is little known. It is believed to intervene in the potentiation of the anti-inflammatory and anti-diabetic effect (Bassaganya-Riera et al., 2010). There are some studies on murine animal models (mice).

5.4.6   OTHER COMPOUNDS AS BIOREGULATORS

Among the PGBs, there are a number of compounds that are not mentioned in the classic treatises on plant biochemistry. Such compounds have been more intensively studied in the last three decades (Kung and Yang, 1998; Buchanan et al., 2000; Bajguz and Tretyn, 2003). Among these are: *brassinosteroids* (**XXVI**), *strigolactones* (**XXVII**), *jasmonates* (**XXVIII**) and *salicylic acid* (**XXIX**) – see Figure 5.17. A brief presentation of them follows.

Brassinosteroids – It represents a family of polyhydroxylated compounds isolated from plants. Their chemical structure reveals their steroidal character, with C_{27} brassinosteroids, C_{28} brassinosteroids, and C_{29} brassinosteroids

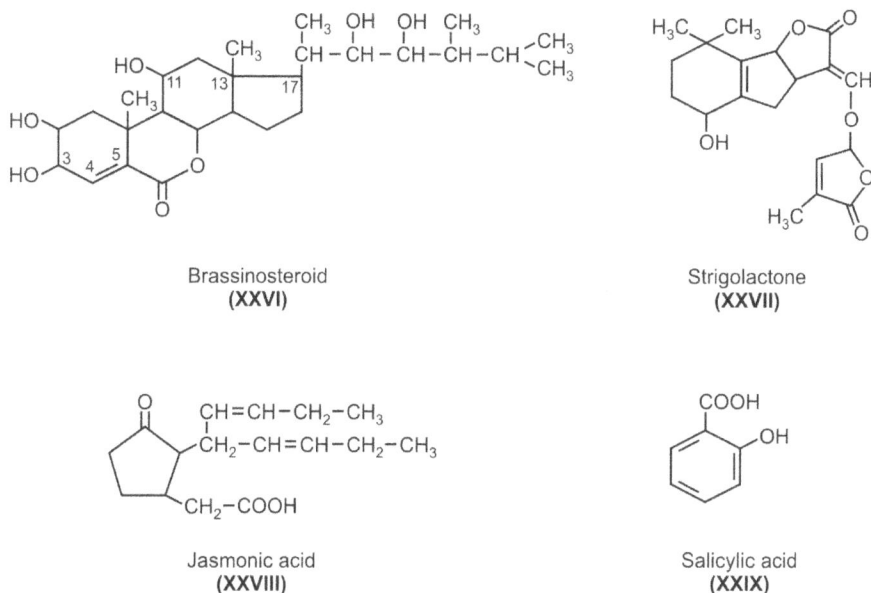

FIGURE 5.17   Various bioregulators – structural formulas.

(Bajguz and Tretyn, 2003). Also included in this group is a steroidal lactone derivative called brassinolide. The first compound from this group was obtained from turnip (*Brassica napus*) pollen. There are currently over 70 compounds isolated from plants. Among them, the most active is brassinolide. In plant physiology, it is known that these compounds intervene in the elongation of cells, involvement in the development of pollen. Brassinosteroids have a receptor present in the plasma membrane (called BRI A), which is activated in the presence of the protein kinase enzyme.

Strigolactones – They are compounds derived from carotenoids and were identified initially in root exudates (primary roots) and later on from other parts of the plant. These substances stimulate the germination of the seeds of parasitic plants from the genera *Striga* (from which they got their names) *Orobanche* and *Phelipanche*. Also, strigolactones were isolated that stimulate the development of different fungi, etc.

Jasmonates – Are compounds containing cyclopentanone. They come from the metabolism of membrane fatty acids. Methyl jasmonate and jasmonic acid are mainly formed.

They are considered multifunctional compounds that intervene in the metabolic processes associated with the development of plants, their survival and implicitly reproduction (flowering, fruiting, senescence). Jasmonates activate the response to pathogenic attacks, to environmental stress, e.g., salinity (increases plant tolerance to environments where Cu and Cd are present); low temperatures. Jasmonates from plants eaten by herbivores cause oral lesions. For this reason, the effects of jasmonates induced the expression of "anti-herbivore chemicals", which contribute to the "systemic defense response" of plants by removing herbivores (Wang et al., 2021). The best investigated compound is jasmonic acid (AJ). It is conjugated with isoleucine (AJ - Ile).

Salicylates – The biosynthesis of salicylates is induced by a pathogenic attack. Salicylates contribute to the acquisition of systemic resistance to pathogens. There are no known receptors for salicylates.

5.4.7 Natural and Synthetized Inhibiting Substances

The activity of growth bioregulators in plants is sometimes limited or even annihilated by other endogenous substances. Such substances, called "natural inhibitors", have been isolated from various plant tissues. Later, such substances were also obtained through organic synthesis, called "synthetized inhibitors".

Sometimes the classification of endogenous compounds with attributes of inhibitors or stimulators is difficult, because the effect is conditioned by the concentration of the substance. Thus, the same compound can act as an inhibitor or as a stimulator in plant growth.

In general, the inhibitors are considered to have distinct anti-gibberellin, anti-auxin and anti-cytokinin action. The succinct presentation of the inhibitors follows.

5.4.7.1 Natural Inhibitors

The separation of compounds with the role of inhibitors was carried out by Linser and Kaindl (1951). Later, their effects on plant metabolism were also investigated.

Structurally, the substances are isolated from various plant tissues and belong to different classes of organic compounds:

- *phenols* – phlorizin (the aglycon is phloretin), quercetin, epicatechin, juglen, etc.
- *carboxylic acids* –caffeic acid, cinnamic acid, chlorogenic acid, etc.
- *lactones* –coumarin, aesculin, scopoletin, etc.

In some specialized works, natural inhibitors are mentioned as a distinct category of compounds, including abscisic acid. Natural inhibitors ensure the maintenance of plants in a state of rest, a circumstance called "dormancy" (in treatises on plant physiology). It also regulates the growth of buds and flowering, favors the abscission of fruits and leaves, prevents the germination of seeds in the fruit, accelerates the senescence (aging) of tissues by decomposition of chlorophyll, etc. Among the natural inhibitors, there are some more well-known compounds: *phlorizin, cinnamic acid* and *coumarin*. The chemical structures of these compounds are shown in Figure 5.18.

Phlorizin – It is a glycoside whose aglycone is phloretin. Its structure highlights the flavonoid nature of the compound.

This compound has an inhibitory role for plants in winter by inhibiting the action of enzymes involved in the growth process. At the end of winter, the amount of phloretin in the buds decreases through enzymatic hydrolysis. This is how growth stimulants can intervene.

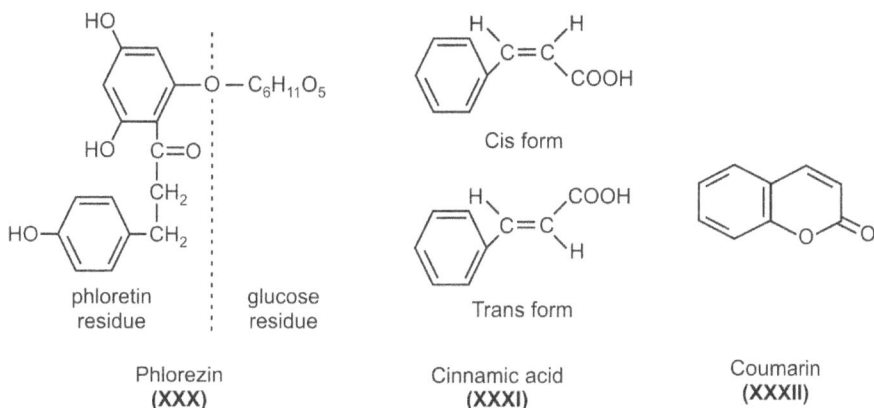

FIGURE 5.18 Growth bioregulators with the role of inhibitory substances.

Cinnamic acid – It is found in plant tissues in the cis and trans forms. The cis form has a synergistic action with heterouxin, and the trans form is antagonistic to the action of indolyl-3-acetic acid.

Coumarin – Structurally, it is ortho hydroxycinnamic acid lactone. In plants it is found in leaves, fruits, seeds. It is present in the form of glucoside.

In plant physiology, it is known as an inhibitor of seed germination, with effects similar to abscisic acid.

5.4.7.2 Synthetized Inhibitors

They are chemical compounds obtained in the laboratory whose synthesis was imposed by the requirements of agricultural practice. From a structural point of view, they differ a lot.

From this group of better known compounds are *maleic hydrazide, actinomycin D, chloramphenicol, puromycin,* etc. Brief data on effects in plant physiology are presented below.

Maleic hydrazide – It intervenes as an inhibitor of cell division at the level of merismatic tissues. This causes a reduction in the length of the shoots in the trees. It also delays the formation of inflorescences, inhibits the sprouting of potato tubers, onion bulbs, etc. From a biochemical point of view, there are disturbances in the enzyme systems from the metabolic reactions.

Actinomycin D – It inhibits the growth of cell membranes and reduces turgorescence. At the biochemical level, there is a blockage of protein synthesis (predilection of RNA synthesis), disturbances of photosynthesis through changes in the biosynthesis of proteins and carbohydrates in the composition of cell membranes. In the medical field, this compound is known as an antibiotic.

Chloramphenicol – In the vegetable kingdom, the action of chloramphenicol (chloromycetin) resides in the disruption of chloroplast morphogenesis (by inhibiting the formation of lamellae from plastids), reducing the amount of chlorophyll and, thereby, photosynthesis. Also, the mitosis process is deregulated and chromosomal breaks occur. In plant physiology, the action of chloramphenicol is discussed as an antagonistic action to gibberellins. From a biochemical point of view, changes occur in the biosynthesis of proteins (metalloproteins, nucleoproteins) and carbohydrates. And this compound is known as an antibiotic in the medical field.

Puromycin – It is a purine mononucleoside produced by fungi. It has an inhibitory action on the activity of enzymes that compete for the formation of anthocyanins from plants.

5.4.8 SYNTHETIZED RETARDANT SUBSTANCES

These substances have been studied alongside PGBs for the reason that they intervene in the process of plant morphogenesis, having an inhibitory character for the processes of cell division, at the same time reducing the elongation of the cells in the stems.

For the reason that they intervene in the metabolic processes, reducing the elongation of the cells and thereby slowing down the height growth of the plants, these compounds have the general name of *retardant substances.*

Analytical methods have been used in plant biochemistry for the isolation of these substances, which involve the fractional extraction of natural compounds, chromatographic separation, investigation by spectrophotometric methods, etc.

The change in the length of the stems occurs especially at the level of the internodes. Thus, there is a shortening of the internodal zones, a thickening of the stems and a stimulation of the lateral ramifications (important in some leguminous plants).

The retardant substances also intervene, stimulating flowering, fruiting and ripening.

The physiological action of the retardants manifests itself in various ways. For example, important changes occur in fruit plantations affecting vegetative growth and fruit production. The increase in the length of the shoots, the numerical increase of the fertile inflorescences, with the stimulation of the growth of the fruits, is noticeable. Premature leaf fall is reduced. As a whole, qualitatively and quantitatively superior harvests are obtained.

It also increases the resistance of plants to abiotic stress – to temperatures (cold/hot), humidity, salinity, but also to biotic stress (diseases and pests).

Through chemical synthesis, various compounds used as retardants were obtained. These were usually called compound I, compound II, etc. Among them, the most active is *2,4-dichlorobenzyl-nicotine chloride* (later called compound I). This causes the beanstalks to elongate. Finally, there is a decrease in the weight of the stems and an increase in the weight of the leaves.

Another retardant is *2-isopropyl-7-trimethyl-ammonium-5-methylphenyl-1-piperidinecarboxylate* (called compound II). This compound, with an ammonium ion, is a retardant for various plants, e.g., beans, cucumbers, chrysanthemums, etc. It is also known for its bactericidal properties.

The number of retarding compounds is greater – their use is of interest for agricultural productions and for some flowers.

5.5 BIOREGULATORS IN FOOD CHEMISTRY AND NUTRITION

PGBs accompany – as biologically active substances – the nutrients present in food products of vegetable origin (Bodea and Enăchescu, 1984; Alais and Linden, 1993; Magnone et al., 2015; Gârban, 2018; Fathy et al., 2022). In the animal and human organisms, these compounds influence the biochemical and physiological processes.

Such influence in humans are discussed in a compilation of data from various scientific works by Mukherjee et al. (2022). In this way, it was found that PGBs intervene in carbohydrate metabolism (glucose homeostasis), in the antioxidant response at cellular level, cell division, inflammatory processes, inhibition of cancer growth, etc.

Auxins – In human biology, it was found that they intervene in the regulation of the cell cycle. Also, they have anti-tumoral properties.

Gibberellins – Were isolated from various food products, especially from vegetable seeds (e.g., beans contain gibberellins A_1 in an amount of 0.2 mg per 100 g of beans). In general, it is estimated that they intervene in the specific phosphorylation reactions for the biosynthesis of protein-enzymes, the biosynthesis of nucleic acids, etc.

They were reported to induce oxidative stress by generating reactive oxygen species. Animal experiments revealed hepatotoxicity in adult albino rats. Gibberellins can induce damage at DNA level, but they have antitumor properties.

Cytokinins – Intervene in the metabolization of proteins (enzymatic biogenesis, biosynthesis of nucleic acids); activates the transport and accumulation in tissues of various inorganic and organic compounds; activates the transmembrane transport of calcium ions, etc. In humans, geronto-modulatory and anti-aging properties were evidenced for cytokinins. Also, they can damage the DNA macromolecule, affecting gene expression, i.e. an intervention at the level of cell cycle. Cytokinins were also isolated in the urine of patients with cancer, derived from tryptophan metabolization.

Ethylene – Is a naturally occurring substance in plants, promoting senescence processes in fruit ripening, stress responses and vegetative development. Ethylene production and response can be genetically and biochemically manipulated to speed up or delay fruit ripening and, in this way, to stabilize foods during their transport to the consumer. Fruits producing ethylene are apples, pears, tomatoes, melons, avocados, apricots, etc.

Physiologically and biochemically, ethylene interferes with the increase of permeability of biomembranes, the biosynthesis of proteins. In the case of ethylene, the applicative effect resides in stimulating the ripening of fruits stored in closed rooms. Premature ripening occurs. Intervines in biosynthesis processes that accompany oxidative stress. One of its derivatives, ethylene dioxide, has genotoxic and carcinogen effects in humans.

Ethylene oxide is banned for use in food and feed since 2011 because can have mutagenic and carcinogenic effects.

Abscisic acid – Is naturally present in fruits and vegetables (e.g., apricot, avocado, banana, carob, soy bean, potato, tomato, pea, cucumber, etc.). It is considered a pro-inflammatory substance in stress conditions. Animal experiments with abscisic acid revealed antidepressant property. Other studies showed that glucose homeostasis is regulated by abscisic acid.

From the point of view of human nutrition, the physiological role of abscisic acid is not elucidated.

5.6 GROWTH REGULATORS IN AGROBIOLOGY

5.6.1 GENERAL CHARACTERISTICS

The evolution of knowledge about PGB started from general data on extraction methods to data on chemical structure and biological activity. This information gradually led to the idea of the need to obtain these substances through organic chemical synthesis (Nickell, 1982; Gergen et al., 1988; Arteca, 1996).

In this sense, the processes carried out in biochemistry and plant physiology were seconded by applications in agrobiology, involving leguminous plants, fruit trees, vines, etc. The existing information gradually led to the synthesis of various compounds with the role of stimulators or inhibitors for plants in agricultural crops (Heller et al., 2004; Jiang and Asami, 2018).

The pleiad of these compounds obtained through organic chemical synthesis has the generic name of "plant growth regulators" (PGR). Thus, the series of substances obtained by chemical extraction from plants, briefly presented in this chapter, was completed. The action of cytokinins on plants can be synergistic or antagonistic with that of auxins and gibberellins.

With reference to synthetic substances (i.e., PGR), it is important to note that their use must be preceded by experimental research to avoid possible risks. In fact, this is a general precept in the field of research regarding compounds of pharmaceutical interest, in this case phytopharmaceuticals.

5.6.2 COMPARATIVE EFFECTS

In intensive agriculture, PGRs have been used. In this way, the increase of economic benefits was pursued. The first applications of PGR in agriculture were carried out in 1930 in the USA. Initially, the effect of ethylene was observed as a natural product – so as a PGB. In these conditions, the extracted ethylene administered during the development of pineapple plants (*Ananas comosus*) was used. Positive effects on flower growth and increased fruit production were observed. The effects on the human body were extremely low. These findings led to the idea of using PGR in agriculture.

In the years that followed, the use of ethylene expanded. This PGR in gaseous form has proven effective in the ripening stage of various fruits, including bananas, apples, pears and watermelons. For this purpose, a compound called ethephon – which has specific effects of ethylene on apples and bananas – was obtained by synthesis. Overall, ethylene proved effective for ripening apples and especially bananas (from one day to the next).

The use of PGR – compounds obtained by synthesis – was also extended in the case cytokinins. In this sense, the use of the compound called for chlorfenuron in the cultivation of watermelons is mentioned. This compound is "suspected" of effects exacerbated by the so-called "watermelon explosion". The compound has been approved for use in the USA for grapes and kiwifruit. Tested on animals (mice), it was found to produce alopecia (hair loss).

By synthesis cytokinin-specific compounds were obtained and used as PGR. Compounds such as benzyladenine, tetrahydropyranyl-benzyladenine and diphenylurea are also mentioned (Figure 5.19). In the structure of the first two compounds,

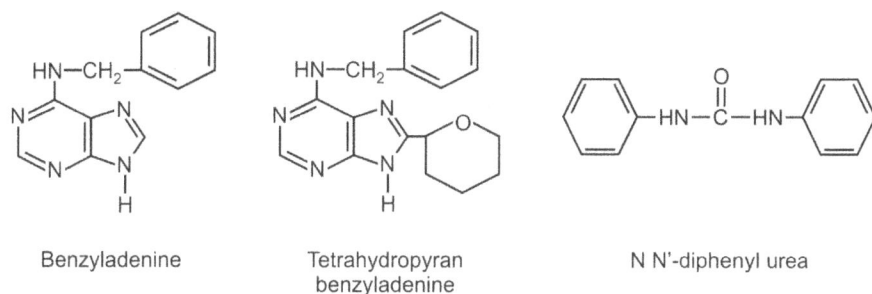

Benzyladenine   Tetrahydropyran   N N'-diphenyl urea
                             benzyladenine

FIGURE 5.19   Other cytokinins – structural formulas.

there are heterocyclic compounds (nitrogens and oxygenates). The third compound contains a urea residue.

These compounds have physiological effects similar to cytokinins. Another group of PGR compounds used in agriculture were auxins. In them, the effect of stimulating the development of the root zone was known. Synthetic auxins have been obtained for applications in agriculture. These include indole-butyric acid and naphthyl-acetic acid. Synthetic auxins have found wide use in preventing fruit drop before ripening, observed in apples.

The problem of the use of synthetic compounds from the PGR group in agriculture is the focus of investigations regarding the so-called "good agricultural practice" (GAP). It is noted that in order to correctly apply the regulations regarding consumer protection in relation to xenobiotics, specific notions are used, e.g., the so-called "good practice". In this framework, there is also the term "good manufacturing practice" – (GMP) used for processed products (in the food industry). For the application of GAP measures, the anticipated registration by the relevant authorities (from each country) of the substances used as PGRs was foreseen.

Only in this situation is the use in agriculture allowed. The measures taken regarding "food safety" led to the establishment of some control rules for an applicative perspective. In this framework, RCPs were classified similar to pesticides. Thus, there are specific legislative provisions for the control of these compounds, similar to those for pesticides.

REFERENCES

Abeles F.B., Morgan P.W., Saltveit M.E. - *Ethylene in Plant Biology*, 2nd edition, San Diego Academic Press, San Diego, 1992.

Addicott F.T., Lyon J.L., Ohkuma K., Thiessen W.E., Carns H.R., Smith O.E., Cornforth J.W., Milborrow B.V., Ryback G., Wareing P.F. - Abscisic acid: a new name for abscisin II (dormin), *Science*. 1968, 159(3822), 1493.

Alais C., Linden G. -*Biochimie Alimentaire*, 2-ème edition, Masson, Paris-Milan-Barcelone-Bonn, 1993.

Arteca R. - *Plant Growth Substances: Principles and Applications*, Chapman and Hall, New York, 1996.

Bajguz A., Tretyn A. - The chemical characteristic and distribution of brassinosteroids in plants, *Phytochemistry*, 2003, 62, 1027–1046.

Bassaganya-Riera J., Skoneczka J., Kingston D.G., Krishnan A., Misyak S.A., Guri A.J., Pereira A., Carter A.B., Minorsky P., Tumarkin R., Hontecillas R. - Mechanisms of action and medicinal applications of abscisic acid, *Curr. Med. Chem.*, 2010, 17(5), 467–478.

Bodea C., Enăchescu G. - *Treatise on Plant Biochemistry, Part II. The Chemical Composition of the Main Cultivated Plants, Vol. V - Vegetables* (in Romanian), Editura Academiei R.S.R., Bucureşti, 1984.

Buchanan B.B., Gruissem W., Jones R.L. - *Biochemistry and Molecular Biology of Plants*. American Society of Plant Physiologists, Rockwill, Maryland, 2000.

Chaves Ana Lúcia S., de Mello-Farias P.C. - Ethylene and fruit ripening: from illumination gas to the control of gene expression, more than a century of discoveries, *Genet. Mol. Biol.*, 2006, 29(3), 508–515.

Cohen J.D., Bandurski R.S. - Chemistry and physiology of the bound auxins, *Ann. Rev. Plant Physiol.*, 1982, 33, 403–430.

Davies P.J. (Ed.) - *Plant Hormones: Physiology, Biochemistry and Molecular Biology*, Kluwer, Dordrecht, 1995.

Eagles C., Wareing P. - Dormancy regulators in woody plants: experimental induction of dormancy in *Betula pubescens*, *Nature*, 1963, 199, 874–875. https://doi.org/10.1038/199874a0.

Fathy M., Saad Eldin S.M., Naseem M., Dandekar T., Othman E.M. - Cytokinins: wide-spread signaling hormones from plants to humans with high medical potential, *Nutrients*, 2022, 14(7), 1495. https://doi.org/10.3390/nu14071495.

Gane R. - Production of ethylene by some ripening fruits, *Nature*, 1934, 134, 1008. https://doi.org/10.1038/1341008a0.

Gârban Z. (Ed.) - *Molecular Biology: Concepts, Methods, Applications* (in Romanian), 6th edition, Editura Solness, Timişoara, 2009.

Gârban Z. - *Biochemistry: Comprehensive Treatise, Vol. II. Biochemical Effectors* (in Romanian), 5th edition, Publishing House of the Romanian Academy, Bucureşti, 2018.

Gergen I., Goian M., Puşcă I., Borza I., Lăzureanu A., Vălceanu R. - *The Use of Bioregulators in Agricultural Production* (in Romanian), Editura Facla, Timişoara, 1988.

Goodwin T.W., Mercer E.I. - *Introduction to Plant Biochemistry*, Pergamon Press Ltd., Oxford-New York-Toronto-Paris-Frankfurt, 1983.

Guleria S., Kumar M., Kaushik R. - Plant hormones physiological role, effect on human health and its analysis, *Journal of Microbiology, Biotechnology and Food Sciences*, 2021, 11(1), e1147. https://doi.org/10.15414/jmbfs.1147.

Haberlandt G. - Zur physiologie der zellteilung, *Sitzber. K. Preuss. Akad. Wiss.*, 1913, 318–345.

Harborne J.B. (Ed.) - *Methods in Plant Biochemistry, Vol. 1*, Academic Press, London, 1989.

Heller R., Esnault R., Lance C. - *Physiologie végétale. Tome 2, Développement*, 6-ème édition, Dunod, Paris, 2004.

Ifrim S. - *Biologically Active Substances* (in Romanian), Editura Tehnică, Bucureşti, 1997.

Jiang K., Asami T. Chemical regulators of plant hormones and their applications in basic research and agriculture, *Biosci. Biotechnol. Biochem.*, 2018, 82(8), 1265–1300. https://doi.org/10.1080/09168451.2018.1462693.

Kögl F., Haage-Smit A.J., Erxleben H. - Über ein neues Auxin (Heteroauxin) aus Harn. XI. Mitteilung über pflanzliche Wachstumsstoffe, *Hoppe-Seyler's Z. Physiol. Chem.*, 1934, 228, 90–103.

Kung S.-D., Yang S.-F. (Eds.) - *Discoveries in Plant Biology, Vol. I*, World Scientific, London-River Edge-Singapore, 1998.

Kurosawa E. - Experimental studies on the nature of the substance secreted by the "bakanae" fungus, *Nat. Hist. Soc. Formosa*, 1926, 16, 213–227.

Letham D.S. - Zeatin, a factor inducing cell division isolated from zea mays, *Life Sci.*, 1963, 2, 569–573.

Linser H., Kaindl K.-The mode of action of growth substances and growth inhibitors. *Science,* 1951, 114, 69-70.

Magnone M., Ameri P., Salis A., Andraghetti G., Emionite L., Murialdo G., De Flora A., Zocchi E. - Microgram amounts of abscisic acid in fruit extracts improve glucose tolerance and reduce insulinemia in rats and in humans, *FASEB J.,* 2015, (12), 4783–4793. https://doi.org/10.1096/fj.15-277731.

Mander L.N. - The chemistry of gibberellins: an overview, *Chem. Rev.,* 1992, 92, 573–612.

McGaw B.A. - Cytokinin biosynthesis and metabolism, pp. 98–117, in *Plant Hormones: Physiology, Biochemistry and Molecular Biology* (Davies P.J., Ed.), Kluwer, Dordrecht, 1995.

Miller C.O., Skoog F., von Saltza M.H., Strong F.M. - Kinetin, a cell division factor from deoxyribonucleic acid, *J. Am. Chem. Soc.,* 1955, 77, 1392.

Mukherjee A., Gaurav A.K., Singh S., Yadav S., Bhowmick S., Abeysinghe S., Verma J.P. - The bioactive potential of phytohormones: a review, *Biotechnol. Rep. (Amst.),* 2022, 8, 35, e00748. https://doi.org/10.1016/j.btre.2022.e00748.

Nickell L.G. - *Plant Growth Regulators - Agricultural Uses,* Springer-Verlag, Berlin, 1982.

Nambara E., Marion-Poll A. - Abscisic acid biosynthesis and catabolism, *Ann. Rev. Plant Biol.,* 2005, 56, 165–185.

Neamţu G.V. - *Ecological Biochemistry* (in Romanian), Editura Dacia, Cluj-Napoca, 1983.

Paál A. - *Uber phototropische Reizleitungen, Ber. d. bot. Ges.,* 1914, 32, 499–502.

Pennazio S. - The discovery of the chemical nature of the plant hormone auxin, *Riv. Biol.,* 2002, 95(2), 289–308.

Simon S., Petrášek P. - Why plants need more than one type of auxin, *Plant Sci.,* 2011, 180(3), 454–460, https://doi.org/10.1016/j.plantsci.2010.12.007.

Sponsel V.M. - Gibberellin biosynthesis and metabolism, pp. 66–97, in *Plant Hormones: Physiology, Biochemistry and Molecular Biology* (Davies P.J., Ed.), Kluwer, Dordrecht, 1995.

Taiz L., Zeiger E. - *Plant Physiology,* 2nd edition, Sinauer Associates, Massachusetts, 1998.

vanOverbeek J., Conklin M.E., Blakeslee A.F. - Factors in coconut milk essential for growth and development of very young Datura embryos, *Science,* 1941, 94(2441), 350–351. https://doi.org/10.1126/science.94.2441.350.

Walton D.C., Li Y. - Abscisic acid biosynthesis and metabolism, pp. 140–157, in *Plant Hormones: Physiology, Biochemistry and Molecular Biology,* Kluwer, Dordrecht, 1995.

Wang Y., Mostafa S., Zeng W., Jin B. - Function and mechanism of jasmonic acid in plant responses to abiotic and biotic stresses, *Int. J. Mol. Sci.,* 2021, 22, 8568. https://doi.org/10.3390/ijms22168568.

Went F.W. - On growth-accelerating substances in the coleoptile of *Avena sativa, Proc. Kon. Akad. Wetensch. Amsterdam,* 1926, 30, 10–19.

Went F.W. - Wuchsstoff und Wachstum, *Extrait Recueil Travaux Botanique Neerl.,* 1928, 25, 1–116.

Went F.W., Thimann K.V. - *Phytohormones,* The MacMillan Co., New York, 1937.

Yabuta T. - Biochemistry of the 'bakanae' fungus of rice, *Agric. Hort. (Tokyo),* 1935, 10, 17–22.

6 Ad Limina Interrelations
Organism – Biologically Active Substances

6.1 INTRODUCTION

Biochemical transformations that continuously take place in the body are specific attributes of living matter, which highlight the so-called "dynamic characteristics of biological processes". Knowing them is important for biochemistry, molecular biology, physiology, nutrition, pharmacology, pathobiochemistry, etc. In general, the "dynamic characteristics of biological processes" define transformations in time and space in a continuous, interrelational, reversible and/or non-reversible manner.

These characteristics usually include: metabolization of nutrients, biotransformation of xenobiotics, chronobiochemistry, homeostasis, allostasis, homeorhesis, bioavailability and pathobiochemistry (Gârban, 2014).

Numerous aspects related to these "dynamic characteristics" refer to biologically active substances of food, pharmacological and agrobiological interest. Next, two important characteristics are presented for understanding the problems related to biologically active substances.

The exposed data are limited to *bioavailability* and *chronobiochemistry*. To have a general picture of the interactions of biologically active substances with the organisms, the issue of their traceability is also discussed. Internal/external traceability monitoring in the case of biologically active substances will provide certainty in ensuring the beneficial effects on the body.

6.2 BIOAVAILABILITY

6.2.1 BASIC CONCEPTS

Bioavailability is generally defined by the relationship established between a biologically active substance (physiological, pharmacological and toxicological) and the organism in which this substance enters. To estimate the bioavailability of a biologically active substance, the amount of it in biological tissues/fluids is assessed, compared to the amount in which it enters (from food, medicine and the environment).

The intensity of the effect and its duration are also evaluated. In this context, various biochemical barriers play a major role. By carefully deepening the concept of bioavailability and its applications, it can be noted that there is an indissoluble connection with biochemistry and xenobiochemistry. In this sense, bioavailability can be approached in relation to the metabolization of nutrients and the biotransformation of xenobiotics.

DOI: 10.1201/9781032702520-6

Obviously, in discussing the issues related to bioavailability (BA), the correlation with the aspects of transit through biochemical barriers is not without importance. It can be concluded that the metabolization of nutrients and the transformation of xenobiotics must be seen as processes conditioned by natural biochemical barriers (Gârban, 1999).

In a more extensive framework, the perspective of specific pathobiochemistry interactions is also considered. The concept of "bioavailability" was used for the first time in relation to the amount of absorption of vitamins from pharmaceutical preparations (Oser et al., 1945- cited by Arnold and Sicé, 1976). It is mentioned that vitamins are both nutrients and medicines, depending on the dose. From this simple example, it can be understood that the problem of bioavailability is common in nutrition, pharmacology and even in toxicology. Along with the concept of "bioavailability", there were those of "physiological availability" and "biological availability".

This reveals the tendency to expand the way of definition and the directions of investigation which – in a wider context – can be applied in the study of various biologically active substances, going beyond the circumscribed framework of substances of pharmaceutical interest. According to the American Pharmaceutical Association (A.P.A.), the notion of bioavailability is defined by "the amount of active principle that reaches the general circulation unchanged after administration in a pharmaceutical form by oral or other means, which requires an absorption process and the speed with which it occurs this process" (A.P.A., 1972).

Seen from the point of view of the US Food and Drug Administration (government body known by the acronym F.D.A.), bioavailability was defined by "the speed and degree with which the active principle is absorbed from a medicinal product and which becomes available at the site of action" (F.D.A., 1977). The same definition criterion can also be used to characterize nutrients (e.g., macro- and micronutrients, biologically active substances, etc.) and chemical xenobiotics that enter the body.

Expanding the scope of reference, it is easy to understand that the problem of bioavailability is not the exclusive prerogative of some pharmaceutical preparations (according to the first definition given by Oser in 1945), equally affecting the nutrients present in the foods for current use or in food supplements (implicitly the biologically active substances).

It is also mentioned that bioavailability can also be discussed in relation to some ecological aspects that refer to biogeochemistry, soil science (pedology), plant physiology and (not last resort) plant biochemistry. If we continue to address the biochemical aspects, we are led to observe continuity in the trophic chain: soil-plant-animal-man.

6.2.2 Interdisciplinary Characteristics

The brief presentation of some particularities of bioavailability, viewed in relation to nutrients, pharmaceutical products and in relation to the environment, draws attention to this dynamic characteristic specific to biological systems.

6.2.2.1 Bioavailability in Nutrition

It represents a more complex subject for the consideration that two environments are "brought into contact": the external environment – through the nutrients present in food; the internal environment – through secretions from the digestive tract and intestinal flora (Gârban, 2000). The bioavailability of a nutrient is given by the proportion in which it is ingested, then absorbed at the level of the digestive mucosa and metabolized to be available for the body's needs. Only a fraction of the nutrient is actually usable.

In nutrition, bioavailability is dependent on the amount of ingested nutrients and the amount of nutrients that overcome biochemical barriers and reach the general (systemic) circulation. The latter is conditioned by the nutrients from the ingested food, the degradation in the stomach and the intestinal lumen, the absorption at the level of the digestive epithelium, but also the intestinal transit (Barrett et al., 2010; Neilson et al., 2017).

In general, in the field of human nutrition, the problem of bioavailability has been extended mainly in the research of micronutrients from the group of mineral compounds and vitamins. The fact is also explained by the deepening of fundamental and applied research with reference to the use of mineral compounds in the form of the so-called, food supplements". The interest in the bioavailability of nutrients also led to the in vitro study for vitamins and minerals (Srinivasan, 2001).

It should be noted that the aspects related to the specific intestinal transit of nutrients (from food/food supplements) are also valid in the case of drugs administered orally. The problem of bioavailability is currently being studied in relation to xenobiotics of food and pharmaceutical interest. Various chemotherapeutics (of extraction and/or synthesis) administered orally and/or parenterally are studied from pharmacological point of view. By studying bioavailability in relation to nutrients and xenobiotics, it was found that there are biochemical, biophysical, physiological and even genetic causes that significantly influence its status.

A special situation exists in the case of compounds that, at the same time, can be considered nutrients – in food or medicine, e.g., vitamin C (ascorbic acid), vitamin D_2 (calciferol) and vitamin B_1 (thiamine). In this case, the terms "catabolism" or "biodegradation" are used correctly regardless of whether the problem of vitamin micronutrients or some pharmaceutical products from the group of vitamins (lipo- or water-soluble) is being addressed. In the case of nutrients, it is interesting to note that bioavailability is dependent not only on the amount of nutrients but also on the combinations of nutrients from the ingested food.

It is also discussed about bioaccessibility which is facilitated by the physicochemical processes in the digestive tract, starting with mastication and continuing with digestion subject to the action of enzymes in the oral cavity, stomach and intestines. For bioavailability in nutrition, there is also an interconditioning of the nutrients present in various foods. A first example is highlighted by lycopene (chromo-protein) from tomato whose bioavailability is increased, during culinary preparation, by the presence of fatty acids (as solvents). The phenomenon is generally valid for carotenoids.

Another example is offered by foods of vegetable origin through the presence of phytic acid (inositol polyphosphate or inositol hexakisphosphate), which leads to the formation of phytates. They decrease the bioavailability of Ca, Mg, Zn, etc., ions.

Fermentation as a technological process (in the food industry) as well as germination as a biological process (in plant physiology) can contribute to decreasing the amount of phytic acid in food, thus increasing the bioavailability of metal nutrients.

An example of the decrease in bioavailability is represented by food phytosterols which, competitively, reduce the absorption of cholesterol (implicitly cholesterol of endogenous origin). It is also exemplified by the fact that proteins, like vitamin C in food, increase the bioavailability of iron. It was also found that the bioavailability for folates present in food (fruits, vegetables and liver) is lower by 20%–70% compared to that for synthetic folic acid (vitamin B_9). This situation explains the need to switch, in some cases, from diet to drug treatment. Bioavailability can also be influenced by the packaging in which food products are placed during the preservation and marketing stage (Gârban, 2018). In relation to bioavailability in nutrition, the term "nutritional inhibitors" was formulated. It refers to certain nutrients from ingested food that compete for absorption and thus change bioavailability. For example, the interactions between Ca and nonheminic Fe are known in this sense. In such cases, in dysmineralosis, it is recommended that foods with minerals (competitors) be consumed at different times of the day.

6.2.2.2   Bioavailability in Pharmacology

Bioavailability, from a pharmacological point of view, quantifies the proportion in which a certain drug is absorbed and is available to produce specific effects at the systemic level or at the loco-regional level (Arnold and Sicé, 1976; Toutain and Bousquet-Melu, 2004). The study of bioavailability and its interrelations with absorption and bioequivalence required attention in pharmacology due to the underdosing of drugs followed by therapeutic failures, but also their overdose, sometimes followed by therapeutic accidents. Studies on bioavailability in pharmacology are presented in various treatises in which the concentration–time relationship is discussed in more detail.

> *Absorption vs. bioavailability* – From a physiological point of view, the terms "absorption" and "bioavailability" are not synonymous or interchangeable. Absorption represents a process, distinct in time, between the moment of administration and the moment when the drug reaches the site of action. Bioavailability is a more complex process that can be evaluated (and quantified) considering the time of administration and the time when the drug produces effects at the site of action. In relation to bioavailability, there are the distinct notions of absolute bioavailability (ABA) and relative bioavailability (RBA).
>
> *ABA vs. RBA* – The presentation of ABA and RBA takes into account the need to know their specific assessment features. Absolute bioavailability (ABA) – is represented by the percentage amount of the administered dose (0%–100%) that is found in the general (systemic) circulation. Its estimation is done comparatively – following the amount of the drug in the body after intravascular administration (intravenous - i.v. or intraarterial - i.a.), considered 100% and extravascular administration carried out by various routes, e.g., per oral (per os), intramuscular (i.m), intraperitoneal (i.p.), sublingual, transdermal, etc.
>
> *Relative bioavailability (RBA)* – is evaluated comparatively between two pharmaceutical formulations of a drug or between two routes of administration for the same formulation of the drug, without referring to intravenous administration. By definition, the pharmaceutical "formulation" of a drug corresponds to the set of substances that make up its composition. In "formulation" two types of substances are included, i.e. the active principle and excipients.

6.2.2.3   Bioavailability in Ecology

Currently, bioavailability and the field of ecology are being discussed, starting from the nutrient and toxic elements in soil and marine waters (Hamelink et al., 1994; Fisher and Reinfelder, 1995). In ecology, bioavailability is not only conditioned by the abundance of an element in the environment, but also by the presence of other elements with which it can be in competition. Regarding ecology in relation to geo-chemistry and soil science, it is considered that bioavailability must be discussed in connection with the ability of the "mineral kingdom" to provide substances neces-sary for plant nutrition.

Thus, in plant physiology, the problem of nutrition is discussed in relation to mineral bioelements (cationic/anionic), with water and with water-soluble gases that enter the plants. It is also discussed in relation to gases in the atmosphere. From a physiological point of view, it is taken into account that mineral elements, water and soluble gases from the soil are taken up by "radicle transit", and gases from the atmosphere are taken up by "foliar transit". In the bioavailability-mineral nutrition relationship in plants, the so-called "mobilization" (solubilization) of the compounds is expressly discussed. In this framework, the way to ensure assimila-bility – an important fact in plant physiology – by bringing it into the ionic form is highlighted. Along with the access to nutrients of vegetable origin, it is also impor-tant to evaluate the presence of "toxic elements", which are studied by ecotoxicol-ogy (Jantunen, 2010; Baveye et al., 2011; Gârban, 2018).

If we consider the ecology in relation to the aquatic environment, it can be noted that in aquatic animal organisms bioconcentration occurs through increased assimilation for metallic elements. It is known that in the marine aquatic environ-ment there is a preliminary bioconcentration in the plankton, then in the fish body (well known is the case of Hg, more precisely methyl mercury). The situation is explained by existence of an acidic environment, which favors bioavailability fol-lowed by tissue retention.

6.2.3   CONDITIONING OF BIOAVAILABILITY

The estimation of bioavailability is conditioned by the so-called intrinsic (physiolog-ical) and extrinsic (nutritional) variables. If the characteristics of these "variables" are analyzed carefully, it will be observed that in fact, bioavailability is conditioned by some particularities of the composition of nutrients/xenobiotics, as well as their metabolization/biotransformation.

The explanation of biochemical homeostasis starting from the intrinsic (physi-ological) and extrinsic (food) variables is the subject of numerous theoretical studies of physiological and even physiopathological interest. Such studies correlate with some applied directions in nutrition, constituting benchmarks in the practice of obtaining and using "food supplements".

In this sense, the obtaining of food supplements with micronutrients and biologi-cally active substances is especially mentioned. This applicative direction has been extended in the last three decades to human nutrition.

6.2.3.1 Intrinsic Variables

The intrinsic variables also called "physiological variables" are characterized by several essential aspects: (i) absorption mechanisms; (ii) interactions between metabolites and (iii) interactions between metabolites and xenobiotics.

Absorption mechanisms – The evaluation of absorption mechanisms highlights differences between the metabolization of macronutrients: proteins, lipids and carbohydrates, as well as micronutrients – from the group of biomineral compounds (cations/anions) and vitamins.

Biologically active substances can also be integrated into the latter. The study of these mechanisms has revealed, for example, the fact that during the neonatal period there is a reduced regulatory capacity for the absorption of trace elements such as: Fe, Zn, Cr and Pb. This fact – explained by the lack of adaptation of the intestinal tract to absorption processes – affects homeostasis. At older ages, there is a "decline in absorption efficiency", which in Biochemical Pathology, as well as in nutrition, can be called "malabsorption". Such aspects were noticed especially in the case of Cu and Zn.

Interactions between metabolites – These interactions are studied in dynamic biochemistry (metabolism). Within metabolic interactions, two particular aspects were noted, namely: (i) metabolic interdependencies and (ii) metabolic interrelations.

Metabolic interdependencies – Such studies – important in dynamic biochemistry - were carried out regarding the storage of mineral elements and various metabolites. It was found that there are phenomena of interdependence between mineral trace elements and other metabolites (protein, lipid) or between trace elements taken as such. In the case of trace elements, biochemistry and physiology studies have shown that there is an interdependence between Fe, Cu and catecholamine metabolization; between Zn and the biosynthesis and biodegradation processes of proteins.

The interdependence was also noted: within non-metallic trace elements, e.g., selenium influences the metabolic processes in which iodine is engaged, an example that reveals a certain synergism.

Metabolic interrelations – The problem of these interrelations is of particular interest in particular cases of loss or reduction, in physiological/physiopathological conditions, of the amount of stored trace elements, e.g., anabolic processes contribute to the fixation of Zn; tissue damage causes Zn loss; intense mental activity causes loss of Cr. Also, within the interrelationships, the issue of the release of the stored elements from the system was addressed, e.g., low levels of Ca trigger the release of Zn from the skeleton; the intensification of catabolism produces the redistribution of Zn, etc.

Interactions between metabolites and xenobiotics – The study of the interaction between metabolites and xenobiotics (represented by chemical contaminants of food and certain pharmaceutical products) is the subject of study of dynamic xenobiochemistry, so the biotransformation of xenobiotics is of interest.

The approach of interactions between metabolites and xenobiotic chemicals is important in xenobiochemistry, also aiming at possible implications in pathobiochemistry. Metabolites are produced by the intake of nutrients (exogenous origin) and/or by metabolic processes specific to morphogenesis (endogenous origin).

Xenobiotics and/or residual xenobioderivatives, formed as a result of biotransformation processes, can react directly with metabolites. Interactions between metabolites and residual xenobiotics/xenobioderivatives occur especially during biotransformation, involving: xenobiodegradation pathways (oxido-reduction reactions and hydrolysis reactions) and xenobiosynthesis pathways (conjugation reactions, adduction reactions).

6.2.3.2   Extrinsic Variables

The considered extrinsic variables and "nutritional variables" are correlated with aspects related to: (i) solubility and dimensions of molecules; (ii) synergistic effects; (iii) antagonistic effects.

Solubility and dimensions of molecules – In evaluating bioavailability, studies on the solubility and dimensions of the molecules are important. The first observations can start from the macro- and micronutrient status of the various chemical compounds present in food.

Obviously, aspects related to the solubility of compounds and their dimensions can also be discussed in relation to medicines. With macronutrients, after digestion, through biodegradation (catabolism), carbohydrate and protein macromolecules are "brought" to molecules with low molecular mass.

Thus, in the case of carbohydrates (e.g. starch, glycogen) we end up with monoglycide molecules, and in the case of proteins, we end up with amino acids. Also, in all categories of nutrients and xenobiotics, polar forms are formed that are very soluble. In the case of micronutrients - represented by mineral and vitamin compounds - solubility issues are discussed in connection with the biologically active specificity and absorption.

It is known that molecules containing trace elements influence absorption at the level of mucous membranes. In this case, the role of Fe oxalate, Cu sulfate and salified silicates with trace elements that are not available in the intestine is exemplified.

The role of phytates (derivatives of phytic acid) is also mentioned. They can fix Fe, Pb, Mg, etc. which influences Ca metabolization.

Synergistic effects – Synergism is defined as an interaction between two or more agents (factors, entities or substances), which results in the increase of the combined effect in relation to their separate effects. In biochemistry/ xenobiochemistry/pharmacology, synergism represents the superior effect given by the associated action of the administered substances.

Regarding synergism, it is mentioned that some substances activate absorption, e.g., citrates and histidine activate Zn absorption; ascorbate modifies the Fe/Cu antagonism generating some synergistic effects.

Other substances, important in synergism, maintain the transport status of trace elements within the framework of systemic mobility (throughout the body). Such an action has transferrins and albumins, which, in general, are considered substances with the role of plasma ligands for trace elements.

Antagonistic effects – Antagonism is defined as the interaction between two or more agents (factors, substances) whose combined effect is less than the sum of the individual effects.

In chemistry/xenobiochemistry/pharmacology, antagonism characterizes the action of two or more substances that, taken together, exert a decrease in relation to their individual effects. The antagonistic effects are noticeable, for example, in the case of the evaluation of the mobility of trace elements.

The antagonism reveals some typical aspects, e.g., (i) decreasing the solubility of elements in the gastrointestinal tract, e.g., Ca/Zn/phytates and Cu associations/sulfides; (ii) achieving a competition between elements in relation to the receptors involved in the supply - flow - retention - elimination relationship; (iii) the highlighting of some competitive effects between cationic mineral compounds, e.g., in the case of Zn/Cu, Zn/Cd associations, etc.

Particular aspects of bioavailability in relation to micronutrients from the group of cationic biomineral compounds can be better revealed under the conditions of the evaluation of intrinsic and extrinsic variables.

An approach to the problem of bioelements reaching the tissues (in metabolically usable forms) is also subject to other factors, among which are mentioned: the physical-chemical characteristics of the nutrients present in food - they can generate interference between bioelements; the biochemical interactions of these elements with synergistic or antagonistic compounds that may occur in the intestine and/or in the tissues; the relationship between the intake and the need of bioelements influences the effectiveness of the absorption of trace elements, their storage and incorporation into the morphological systems in whose composition and functionality they are necessary.

Seen in a wider context, the problem of bioavailability raises new and interesting aspects in relation to mineral nutrition and, especially, in relation to essential mineral micronutrients, e.g., Zn, Cu, Mo, etc. – of metallic nature and Se, I – of non-metallic nature. An interesting aspect in terms of bioavailability in relation to biochemical homeostasis is highlighted by alcohol consumption and its effects on embryo-fetal metabolism in laboratory animals (Gârban et al., 1993). And in this case, the antagonistic effects are discussed in relation to the eredopathology. The existing data in the specialized literature also reveal correlations between alcohol consumption and changes in maternal and fetal metabolism, especially lipid and hydroelectrolytic metabolism. In the study of the bioavailability of biomineral compounds undertaken on laboratory animals as part of experimental research, the status of biochemical homeostasis is monitored.

Such studies are of predictive interest for the medical clinic, finding the existence of influences due to intrinsic and extrinsic variables (e.g. in dysmineralosis). In the field of nutrition - in relation to bioavailability - the issues of bioequivalence/bioinequivalence can also be discussed. One can exemplify - in this context - the existence of bioinequivalence in the case of micronutrients from the group of mineral compounds containing Ca and Mg, which are poorly absorbed if associated with an increased consumption of bakery products.

The phytic acid, present in these products, will retain part of the calcium and magnesium so that a smaller quantity will reach the body. Hence an apparent depletion with the modification of biochemical homeostasis.

6.2.4   The Relationship Ligands – Receptors and Bioavailability

In addressing the problems related to bioavailability, it is important to consider the ligand-receptor relationship (Dollery, 1991; Deshpande, 2002; Landis and Yu, 2004). In this framework, it is necessary to know the aspects of chemical structure, interaction mechanisms, agonistic/antagonistic effects, biochemical and physiological implications, etc. These aspects are also addressed in Molecular Biology.

6.2.4.1   Ligands and Receptors – Characterization

6.2.4.1.1 Ligands

The nature of the ligands – It was designated by the term ligand in biochemistry "a substance capable of binding to a receptor". Such a receptor is activated in relation to a neurotransmitter or a hormone. Activation of the receptor in relation to drugs or other substances is also possible. Based on these considerations, a classification of ligands reveals that there are:

  i. *natural ligands* – originating from nutrients and/or metabolites (e.g., neurotransmitters, hormones, etc.);
 ii. *artificial ligands* – derived from chemical xenobiotics of food, pharmaceutical or strictly toxicological interest. In part, the chemical structure of the ligands is known, so that a number of ligands have been obtained by synthesis.

Ligand isomerism – Some of the ligands have isomeric structures. Several types of isomerism are known that are the basis of biological activity.

It is mentioned in this sense: (i) spatial configuration isomerism - represented by dextrorotatory (+) and levorotatory (−) enantiomers, their mixture being called racemic (inactive form by compensation); (ii) D-L stereoisomerism; (iii) R-S stereoisomerism (rectus -sinister) and (iv) α–β anomeric, specific to carbohydrates.

Ligand interactions – Specific aspects of the stereoselectivity of some ligands can be exemplified; e.g., levorotatory adrenaline is pharmacologically active; antimalarials and dextrorotatory quinidine - are also pharmacologically active. Stereoselectivity is also influenced by the number of asymmetry centers.

A first example, molecules that have two centers of asymmetry (two asymmetric C* carbon atoms) show four isomeric forms. Another example is represented by thalidomide which has two isomers with hypnosedative effects. Of these, only thalidomide levogyrate - is teratogenic in certain animal species. The isomeric structures are also important because they intervene in the processes of biotransformation and elimination. A typical example is prilocaine, a local anesthetic in which the R (−) form hydrolyzes rapidly and intervenes in the production of methemoglobinemia. In contrast, the S (+) form hydrolyzes slowly and does not lead to methemoglobinemia (Lechat, 1971).

6.2.4.1.2 Receptors

The nature of receptors – In general, formations with receptor properties are represented by proteins with a macromolecular structure. Better known are the receptors represented by soluble proteins with a molecular mass of 10,000–500,000 Da. However, there are also receptors with glycoprotein structures.

In this case, part of the receptor site is made up of polysaccharides. Also, the receptors include deoxyribonucleic acid (DNA) which can interact – at the level of heterocyclic nitrogenous bases – with the ligand (natural or artificial).

Identification of receptors – It can be done in vitro and in vivo. In vitro identification can be done by methods of evaluating the ligand-receptor binding with labeled compounds (in which there are radioisotopes). In vivo receptor identification is possible by positron emission tomography. In this way, the maps of brain receptors of dopamine, benzodiazepamin, etc. can be visualized.

Location of receptors – The study of the localization of the receptors led to the observation that they are located in: (i) the cell membrane, (ii) some cell organelles (e.g., ribosomes, mitochondria), and (iii) the cell nucleus.

Distribution of receptors – There are receptors with a wide distribution in the cells of various tissues and receptors with a limited distribution only in certain tissues (typical example the uterus, e.g., the response to oxytocin). In a certain organ, there may be receptors whose response to stimulation triggers various physiological effects, conditioned by the dose of the substance. There are cases when the response to certain doses is weak, a response is obtained only at increased doses. Such situations are encountered in the experiments with xenobiotics of pharmaceutical interest carried out on animals, examples are: (i) at increased doses of the same drug, seemingly opposite effects may occur; (ii) the stimulation produced by two drugs that bind to two types of receptors can induce opposite effects but of unequal importance in relation to the dose.

6.2.4.2 Interactions between Ligands and Receptors

The interaction between ligands and receptors is primarily dependent on the composition of the ligands (represented by nutrients, pharmaceutical products and xenobiotics) but also on their concentration. For example, we will consider that there is a certain ligand with xenobiotic attributes denoted in the general case by X_i, it can bind to a receptor R. The substance with xenobiotic attribute (X_i) and with the role of ligand can be:

- xenobiotic of food interest (X_a) e.g., pollutants from the class of mycotoxins (aflatoxin), additives from the class of preservatives (sodium benzoate E 211);
- xenobiotic of pharmaceutical interest (X_f), e.g., paracetamol, cisplatin;

- xenobiotic of strictly toxicological interest (X_t) used as a pesticide: organo-chlorine products (e.g., DDT); organophosphorus products (e.g., parathion). In the situation where there is an interaction between the xenobiotic (X_i) and the receptor (R), two types of interactions can be observed – depending on the nature of the xenobiotic: (i) xenobiotic-agonist and (ii) xenobiotic-antagonist.

6.2.4.2.1 Xenobiotic-Agonist Ligand $X_{i(1)}$

The ligand can interact with the R_x receptor forming the $X_{i(1)}R_x$ complex, which can be biologically active. In this complex, the R_x receptor has the role of an agonist.

The interaction is as follows:

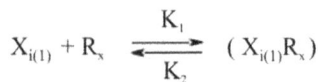

$$X_{i(1)} + R_x \underset{K_2}{\overset{K_1}{\rightleftarrows}} (X_{i(1)}R_x)$$

biologically - active

where
K_1 - association speed
K_2 - dissociation speed

The specification of biologically active, depending on the nature of the xenobiotic, can be: physiologically active in the case of X_a, pharmacologically active in the case of X_f, and respectively toxicologically active in the case of X_t.

6.2.4.2.2 Xenobiotic-Antagonist Ligand $X_{i(2)}$

It is highlighted that the ligand interacts with the R_y receptor, fixing on it. The formed complex $X_{i(2)}R_y$ is biologically inactive:

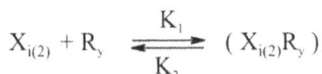

$$X_{i(2)} + R_y \underset{K_2}{\overset{K_1}{\rightleftarrows}} (X_{i(2)}R_y)$$

biologically - inactive

In this situation, the $X_{i(2)}$ type ligand is considered a competitive antagonist in relation to the $X_{i(1)}$ ligand. The competition takes place between the ligands, in relation to the fixation on the receptors. The interaction is dependent on the number of xenobiotics and especially on the affinity to the R_y receptor.

6.2.5 BIOAVAILABILITY AND BIOLOGICALLY ACTIVE SUBSTANCES

Biologically active substances of food and pharmaceutic interest derive mainly from plant sources and in a lesser extent from animal sources. It was found that the bioactive substances that reach the organism by food intake or used as pharmaceutical extracts have a beneficial effect on the health status and reduce the risk of certain chronic conditions as well as of degenerative diseases.

Fruits, vegetables, cereals, dairy products, a.o. are considered sources of biologically active substances of nutritional interest. For example, it is known that plants are rich in carotenoid compounds, while some aquatic animals provide polyunsaturated fatty acids.

Approaching the problems related to the bioavailability of biologically active substances from foods, must take into account *bioaccessibility* (Li et al., 2012; Galanakis, 2017).

In the case of plant-based foods, it can be defined as the presence of the biologically active substance fraction released from the ingested food into the gastrointestinal tract and became available for intestinal absorption.

Bioaccessibility of biologically active substances from vegetal food sources depends on the processing (the resistant cell walls of vegetable limit the breakdown in the small intestine), components of the ingested foods, etc. Once released, the biologically active substances can be absorbed and are subject to many transformations in the gastrointestinal tract. During metabolic reactions, these substances interact in the large intestine with microbiota, too.

In the case of pharmaceutical products bioaccessibility is partly resolved because these are as plant or animal extracts and can be administered without food consumption. Aspects related to the absorption of pharmaceuticals with biologically active compounds and their bioaccessibility are of interest in the pharmaceutical industry.

There are differences and similarities in the relationship between biologically active substances of plant food origin and pharmaceutical products in terms of bioavailability. Both the bioactive substances present in food and those in medicines are subject to physical and chemical transformations before they pass into the systemic blood circulation.

It is known that along with nutrients in food can be also chemical xenobiotics. In all cases there are stages of metabolization/biotransformation in which they are subjected to oxidation and reduction reactions, and subsequently to conjugation reactions.

Bioavailability of biologically active substances depends on their sources, chemical structure (knowing the large heterogeneity of the composition), molecular mass, processing techniques, etc. and the receptor organism: its physiological and health status; food components that access in the organism (implicitly the possible presence of xenobiotics).

More details and examples regarding the biodegradation of biologically active substances present in some foods, some medicines (e.g. polyphenols, polyunsaturated fatty acids) and of xenobiotics are found in the literature (Rein et al., 2013; Garban, 2018). Changes at the level of hepatocytes and enterocytes are especially emphasized. The elimination of residual substances at the level of the digestive tract is by feces, and at the circulatory level by urinary excretion. In the transit of food nutrients, there is the following sequence of phases: absorption-distribution-metabolization-elimination (ADME). In the case of chemical xenobiotics in food, the sequence of phases is: absorption-distribution-biotransformation-elimination (ADBE) – see also Chapter 1.

6.3 CHRONOBIOCHEMISTRY

6.3.1 Basic Concepts

The biochemical interactions specific to metabolic processes take place in time and space, which gives biological systems the possibility of correlation in terms of chronobiochemistry and topobiochemistry (Halberg, 1983; Haus and Kabat, 1984; Garban, 1997; Dunlap, 1999; Reinberg, 2003; Schwob, 2007).

Chronobiochemical metabolic processes (chronos-time; bios-life) that take place in the body maintain the status of various functions (e.g., respiration, circulation, etc.) and the relatively constant level over time of certain metabolites (e.g., metabolites detected in the blood: calcium, phosphorus, glucose, etc.).

The first observations of chronobiology were made by Mairan in 1729, who noticed that mimosa plants (*Mimosa* sp.) open during the day and close during the night. The fact that this phenomenon is dependent on the light-dark relationship has led to the idea that the rhythm is endogenous. After almost two centuries, Johnson, in 1926, noticed that there were changes in the circadian rhythm in animals as well (observations made on mice).

Based on this remark, the author postulated that there is an "internal physiological clock" independent of fluctuations in medium. Currently, it is believed that there are *clock-genes* that control the existence of molecular mechanisms of chronobiochemical processes. Studies on the existence of "clock-genes" were performed by Konopka and Benzer (1971) on *Drosophila melanogaster*. The authors isolated three mutants that drastically alter circadian biorhythm in Drosophila, i.e., a mutant that induces dysrhythmia, a mutant that shortens the circadian biorhythm to approx. 19 hours and another that increases this biorhythm to over 24 hours.

These mutants intervene in the activity of *putative clock-genes*. The mutants intervene in the circadian rhythm of insect populations and in individual locomotor activity.

In nature, it has been observed that for the same biological process, there are periodic alternations with "times in which they decrease" and "times in which they increase" various values.

Chronobiochemical variations are manifested in a regular, periodic and predictable manner, having the features of a rhythm. This observation has led to the accreditation and consecration of the notion of *biological rhythm*.

Chronobiochemistry – is defined by the variation of the amount of metabolites given by biochemical changes, statistically validated, which have an undulating and reproducible character in certain frequency ranges.

The biochemical changes that determine the amount of metabolites are genetically conditioned. Thus, it can be concluded that there is a *genetic programming* (or more commonly a "gene coding") in relation to chronobiochemistry. Obviously, this observation can be extended to all the dynamic characteristics of biological processes.

The existence of "gene coding" makes it possible to explain the situations in which different species of animals have similar rhythms, without being absolutely identical. Although there is general recognition of the role of "gene coding" in biorhythms, there are no systematic studies of specific genes in the transmission of these traits in relation to chronobiochemistry.

Chronobiochemistry – and chronobiology – in general – addresses the issue of the role of biological rhythms (biorhythms) and their implications for the integration of biological systems. Biorhythms are burdened by biochemical and physiological changes, which vary over time.

Viewed in a broader context, chronobiochemistry – and chronobiology in general – addresses the issue of biological rhythms (biorhythms) and their implications for the integration of biological systems. Biorhythm is burdened by biochemical and physiological changes, which vary over time.

6.3.2   BIORHYTHM CONDITIONING

The manifestation of biorhythms and their specific chronobiochemical changes are conditioned by the phenomenological origin (intrinsic and extrinsic) and are dependent on metabolic processes and environment. The *intrinsic* component – often called "endogenous" – attests to the fact that a certain innate biorhythm is present in the body's genetic code and is passed from one generation to another. In this case, it is accepted – *in eodem tempore* – the existence of a transmission in the sequence of replication-transcription-translation processes.

Thus, biological information is transmitted from genomic DNA to proteins (Karahalil, 2010). The proteins involved in maintaining biorhythms influence the rhythmicity of biological processes.

Studies in this field of molecular biology are becoming more numerous, in the last two decades being oriented mainly toward certain correlations between cellular ultrastructure and physiological processes, e.g. oxidative stress in relation to DNA, effects induced by xenobiotics, etc. (Kristal and Yu, 1992).

Details regarding mitochondrial ultrastructure and the senescence process were also investigated, e.g., electron microscopic characteristics and topobiochemical aspects of mitochondrial DNA, commonly referred to as mtDNA (Katsumata et al., 1994).

The *extrinsic* component - also called "exogenous" – is dependent on environmental factors, e.g., temperature, humidity, atmospheric pressure, irradiation threshold, etc. These factors can greatly influence the biorhythms and especially their periodicity. The chronobiochemical effects induced by: thermal stress, both in the plant and in the animal kingdom, water stress and baric stress (determined by changes in atmospheric pressure) are known in this sense.

The listed effects are known and increasingly studied in physiology and pathophysiology (e.g., respiratory diseases, cardiovascular diseases, etc.). The conditioning of the biorhythms of the extrinsic component is, to a certain extent, influenced by man through the interventions (possible in part) on the environment.

6.3.3   PERIODICITY OF BIORHYTHM

Bioperiodicity appears as a ubiquitous phenomenon, which concerns all aspects of life. Thus, a morphophysiological bioperiodicity is distinguished, interesting: *biochemical* interactions – e.g. bioconstituents, micronutrient biochemical mediators (implicitly biologically active substances), etc.; *biophysical* variations – e.g. membrane potential, *physiological* processes – e.g. cardiac chronotropism; *morphological* changes, e.g., the heart in the systole-diastole alternation.

Biorhythms could be detected at various levels of biological organization: organisms, organ systems, tissues, cells and even cellular ultrastructural formations. They can extend over an extremely long period of time, from 10^{-3} to 10^8 seconds.

Their grouping is done in: rhythms with microintervals – with periods of the order of seconds or minutes; rhythms with macro-intervals – whose period is of the order of hours, days, months or years (Gârban, 2014).

In current studies of chronobiology and chronobiochemistry, the grouping of biorhythms is preferred according to the time interval. Thus, several types of biorhythms were detected:

- *ultradian* – which looks at oscillations with shorter periods: from a few fractions of a second, to minutes and even hours;
- *circadian* – which lasts about 24 hours (in principle it is between 21 and 28 hours);
- *infradian* – in which the period is several days, e.g., 7 days – weekly rhythms, about a month (circamensal or lunar rhythms), about a year (annual rhythms), several years (multiannual rhythms).

Addiction to astronomical phenomena characterizes biorhythms with long time intervals, extending from one season (summer and winter) to multiannual intervals. These biorhythms can be seasonal, annual and multiannual with a duration of 11 years (solar).

6.3.4 BIORHYTHM TYPES AND PARAMETERS

6.3.4.1 Ultradian Biorhythms

Estimating biorhythms of some physiological processes - on the basis of which the biochemical and biophysical mechanisms are based – a great variation of their periodicity is noticed.

Thus, it was found that in humans the periodicity of ultradian biorhythms (expressed in seconds) is: in *neural processes* 10^{-3} to 10^{0} s (e.g., on encephalogram 10^{-2} s); in *respiratory* movements approx. 0.27×10^{0} s; in peristaltic movements 10^{1} s; in *bloodstream* 10^{1} to $10^{2.5}$ s.

From the above mentions, it can be seen that ultradian biorhythms originate from the biochemical interactions that take place in the body and characterize par excellence the physiological processes in the nervous system, respiratory, digestive and circulatory systems.

6.3.4.2 Circadian Biorhythms

Circadian biorhythms (also called *nyctemeral* or *daily*) last about 24 hours. Their limits – given the phase lag – are between 21 and 28 hours. These variations are due to "biochemical individuality", included in the form of a program in the genetic heritage of each organism.

Mathematical calculations that have followed the expression of the periodicity of circadian rhythms in the second led to the conclusion that their average duration is approx. $0.864. 10^{5}$ s. It is generally estimated that circadian biorhythms, compared to other classes of biorhythms, are characteristic for humans, animals and plants.

The animal kingdom – Circadian chronobiochemical variations with respect to metabolism, in animals and humans, show that the night is dominated by anabolism and the day by catabolism. Taken as a whole – within the circadian biorhythm – there is a morning exacerbation and an evening remission of physiological parameters (e.g., body temperature) and biochemical parameters (e.g., the amount of nucleoproteins in tissues, metals in the blood, etc.) – see Gârban (2014).

Regarding the dual phases of catabolism-anabolism in the day-night relationship, it is mentioned that the animals whose activity takes place at night are an exception to this correlation, e.g., the field mouse, the snake, the bat, etc. Endocrinology studies have shown that in humans there are chronobiochemical changes specific to the circadian biorhythm whose intensity increases at night, e.g., biogenesis of growth hormone, prolactin and melatonin (Cajochen et al., 2003).

Chronobiochemistry studies performed by Reinberg (1974) on specific parameters of material metabolism (e.g., serum and urinary metabolites) revealed notable changes in day/night alternation, noting the existence of different maximum values (acrophase) from case to case, within 24 hours.

The plant kingdom – In the case of plants, there are metabolic processes that show variations with environmental specificity. A conclusive example of circadian biorhythm is *photosynthesis* and *nitrogen retention.*

These aspects are particularly important for *agrobiology.* It is interesting to note that the intensity of these processes is antithetical and is maintained by the alternation day/ night. Photosynthesis – as it is known – is the process of forming living matter in the presence of light. Nitrogen retention is achieved during the night (in the absence of light).

Biosynthesis processes followed by nitrogen fixation – studied especially in cyanobacteria – are inactivated by the presence of oxygen that is released during photosynthesis.

6.3.4.3   Infradian Biorhythms

As shown, infradian biorhythms have longer periods and are characteristic of: (i) *weekly,* (ii) *circamensal,* (iii) *seasonal,* (iv) *annual* and (v) *multiannual rhythms.* A brief presentation of them is given below.

a. *Weekly biorhythm* – Interesting findings have been made on the weekly biorhythm. In the case of this biorhythm in humans, the variation of the status of some blood biochemical parameters (glycemia, lipemia, bioelectrolytes, etc.) was detected. Changes in hormonal status and changes in the amount of chemical mediators of nerve influx have also been observed. The characteristics of this biorhythm have been studied more in the last three decades. The biorhythm is correlated with the period of human professional activity and the rest period, respectively.

b. *Circamensal biorhythm* – It also has the names of "monthly" or "lunar" biorhythm. This biorhythm is well studied in women, correlating with the "estrous cycle" (average 28 days), being approached in more detail in the treatises on physiology and endocrinology. Remarkable is the fact that a "quasi-monthly biorhythm" was also found in men, related to the excretion of thyroid hormone T3 (trit) with a periodicity of approx. 29 days and 5 hours. The monthly biorhythms do not have a strict temporal delimitation, there are variations of the order of approx. 2 days, conditioned by the activity regime, geomagnetic phenomena, etc.

    c. *Seasonal biorhythm* – It has both intrinsic (hereditary) and extrinsic (seasonally dependent) conditioning. In the *animal kingdom*, this biorhythm can be exemplified by hibernation to animals - with the reduction of metabolic functions and body temperature, etc., in the autumn-winter period and the resumption of activity in the spring-summer period. It is also possible to explain the migration of birds, which has a seasonal conditioning in relation to the season. The extrinsic conditioning of this biorhythm is dependent on factors of the existential area.

        In the *vegetable kingdom*, the existence of seasonal biorhythm can be exemplified by the existence of phenophases in plants. Thus, in the spring-summer period, there are distinct phenophases, e.g., growth, development, maturation and seed production. During the autumn-winter period, the essential phenophase lies in the preservation of the morphofunctional attributes of the seeds.

    d. *Annual biorhythm* – It has an intrinsic conditioning inscribed in the "genetic program" of each organism (so-called "gene coding"). For the animal kingdom, a typical example of an annual biorhythm is reproduction in some wild animals, in which the birth of chicks takes place in a certain period limited to approx. 1 month during the year. This biorhythm is also characteristic of the plant kingdom, knowing that there are annual and perennial plants depending on how they produce seeds. In humans, the annual biorhythm has been studied mainly from the endocrinologist's point of view, finding that there is an increased status of the amount of androgen and estrogen hormones in the summer-autumn period, which is also confirmed by biomedical statistics that show an increase in birth rates, periods of gonadal endocrine hyperfunction.

    e. *Multiannual biorhythm* – It is generally less known because there are fewer studies in this regard. However, it is known that depending on the solar activity – which has a periodicity of 11 years – there are chronobiochemical changes in blood metabolites, the composition of biological fluids and hematological changes – characterized by a decrease in the number of leukocytes. It is interesting to note that this biorhythm is correlated with the increase in the incidence of infectious diseases, neuropsychiatric diseases, strokes (heart attacks, emboli) and cases of hypertension. It is mentioned that the multiannual biorhythm simultaneously influences the other biorhythms through the magnitude of the cosmic phenomena, the increase of the cosmic radiation threshold, the magnetic effects, etc. A calculation of the duration of some infradian biorhythms, performed in seconds – a procedure applied to make the comparison possible – shows that the lunar rhythms have a duration of $2,593 \times 106$ s, and the circannual rhythms of 3×10^7 s.

Biorhythmology also reveals important changes during ontogenesis. In this sense, the chronobiochemical changes of endocrinological interest of steroid hormones with important differences from the juvenile, adult and senescence periods are exemplified.

    The approach of the problems of biochemistry and molecular biology of the endocrine system in humans also revealed the role of gonosomes (sex-specific chromosomes), in women XX and men XY. There were changes in women around the age of

45–50 years and in men at 55–60 years. In general, in connection with the endocrine system and senescence, it is stated that hormones do not cause, but participate in the process of senescence. This statement correlates with the fact that estrogen (female) and androgen (male) hormones are not only *sex* hormones but also anabolic hormones.

6.3.4.4 Characteristic Parameters

A regular, cyclic variation – specific to biological rhythms – can be assimilated to a *sinusoidal function*, which allows a mathematical approach. Thus, it is possible to investigate the sinusoidal function of some quantities (i.e. chronobiochemical values) determined experimentally in a time series in order to allow the estimation of some parameters and even their biostatistical evaluation.

In this context, Halberg (1983) reveals the existence of so-called characteristic parameters for each biorhythm. Defining the characteristic parameters for the chronobiochemical changes is of essential importance, among which the circadian changes are exemplified (Figure 6.1).

Their definition is as follows:

- *period* – characterizes the duration of a complete cycle of variation.
- *acrophase* – corresponds to a maximum peak of the chronobiochemical quantity in the considered period. It is usually rendered in units of time (hours, minutes…) or weight (mg, µg…) in relation to a phase arbitrarily chosen reference time (e.g. local time for daily rhythmic variations).
- *amplitude* – is defined by half the total variation of a chronobiochemical quantity (studied) for a given period. Overall, there is a variation of the function, which is performed on both sides of a rhythm level. The tracing of the sinusoidal curve - specific to the biorhythm - is done integratively based on experimental measurements. The effects of chronobiochemical changes are not limited to circadian variations, extending to the selenium level (especially in the endocrine aspect) and even to the seasonal level (more noticeable in the electrolytes).

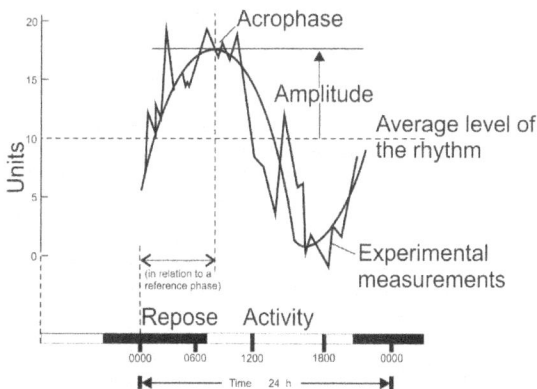

FIGURE 6.1   Circadian chronobiological changes.

6.3.5 Autonomy and Synchronization of Biorhythms

Bioperiodicity is presented as a specific phenomenon to organisms. It has an endogenous (genetic) component, which is at the origin of the "autonomy" of biorhythms. However, there is also the possibility of "synchronization" in relation to the existential area. This aspect is due to exogenous factors - it has been studied more in circadian biorhythms.

6.3.5.1 Biorhythms Autonomy

The question of the *autonomy* of biorhythms can be discussed in relation to the existence of a "genetic programming". So, it has an endogenous origin. The genetic factor explains both the qualitative and/or quantitative chronobiochemical changes of some intrinsic biochemical effectors (e.g. hormones, electrolytes) and the existence of a biofeedback in the neuroendocrine system - a problem addressed more in physiology and neurobiochemistry. A lesser-known aspect of biorhythm autonomy is related to "*psycho-physical* biorhythms". In this sense, the studies undertaken show the existence of three characteristic biorhythms in humans: (i) *physical*; (ii) *emotional* and (iii) *intellectual*. Brief information about them is given below.

a. *Physical biorhythm* – lasting 23 days. This biorhythm actually consists of two circumscribed stages, each at 11 days and 12 hours. Specific to this biorhythm is the fact that: the first stage – is characterized by an increased resistance of the body to physical exertion; the second stage - of identical duration, is characterized by a remission of physical capacity.

b. *Emotional biorhythm* – extends over a period of 28 days, has two stages, each lasting 14 days. In the case of this biorhythm, the first stage is characterized by a tonic state and a calm attitude; the second stage is marked by nervousness and even irritability.

c. *Intellectual biorhythm* – lasting 33 days. It is also characterized by the existence of two distinct stages lasting 16 days and 12 hours each: the first stage is defined by more intense brain activity, increased memory, spontaneity; the second stage is characterized by reduced ability to focus on the topics addressed, increased "switchability of attention" (in psychology, often studying this issue in relation to the possibility of maintaining attention on a topic, it was found that it is between 2–7 minutes).

Regarding the psycho-physical biorhythms, various studies have been done concerning the activity of the air-navigation personnel, the activity of the drivers and the related measures to prevent accidents. These aspects were correlated with physical and emotional biorhythms.

Also, studies of biomedical interest were performed, following the surgical practice in relation to the physical biorhythm. In the field of forensic medicine, scientific investigations have been carried out on the frequency of deaths, noting that they occur mainly in the negative phases of psycho-physical biorhythms. Important observations were made on the basis of which were the "biorhythmograms" developed with the help of the computer and which followed the parallel evolution of all psycho-physical biorhythms.

6.3.5.2 Biorhythms Synchronization

The problem of biorhythm *synchronization* reveals that the body has the ability to adjust its rhythm (e.g., metabolic processes) according to external periodicity, using various signals from the environment. Thus, for example, variations in light intensity (e.g., night, day, twilight, dawn and dusk); temperature variations (natural or artificial); pressure changes (so-called "bar variations"); changes in noise levels, etc., may be external environmental factors that act as stimuli. External environmental factors, natural or artificial, that act on a biorhythm are termed with a generic term *synchronizers* (fr. donneur de temps; germ. zeitgeber).

An overview of synchronizers indicates that they are environmental-external factors that intervene in maintaining or modifying chronobiological processes and implicitly their substrate: variation of chronobiochemical parameters (e.g. change in melatonin and secretory activity of the epiphyseal gland in relation to the conditions of wakefulness/sleep).

Synchronizers include:

a. *physical factors*, e.g., alternating light/dark (with the possible intervention of artificial light), temperature and humidity fluctuations. These factors have applications in the plant kingdom (e.g., experimental vegetation houses) and in the animal kingdom (e.g., constructions suitable for intensive growth);

b. *chemical factors*, e.g., feeding program to animals, administration of fertilizers and water (irrigation) to plants;

c. *biological factors*, e.g., integration of species in a certain ecosystem;

d. *social factors*, e.g., alternation of periods of activity (physical/intellectual) with periods of rest;

e. *geophysical factors*, e.g., dependent on atmospheric ionization; the rotational motion of the Earth around its own axis; the rotational motion of the Earth around the Sun, etc. In human existence, for example, the issue of circadian synchronization of chronobiological manifestations (and obviously their chronobiochemical substrate) is important. Mention is made in this regard of trans-meridian movements, such as air travel by exceeding many time zones, changing working hours (alternating hours day/night) and even changing the official time (transit from winter to summer and vice versa).

Synchronization is important both in the animal kingdom (e.g., bird migration, transhumance grazing) and in the plant kingdom (e.g., the process of photosynthesis in the diurnal cycle, etc.).

In terms of chronobiochemistry, synchronization correlates with changes in material metabolism (protein, lipid, carbohydrate, hydroelectrolytic) and energy metabolism. Thus, chromoproteins (hemoglobin in animals, chlorophyll in plants); nucleoproteins (especially DNA and RNA nucleic acids); lipids (glycerophospholipids, steroids); carbohydrates (glycogen – in animals; cellulose, starch – in plants) reveal important changes in the amount over time.

For example, the observation that hepatic DNA has a higher amount in the morning (Mayersbach, 1967; Gârban et al., 1984) may gain applications in chronopharmacology-based chemotherapy, in "synchronizing" chemotherapy with chronobiochemical changes in tissue DNA (e.g. hepatic DNA). A particular aspect of synchronization is the investigations that follow "cell cycle synchronization" in eukaryotes. Such studies have been performed to evaluate the effects of cytostatic therapy in correlation with the stage of DNA synthesis in the cell cycle.

Viewed in a broader context, the study of chronobiology problems – with their molecular basis chronobiochemistry – has made it possible to find that biorhythms are the consequence of existing interactions in cells. These are received at the level of: tissues, organs, organ systems or the population (e.g., migration to birds, hibernation to mammals).

The extension of chronobiochemistry studies in humans has led to the elucidation of many problems and interdisciplinary connections, leading to the emergence and circumscription of new fields such as chronophysiology, chronopathology, chronopharmacology, chronotoxicology (Halberg, 1983; Hayes et al., 1990; Reinberg, 2003). The existence of biological rhythms, as an expression of the periodic movements in and/or from organisms, allowed the comparison – on a universal scale – with the periodicity of the characteristic movements, both to the macrocosm (galaxies) and to the microcosm (atomic nucleus).

6.3.6 Chronobiochemistry and Phenology Related to the Biologically Active Substances Biogenesis

Chronobiology represents a subsequent field of general biology, which has "biological rhythms" as its object of study. In this field, chronobiochemistry was evolved, which allowed an accurate approach to the problems of medicine and agrobiology.

The concept of "biorhythms" was outlined in 1729 by Jean-Jacques d'Ortous de Mairan (1678–1771) – French geographer, astronomer and biologist. His observations and experiments were the basis for the initiation of research on the "circadian rhythm" in plants. The experiments mainly followed the rhythm of closing and opening of flowers and leaves. He concluded that they have a circadian periodicity, being conditioned by light and heat.

Chronobiochemistry – by extending biochemical analytics data – has made an important contribution to the development of phenological applications – a new field of "applied biology" in plants and animals – with important economic consequences.

The concept of *phenology* was introduced in 1853 by Charles Francois Antoine Morren (1807–1858) – Belgian botanist at the University of Liège, in an attempt to explain the occurrence of events in the life of plants and animals (Demarée and Chuine, 2006).

Thus, phenology becomes a subsequent field of plant and animal chronobiology. Subsequently, chronobiochemistry contributed to the explanation of molecular mechanisms specific to plant physiology and animal physiology. Phenology (gr. phainestain-appearance; logos-science) has as its object of study the periodic phenomena (chronobiochemical) that appear in the life of plants and animals as a consequence of the interaction with environmental factors.

Afterward, chronobiochemical and phenological problems were correlated with the formation of biologically active substances in the tissues of plant and animal organisms.

To evaluate these phenomena, with chronobiochemical evolution based on the existence of specific phases, the notion of *phenological phases* or *phenophases* was introduced. Finally, phenophase has been defined as a stage in the life cycle of plant or animal species in which morphologically, physiologically and obviously biochemically detectable biological changes can be distinguished.

6.3.6.1 Particularities in the Plant Kingdom

Plant physiology studies have shown specific phenophases, i.e. germination, budding, leafing, flowering, fruiting, ripening, etc. During the ripening period, for example, the phenophases reveal distinctive aspects: in the plants of the spontaneous flora: the dissemination of the seeds, the loss of the foliar apparatus, etc.; in plants from agricultural/fruit/flower crops, harvesting, defoliation, etc. appear as different phenophases. In each phenophase, the biosynthesis of nutrients – especially of biologically active substances – is dissimilar and therefore the sampling of tissues for the obtainment of bioactive substances must be made in the moment of their maximum amount.

Scientific research in the field of plant physiology has followed in crop plants through experiments on various varieties with rigorous integration of data on phenological phases, e.g., flowering phenophase (with specific details on the beginning, maximum, end), ripening phenophase.

Regarding the biologically active substances, more in-depth chronobiological/chronobiochemical studies were performed on polyphenol compounds from cultivated plants and plants from the spontaneous flora (Arola-Arnal et al., 2019; Ávila-Román et al., 2021)

Such data are suitable for use in plant *genetics* and *breeding*. In applied agrobiology, phenology temperature. Phenograms can be established even with graphical tracing (histograms) for various phenophases, e.g., flowering phenophase, fruiting phenophase, maturation and ripening phenophase. Such studies can be carried out using the data of a meteorological station and agricultural research stations with rigorous records. If the issue of the involvement of plant chronobiochemistry in phenology studies is discussed, it is important to point out that it can be correlated with the assessment of the amount of nutrients in tissues, e.g., accumulation of starch, gluten, proteins, lipids in plant tissues in the phenophase of ripening and maturation.

Such data are important for agricultural production (cereal plants, technical plants). The same model of applied research can be used for *biologically active substances in plant physiology*. In their case, the *specific phenophase* is followed, in which, in cells and tissues, larger amounts of substances accumulate. It is recommended for harvesting plants.

Extremely important is plant chronobiochemistry and phenology applied in investigations of cultivated medicinal plants and medicinal plants from spontaneous flora. Such studies are concerned with the (gradual) accumulation of *biologically active substances* in various organs of plants. These correlate with economic issues.

6.3.6.2 Particularities in the Animal Kingdom

Phenophases have other distinct characteristics in animals. For example, in the case of wild animals and birds, the phenophases follow:

a. the period of *pro-migration* in spring from the southern areas to the northern areas, construction of nests, spawning, hatching;

b. the autumn *retro-migration* period. Each period is well characterized by physiological features and, obviously, by changing some chronobiochemical "parameters".

For terrestrial wild animals (some mammals) the phenophases follow the period of reproduction, hibernation, etc. Such data are important for ethology in mountain and hilly wilderness areas. In the case of domestic animals, by studying the comparative animal physiology, the application aspects can be evaluated. For this purpose, the *phenophases related* to animal production (e.g., meat, milk, honey) are followed. Obviously, these correlate with metabolism, in particular with the formation of biologically active substances.

It exemplifies the correlation of body mass – age – maximum production. This is an example of the use of modern zooeconomic principles.

Although there are relatively few scientific works on chronobiochemistry and biologically active substances in the animal kingdom, this problem is tangentially mentioned in specialized works (Anton et al., 2005; Möller et al., 2008; Cardinault et al., 2012).

The same can be said for aquatic animals. For example, in the case of fish following the extraction of fish oil in the "specific phenophase" correlated with a maximum of the synthesis of lipids (fatty acids) depending on body weight - age. In fact, the biologically active substances extracted from them are tracked, i.e., eicosapentaenoic acid (EPA) and docosahexaenoic acid (DHA).

6.4   TRACEABILITY OF PRODUCTS OF FOOD AND PHARMACEUTIC INTEREST

6.4.1   BASIC CONCEPTS

The issue of traceability of products of food, pharmaceutical, cosmetic and agrobiological interest has acquired a major importance with the globalization of agriculture and industries in these fields. In the field of food, there are data on the origin of the raw material, its processing, distribution and location (Cunningham and Meghen, 2001; Pouliot and Sumner, 2010; Gârban, 2018). In the pharmaceutical field, both products obtained by extraction (phyto- and zoochemical) and products obtained by chemical synthesis are of interest. The implementation of traceability systems to improve the ability to verify the safety and quality of products is aimed primarily at products of food interest (e.g., raw materials and processed foods), pharmaceuticals (e.g., drugs obtained by extraction or synthesis) and agrobiological (e.g., plant growth bioregulators, especially synthetic compounds). From a historical point of view, more conclusive data on traceability were provided by the International Organization for Standardization (ISO) and others. The first mentions (definitions) include ISO 8402 from 1994 in which traceability was defined as "the ability to track the history, application or location of an entity through the means of recording identifications". It was subsequently revised by ISO 9000: 2000 and ISO 22005: 2007. In fact, the ISO abbreviated name is not an acronym for International Organization for Standardization.

In this sense, it is mentioned that "ISO" comes from the Greek word *isos* (gr., equal, equivalent). The notion is used to define various legislative norms with various areas of technological applicability.

Traceability has been defined in the Codex Alimentarius as "the ability to track the movement of food at various stages of production, processing and distribution" (http://www.fao.org/fao-who-codexalimentarius). There are also definitions agreed by various other international institutions, e.g., FAO, WHO, EU, IUFoST, EFSA and others.

Considered a matter of excellence, traceability has also been of concern to the *International Union of Food Science and Technology* (IUFoST), which has addressed historical issues, theoretical issues and, in particular, application issues with specific features at Earth level. (IUFoST Scientific Bulletin, 2012). This report emphasizes, among other things, the importance of identifying the source of raw materials as the main basis for consumer protection.

In acceptance of Regulation (EC) No 178/2002, for example, traceability means the ability to detect and track certain foodstuffs for human consumption, animal feed, food-producing animals, certain substances for incorporation into various food products during all stages of production, processing and distribution".

The use, reproduction and dissemination of traceability information were encouraged by the FAO. According to the FAO Guide (2017), traceability is defined as "the ability to discern, identify and track the movement of food or substances to be or are incorporated into a food during the stages of production, processing and distribution". We limit the exposure to various definitions by referring to Webster's Online Dictionary (WOD), which states: "Traceability is the ability to track history, applications or location of an article, or test data, using recorded documents."

To track the traceability of various items (articles), there are computerized systems that take over and transmit the information. Globally, traceability issues are specified by Codex Alimentarius and are "extrapolated" through the use of Information Technology and Communication (ICT).

In this way, a *Global Traceability Standard* (GRS) could be established. Acronyms are suitable for spreading information on the web (so-called webography) by interested parties.

In a broader context, it is important to follow the traceability of raw material sources. In this sense, the importance of knowing that xenobiotics (physical, chemical and biological agents) can contribute to the sanitation of the environment with harmful effects in time and space (Ceccarelli, 1996; Sherr and McNeely, 2008; Gârban 2020).

Regarding consumers, traceability makes it possible to know the specifics of foods, the nature of food ingredients and the possible risk, e.g., allergies, food intolerances, etc.

This can increase food security and nutritional security. In special cases, traceability components can be tracked, e.g., traceability of the supplier, the process and even the customer (Banu et al., 1982; Pouliot and Sumner, 2010). The following will briefly present data on the main types of traceability and food identification systems.

6.4.2 TRACEABILITY TYPES

Taken as a whole, traceability is a system that presents as a strategy aimed at prevention in the management of consumer products. Tracking traceability ensures quality requirements and consumer safety (Zhou et al., 2007; Pouliot and Summer, 2010; Vermeer et al., 2013, Tudora and Tîrziu, 2019). Obviously, the food, pharmaceutical and agrobiological products are of special interest. Traceability can be approached in two distinct ways: "in an *integrated system*" and "in a *differentiated system*". The specifics of these systems will be briefly discussed below.

a. *Traceability in an integrated system* – This system involves tracking a particular product in the chain: raw material-transport-processing-storage-distribution-marketing-consumer (Figure 6.2).

   In the integrated traceability system, the product can be tracked in two directions (actually in reverse directions). These are represented by:

   i. *Descending traceability* – It consists of "forward-looking" and assuring the possibility of locating a product taking into account specific criteria – regardless of where it is in the distribution chain.

   ii. *Ascending traceability* – It involves "going back" – circumstance in which the origin and characteristics of a product can be identified taking into account certain criteria established for distribution points. In the case of food, to better understand the importance of reducing the distribution chain, the laconic expression is used, specific to the French language "de la ferme à la fourchette" and "from farm to fork". For more detailed information on traceability systems, one can follow data from the literature that is much more extensive in the case of food and drugs and more sparingly in the case of other products.

b. *Traceability in a differentiated system* – This system covers more limited areas of logistics management in the food chain. Within this system, internal traceability and external traceability are distinguished. These types are

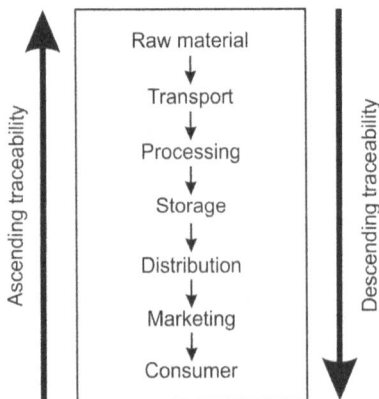

FIGURE 6.2 Traceability in an integrated system (details in the text).

integrated into the "food chain". The expression is frequently used in food law "the chain of economic operators in the food field".

i. *Internal traceability* – In the case of this type of traceability, a sum of information is considered which interests a certain enterprise/company in the evolution of obtaining the food/products. So, it starts from the reception of the raw material, followed by the succession of the stages of processing, storage (implicitly the destruction of waste), marketing.

ii. *External traceability* – It considers the pursuit of a product along a food chain segment, which starts from the finished product obtained by the processor (enterprise/company) to the consumer. According to some authors, this type of traceability covers the entire "food chain". However, the name "external" is not explained.

Please note that where food is distributed in different markets, the measure must take into account the legislative differences in each market (e.g. community market, national market, etc.). Overall, it can be stated that traceability systems (integrated and differentiated) are of interest not only to raw material producers, processors, distributors and consumers but also to the executive power of the state. In this way, coherent measures of socioeconomic and medical interest in relation to human nutrition can be ensured.

6.4.3   Systems of Product Identification

Different systems of identification can be used to know the food traceability data. Among these more important proved to be: (i) *bar codes*; (ii) *landmarks based on radio frequency*; (iii) *biological and biochemical tests* and (iv) *biodegradable markings*.

A brief description of them is of theoretical and applied interest (some in the perspective of expanding use).

6.4.3.1   Barcodes

This means of identification encodes the information on the basis of figures represented by a sequence of black and white bars of various sizes. Barcode decryption is done with a scanner.

In practice, the introduction of this system has been made in the USA and Canada since 1973 and has been called the "Uniform Product Code" (UPC). Subsequently in Europe, since 1979, the code has been used at the suggestion of the European Article Numbering (EAN) Association. It has gradually spread to all continents. Representative figures for a particular code are integrated based on the "labeling program" entered into the scanner and can be used as a verification tool.

In practice, depending on the importance of the marketed product, 8–14 digit codes are used for identification (indicating the country, manufacturer, product, etc.).

Bar codes also allow the integration of a particular item in global trade. These codes can be used online. Mobile phone scanning is also possible.

6.4.3.2   Landmarks Based on Radio Frequencies

In this case, the so-called "identification by radio frequency devices" – Radio Frequency Identification Devices (RFIDs) are used. For this purpose, food data are stored in "electronic circuits" or in "microchips" embedded in plastic material, constituting the so-called *electronic label*.

Various devices that operate at various radio frequencies, in the ranges: 100 kHz–2 GHz, are used for identification. These "tags" allow remote data reading. The data can also be entered in "menus" that are suitable for input on the touch screen.

6.4.3.3   Biological and Biochemical Tests

Biological and biochemical tests used for identification draw attention to the performance of histology, biochemistry, and molecular biology (Alford and Caskey, 1994; Cunningham and Meghen, 2001; Gârban, 2009).

Among the specific methods of this system are mentioned: (i) identification of the *retinal image*; (ii) identification of the *genetic imprint* – also known as "*DNA imprint*" (deoxyribonucleic acid) or simply "DNA test". The application of these methods, although accurate, is limited due to high costs. One can mention their specificity.

    a. *Identification of the retinal image.* The method is specific to biology (histology), being based on recording with special digital cameras the "retinal vascular aspect". This is an attribute of individuality (especially in animals), which is maintained throughout life. This method can be applied to live animals transported for slaughter (elsewhere in the world).

    The method is also suitable for use in "modern zooculture" for the surveillance of live animals as well as animals intended for use in breeding to breed improvements (sent to various locations around the world). One of the applications of the method - given the rigor of the information – concerns zoo parks and nature reservations for animals protected by law.

    b. *Identification of the DNA imprint.* The method is specific to biochemistry and molecular biology. It is suitable for application to live animals, by taking samples from blood, hair, saliva, etc., but also from animal products (animal carcasses). This method is based on specific analyses to molecular biology applied in genetics. In the case of meat and meat products, for example, analyzes based on "DNA imprints" can be compared with data on animals in slaughtered lots. Details on these applications (Cunningham and Meghen, 2001) reveal the importance of using biomarkers/markers in this field.

    In general, there is a great ability to discriminate methods based on molecular biology (Gârban, 2009). It is reiterated that the method has been very useful in forensic medicine, pathology and animal and plant identification studies (Alford and Caskey, 1994).

    From the above data, it is noted that tests based on biology (histology) and molecular biology are in fact specific biomarkers/markers that provide accurate information on the traceability of animals/products of animal origin.

6.4.3.4   Biodegradable Markings

These markings are also known as *edible markings* because they are placed directly on the food. They are invisible and made of an edible substance, e.g., cellulose derivatives.

The compound used for labeling is mixed with a certain food ingredient (usually additives). They are "fixed" due to the physico-chemical effects of food constituents (electrostatic forces, non-destructive interactions with protein compounds, lipids, etc.). The size of such markings is about 200 μm^2 readable area for a barcode.

6.4.3.5 Markings Based on Geospatial Technology

These markings include: (i) *Geographic Information System* - GIS; (ii) *Global Positioning System* - GPS. The latter is a satellite-based radio positioning system, which contains information and the GPS receiver indicates the location in the field.

For information, it is mentioned that for agricultural activities it is possible to collect, analyze and present data. It even becomes possible to *mapping information* to certain regions of the Earth. Marks that use geospatial technology may include the "Quick Response Code", commonly known as the QR code.

QR codes arranged in squares (so two-dimensional) were initially used only in Japan (1995). The two-dimensional specificity originates from the barcode system. Such codes are printed on packaging, posters, billboards, online advertisements, various types of advertising, labels and even business cards. Usually, the QR code is a means of storing information in a visual tag, which can be read by a device (even a smartphone).

Such a code integrates black dot and white space templates, arranged in a square grid. QR codes have an extremely wide use, e.g., postal addresses, telephone numbers, email addresses, websites or web pages, company inventory labels, etc.

In current practice, there are online applications that allow the generation of QR codes. Consumers can access accurate information and valid traceability as well as food safety information only by accessing the QR code printed on the product label. A general observation regarding geospatial technologies draws attention to the difficulties related to the high economic costs of implementing the system.

6.4.4 Traceability of Biologically Active Substances

To ensure the expected effects on the body, it is recommended to ensure a rigorous documentation on the raw material from which products for food and pharmaceutical use are obtained. For this purpose, traceability for biologically active substances can also be determined. For this purpose, additional documentation with reference to the raw material is sought. For this, it is recommended to know: (i) *natural sources*; (ii) *manufacturing procedures*; (iii) *control methods* (chemical and microbiological) and (iv) *distribution chain*.

Documentation is done by evaluating product traceability. In the case of products of food and pharmaceutical interest, it is essential to investigate traceability so as not to affect the health of consumers / patients (Vermeer et al., 2013; Klein and Stolk, 2018).

In the case of the traceability of biologically active substances, the measures regarding nutrivigilance, respectively pharmacovigilance are important (Cutroneo et al., 2014; Gârban, 2020).

Physicochemical methods that can provide rigorous data on biologically active substances that are difficult to detect are often used to investigate traceability. Such a study by Zhou et al. (2007) followed the geographical traceability of samples of propolis from various provinces in China.

The determinations were made by the HPLC method and the HPLC-UV methodological tandem. Chromatograms obtained by HPLC allowed the identification of various compounds such as rutin, myricetin, quercetin, kaempferol, apigenin, pinocembrin, chrysin and galanin. The data obtained made it possible to establish the presence or absence of comparatively evaluated compounds for propolis of various origins.

Investigation of traceability of biologically active substances is of particular interest for laboratories in enterprises where food products (e.g., food supplements, fortified foods) and pharmaceuticals (e.g., extracts from natural products) are processed. These laboratories typically use standardized methods that are integrated into the technology (i.e., production flow control).

In particular, traceability is the subject of activity by institutions that monitor the action/effects of xenobiotics. In this framework, the action is recognized: (i) *chemical xenobiotics* – organic and inorganic compounds present in the environment of natural and/or anthropogenic origin, resulting from agriculture, industry and household waste; (ii) *physical xenobiotics* – represented by radiation emitted by installations used in industry and by nuclear power plants but also by natural radiation; (iii) *biological xenobiotics* - represented mainly by viruses and bacteria, fungi, often with modified genome (which makes analytical investigation difficult).

Another important aspect of traceability investigation is the opportunity to detect counterfeiting (adulteration) of food and drugs (Olsen and Borit, 2013; Cutroneo et al., 2014; Klein and Stolk, 2018).

Traceability in the case of biologically active substances has in view two essential aspects: their quantity after plant harvesting and processing. Such studies were performed following polyphenolic compounds (flavanones, flavonones, anthocyanins, etc.) in cocoa (Gil et al., 2021).

There is research which follows the concentration of ascorbic acid, carotenoids, tannins, minerals, etc. at the beginning and end time of physiological processes. Other studies revealed a decrease in their quantity toward the end of the maturation period, especially in phenols, tannins and flavonoids (Marietti et al., 2012).

Investigations regarding the traceability of biologically active substances are, generally, fewer due to their reduced amount and because it requires high-performance equipment and rigorous periodic checks to notify the evolution/involution of their amount in natural or processed food products.

In the case of traceability investigations, various "product identification systems" (mentioned above) are used. The traceability study becomes important in case of side effects due to the consumption of products of food and/or pharmaceutical interest. If you are interested in nutrivigilance or pharmacovigilance data, these are entered in "databases" that can be used in certain situations.

Addressing the issues of traceability to biologically active substances of food and pharmaceutical interest involves, among others: the conservation of biodiversity in natural habitats; maintaining the balance of ecosystems; the continuous limitation of the sources of contamination of the raw material (of vegetal and animal origin) destined to the production of processed foods and drugs obtained by extraction from natural products; promoting non-polluting processing technologies; promoting education and information on the involvement of xenobiotics in possible environmental contamination.

REFERENCES

Alford R.L., Caskey C.T. - DNA analysis in forensics, disease and animal/plant identification, *Curr. Opin. Biotechnol.*, 1994, 5(1), 29–33.

Anton M., Nau F., Nys Y. - Bioactive egg components and their potential use, pp. 237–244, in *Proceedings of the XIth European Symposium on the Quality of Eggs and Egg Products*, Doorwerth, The Netherlands, 23–26 May, 2005.

Arnold D.J., Sicé J. - Biologic availability and therapeutical equivalence, *J. Clin. Pharmacol.*, 1976, 16, 546–549.

Arola-Arnal A., Cruz-Carrión Á., Torres-Fuentes C., Ávila-Román J., Aragonès G., Mulero M., Bravo F.I., Muguerza B., Arola L., Suárez M. - Chrononutrition and polyphenols: roles and diseases. *Nutrients.* 2019, 11(11):2602. https://doi.org/10.3390/nu11112602.

Ávila-Román J., Soliz-Rueda R.J., Bravo Francisca I., Aragonès G., Suárez M., Arola-Arnal A., Mulero M., Salvadó M. -J., Lluís Arola L., Torres-Fuentes C., Muguerza B. - Phenolic compounds and biological rhythms: who takes the lead?, *Trends Food Sci. Technol.*, 2021, 113, 77–85. https://doi.org/10.1016/j.tifs.2021.04.050.

Barrett E.K., Brooks H., Boitano S., Barman M.S. - *Ganong's Review of Medical Physiology*, 23rd edition, McGraw Hill Co., New York-San Francisco-Lisbon-London-Toronto, 2010.

Baveye P., Block J.C., Goncharuk V.V. - *Bioavailability of Organic Xenobiotics in the Environment: Practical Consequences for the Environment*, Springer, London, 2011.

Banu C., Preda N., Vasu S.S. - *Food Products and their Innocuity* (in Romanian), Editura Tehnică, Bucureşti, 1982.

Cajochen C., Krauchi K., Wirz-Justice A. - Role of melatonin in the regulation of human circadian rhythms and sleep, *J. Neuroendocrinol.*, 2003, 15, 432–437.

Cardinault N., Cayeux M.-O., Percie du Sert P. - La propolis: origine, composition et propriétés, *Phytothérapie*, 2012, 10, 298–304.

Ceccarelli S. - *Positive Interpretation of Genotype by Environment Interactions in Relation to Sustainability and Biodiversity*, The International Center for Agricultural Research in the Dry Areas, Aleppo, Syria, 1996.

Cunningham E.P., Meghen C.M. - Biological identification systems: genetic markers, *Rev. Sci. Tech. Off. Int. Epiz.*, 2001, 20(2), 491–499.

Cutroneo M.P., Isgrò V., Russo A., Ientile V., Sottosanti L., Pimpinella G., Conforti A., Moretti U., Caputi P.A., Trifirò G. - Safety profile of biological medicines as compared with non-biologicals: an analysis of the Italian spontaneous reporting system database, *Drug Safety*, 2014, 37, 961–970.

Demarée G.R., Chuine I. - A concise history of the phenological observations at the Royal Meteorological Institute of Belgium, pp. 815–824, in *Proceedings of the HAICTA 2006 Conference, Vol. III*, Volos, Greece, 20–23 September, 2006.

Deshpande S.S. - *Handbook of Food Toxicology*, Marcel Dekker Inc., New York, 2002.

Dollery C. (Ed.) - *Therapeutic Drugs, Vol. I–II*, Churchill Livingstone, Edinburgh-London-Melbourne, 1991.

Dunlap J.C. - Molecular bases for circadian clocks, *Cell*, 1999, 96(2), 271–290.

Fisher N.S., Reinfelder J.R. - The trophic transfer of metals in marine systems, pp. 363–406, in *Metal Speciation and Bioavailability in Aquatic Systems* (Tessier A., Turner D. R., Eds.), Wiley, Chichester, UK, 1995.

Galanakis M.C. - Factors affecting the bioaccessibility and bioavailability of bioactive compounds, *Food Environ.*, 2017, https://www.linkedin.com/pulse/facors-affecting-bioaccessibility-bioavailability-galanakis/.

Gârban Z., Eremia I., Daranyi G. - Chronobiochemical aspects of the hepatic DNA biosynthesis in experimental animals under the action of some metals, *J. Embryol. Exp. Morph. Suppl. Cambridge*, 1984, 1, 6.

Gârban Z., Aumüller C., Daranyi G., Riviş A., Precob V. - Implications of the chronobiochemistry-metabolism relationship in the induction of homeostasis changes. IV. The action of Cu^{2+} and Mn^{2+} on hepatic DNA biosynthesis and on serum proteins, pp. 164–167, in *Proceedings of the 8th International Symposium on Trace Elements in Man and Animals*, Dresden-Germany, May 16–21, 1993 (Anke M., Meissner D., Mills C.F., Eds.), Verlag Media Touristik, Gersdorf, 1993.

Gârban Z. - Implications of the chronobiochemistry metabolism relationship in the induction of homeostasis changes: the action of the Mg^{2+} and Zn^{2+} ions on DNA. Chapter 71, pp. 482–489, in, *Advances in Magnesium Research* (Smetana R., Ed.), J. Libbey, London Paris Rome Sidney, 1997.

Gârban Z. - The concept of bioavailability: significance and applications. I. Nutritional and pharmacological aspects (in Romanian), pp. 125–130, Symposium, *Agro-food sciences, processes and technologies*, Editura Mirton, Timişoara, 1999.

Gârban Z. - *Biochemstry: Comprehensive Treatise, Vol. I. Basics of Biochemistry* (in Romanian), 5th edition, Publishing House of the Romanian Academy, Bucureşti, 1999.

Gârban Z. - *Human Nutrition, Vol. I* (in Romanian), 2nd edition, Editura Didactică şi Pedagogică R.A., Bucureşti, 2000.

Gârban Z. - *Molecular Biology: Concepts, Methods, Applications* (in Romanian), 6th edition, Editura Solness, Timişoara, 2009.

Gârban Z. - *Biochemistry: Comprehensive treatise, Vol.I. Basics of biochemistry* (in Romanian), 5th edition, Publishing House of the Romanian Academy, Bucureşti, 2014.

Gârban Z. - *Quo Vadis Food Xenobiochemistry*, 3rd edition, Publishing House of the Romanian Academy, Bucharest, 2018 (www.zeno-garban.eu).

Gârban Z. - Nutrivigilance a domain of excellence in food science. Note I. Conceptual and applicative problems, *Journal of Agroaliment. Proc. Technol.*, 2020, 26(2), 47–54.

Gil M., Uribe D., Gallego V., Bedoya C., Arango-Varela S. - Traceability of polyphenols in cocoa during the postharvest and industrialization processes and their biological antioxidant potential. *Heliyon*, 2021, 7(8), e07738. https://doi.org/10.1016/j.heliyon.2021.e07738.

Halberg F. - Quo vadis basic and clinical chronobiology: promise for health maintenance, *Am. J. Anat.*, 1983, 168(4), 543–594.

Hamelink J.L., Landrum P.F., Bergman H.L., Benson W.H. (Eds.) - *Bioavailability: Physical, Chemical and Biological Interactions*, CRC Press Ltd., Boca Raton, FL, 1994.

Haus E., Kabat H. (Eds.) - *Chronobiology 1982–1983*, S. Karger, Basel, 1984.

Hayes D.K., Pauli J.E., Reiter R.J. (Eds.) - *Chronobiology: It's Role in Clinical Medicine, General Biology, and Agriculture, Part B*. Wiley-Liss, New York, 1990.

Jantunen P. - *The Role of Sorption in the Ecological Risk Assessment of Xenobiotics*, University of Eastern Finland, Dissertations in Forestry and Natural Sciences, Kopijyvä Oy, Joensuu, 2010.

Karahalil B. - Pharmacogenomics and toxicogenomics in food chemicals, pp. 477–496, in *Advances in Food Biochemistry* (Fatih Y., Ed.), CRC Press, Taylor and Francis Group, Boca Raton, London - New York, 2010.

Katsumata K., Hayakawa M., Tanaka M., Sugiyama S., Ozawa T. - Fragmentation of human heart mitochondrial DNA associated with premature aging. *Biochem. Biophys. Res. Commun.*, 1994, 202(1):102–110. https://doi.org/10.1006/bbrc.1994.1899.

Klein K., Stolk P. - Challenges and opportunities for the traceability of (biological) medicinal products, *Drug Safety*, 2018, 41, 911–918.

Konopka R. J., Benzer S. - Clock mutants of Drosophila melanogaster. *Proc. Natl. Acad. Sci. U S A*, 1971, 68, 2112–116.

Kristal B.S., Yu B.P. - An emerging hypothesis: synergistic induction of aging by free radicals and Maillard reactions. *J. Gerontol.*, 1992, 47(4), B107–B114.

Landis W.G., Yu M.-H. - *Introduction to Environmental Toxicology*, 3rd edition, Lewis Press, Boca Raton, FL, 2004.

Lechat P. (Ed.) - *Local Anesthetics, Vol. 1. International Encyclopedia of Pharmacology and Therapeutics*, Pergamon Press Ltd., Oxford, 1971.

Li H., Tsao R., Deng Z. - Factors affecting the antioxidant potential and health benefits of plant foods, *Can. J. Plant Sci.*, 2012, 92, 1101–1111 https://doi.org/10.4141/CJPS2011-239.

Maietti A., Tedeschi P., Stagno C., Bordiga M., Travaglia F., Locatelli M., Arlorio M., Brandolini V. - Analytical traceability of melon (*Cucumis melo* var *Reticulatus*): proximate composition, bioactive compounds, and antioxidant capacity in relation to cultivar, plant physiology state, and seasonal variability. *J. Food Sci.*, 2012, 77(6), C646–C652. https://doi.org/10.1111/j.1750-3841.2012.02712.x.

Mayersbach H. - Seasonal influences on biological rhythms of standardized laboratory animals, pp. 87–99, in *The Cellular Aspects of Biorhythms*, Springer Verlag, Berlin-Heidelberg-New York, 1967.

Möller P.N., Scholz-Ahrens E.K., Schrezenmeir J. - Bioactive peptides and proteins from foods: indication for health effects, *Eur. J. Nutr.*, 2008, 47(4), 171–182.

Neilson P.A., Goodrich M.K., Feruzzi G.M. - Bioavailability and Metabolism of Bioactive Compounds From Foods, Chap. 15, pp. 301–319, in *Nutrition in the Prevention and Treatment of Disease* (Coulston M.A., Boushey J.C., Ferruzzi G.M., Linda M., Delahanty M.L., Eds.), 4th edition, Academic Press, Cambridge, MA, 2017.

Olsen P., Borit M. - How to define traceability, *Trends Food Sci. Technol.*, 2013, 1–9.

Pouliot S., Sumner A.D. - *Traceability, Product Recalls, Industry Reputation and Food Safety*, Agricultural Issues Center, University of California, Davis, 2010.

Rein M.J., Renouf M., Cruz-Hernandez C., Actis-Goretta L., Thakkar S.K., da Silva Pinto M. - Bioavailability of bioactive food compounds: a challenging journey to bioefficacy. *Br. J. Clin. Pharmacol.*, 2013, 75(3), 588–602. https://doi.org/10.1111/j.1365-2125.2012.04425.x.

Reinberg A. - *Des Rythmes Biologiques à la Chronobiologie*, 3-ème édition, Gaultier- Villars, Paris, 1974.

Reinberg A. - *Chronobiologie médicale, chronothérapeutique*, 2-ème édition, Flammarion-Coll, Médecine Sciences, Paris, 2003.

Schwob M. - *Les rythmes du corps: Chronobiologie de l'alimentation, du sommeil, de la santé*, Odile Jacob, Paris, 2007 https://rhuthmos.eu/spip.php?article160.

Sherr S., McNeely J.A. - Biodiversity conservation and agricultural sustainability : towards a new paradigm of ecoagriculture landscapes, *Philos. Trans. R. Soc. London, Ser. B*, 2008, 363(1491), 477–494.

Srinivasan V.S. - Bioavailability of nutrients: a practical approach to in vitro demonstration of the availability of nutrients in multivitamin-mineral combination products, *J. Nutr.*, 2001, 131(4), 1349S–1350S.

Toutain P.L., Bousquet-Mélu A. - Bioavailability and its assessment, *J. Vet. Pharmacol. Therap.*, 2004, 27, 455–466.

Tudora E., Tîrziu E. - Traceability technologies in the agri-food sector, *Rom. J. Inform. Technol. Autom. Control*, 2019, 29(2), 101–112.

Vermeer N.S., Straus S.M.J.M., Mantel-Teeuwisse A.K., Domergue F., Egberts T.C.G., Leufkens H. G. M., De Bruin M. L. - Traceability of biopharmaceuticals in spontaneous reporting systems: a cross-sectional study in the FDA Adverse Event Reporting System (FAERS) and EudraVigilance databases. *Drug Safety*, 2013, 36, 617–625.

Zhou J., Li Y., Zhao J., Xue X., Wu L., Chen F. - Geographical traceability of propolis by high-performance liquid-chromatography fingerprints, *Food Chem.*, 2007, 108(2), 749–759.

Index

For Product Safety Concerns and Information please contact our EU
representative GPSR@taylorandfrancis.com
Taylor & Francis Verlag GmbH, Kaufingerstraße 24, 80331 München, Germany

9 781032 702513